AF710092

CLIMATE CHANGE VULNERABILITY AND ADAPTATION IN ASIA AND THE PACIFIC

CLIMATE CHANGE VULNERABILITY AND ADAPTATION IN ASIA AND THE PACIFIC

Manila, Philippines, 15–19 January 1996

Edited by

LIN ERDA, WILLIAM C. BOLHOFER,
SALEEMUL HUQ, STEPHANIE LENHART,
SHISHIR K. MUKHERJEE, JOEL B. SMITH,
JOE WISNIEWSKI

with editorial assistance of

Billy M. McCormac
Editor-in-Chief, WASP

Reprinted from *Water, Air, and Soil Pollution*
Volume 92, Nos. 1–2, November 1996

KLUWER ACADEMIC PUBLISHERS
DORDRECHT / BOSTON / LONDON

Library of Congress Cataloging-in-Publication Data

Climate change vulnerability and adaptation in Asia and the Pacific / edited by Lin Erda ... [et al.].
p. cm.
"Regional Workshop on Climate Change Vulnerability and Adaptation in Asia and the Pacific, held in Manila on 15-19 January 1996."--Foreword.
Includes indexes.
ISBN 0-7923-4204-6
1. Climatic changes--Environmental aspects--Asia--Congresses. 2. Climatic changes--Environmental aspects--Pacific Area--Congresses. I. Erda, Lin. II. Regional Workshop on Climate Change Vulnerability and Adaptation in Asia and the Pacific (1996 : Manila, Philippines)
QC981.8.C5C524 1996
551.695--dc20 96-38429

ISBN 0-7923-4204-6

Published by Kluwer Academic Publishers,
P.O. Box 17, 3300 AA Dordrecht, The Netherlands.

Kluwer Academic Publishers incorporates
the publishing programmes of
D. Reidel, Martinus Nijhoff, Dr W. Junk and MTP Press.

Sold and distributed in the U.S.A. and Canada
by Kluwer Academic Publishers,
101 Philip Drive, Norwell, MA 02061, U.S.A.

In all other countries, sold and distributed
by Kluwer Academic Publishers Group,
P.O. Box 322, 3300 AH Dordrecht, The Netherlands.

Printed on acid-free paper

All Rights Reserved
© 1996 Kluwer Academic Publishers
No part of the material protected by this copyright notice may be reproduced or
utilized in any form or by any means, electronic or mechanical,
including photocopying, recording or by any information storage and
retrieval system, without written permission from the copyright owner.

Printed in the Netherlands

WATER, AIR AND SOIL POLLUTION / *Volume 92 Nos. 1/2 November 1996*

FOREWORD vii

PREFACE ix

ACKNOWLEDGMENTS xi

L. AMADORE, W.C. BOLHOFER, R.V. CRUZ, R.B. FEIR, C.A. FREYSINGER, S. GUILL, K.F. JALAL, A. IGLESIAS, A. JOSE, S. LEATHERMAN, S. LENHART, S. MUKHERJEE, J.B. SMITH, and J. WISNIEWSKI / Climate Change Vulnerability and Adaption in Asia and the Pacific: Workshop Summary 1–12

A. IGLESIAS, LIN ERDA, and C. ROSENZWEIG / Climate Change in Asia: A Review of the Vulnerability and Adaption of Crop Production 13–27

I. AMIEN, P. REJEKININGRUM, A. PRAMUDIA, and E. SUSANTI / Effects of Interannual Climate Variability and Climate Change on Rice Yield in Java, Indonesia 29–39

R.D. BUAN, A.R. MAGLINAO, P.P. EVANGELISTA, and B.G. PAJUELAS / Vulnerability of Rice and Corn to Climate Change in the Philippines 41–51

Z. KARIM, S.G. HUSSAIN, and M. AHMED / Assessing Impacts of Climatic Variations on Foodgrain Production in Bangladesh 53–62

LIN ERDA / Agricultural Vulnerability and Adaptation to Global Warming in China 63–73

WANG JINGHUA and LIN ERDA / The Impacts of Potential Climate Change and Climate Variability on Simulated Maize Production in China 75–85

M.A. WIJERATNE / Vulnerability of Sri Lanka Tea Production to Global Climate Change 87–94

B. BOLORTSETSEG and G. TUVAANSUREN / The Potential Impacts of Climate Change on Pasture and Cattle Production in Mongolia 95–105

K. BOONPRAGOB and J. SANTISIRISOMBOON / Modeling Potential Changes of Forest Area in Thailand under Climate Change 107–117

K.I. KOBAK, I.YE. TURCHINOVICH, N.YU. KONDRASHEVA, E.-D. SCHULZE, W. SCHULZE, H. KOCH, and N.N. VYGODSKAYA / Vulnerability and Adaptation of the Larch Forest in Eastern Siberia to Climate Change 119–127

S. SOMARATNE and A.H. DHANAPALA / Potential Impact of Global Climate Change on Forest Distribution in Sri Lanka 129–135

R.T. PEREZ, R.B. FEIR, E. CARANDANG, and E.B. GONZALEZ / Potential Impacts of Sea Level Rise on the Coastal Resources of Manila Bay: A Preliminary Vulnerability Assessment 137–147

S.P. LEATHERMAN / Shoreline Stabilization Approaches in Response to Sea Level Rise: U.S. Experience and Implications for Pacific Island and Asian Nations 149–157

A.B. PITTOCK, K. WALSH, and K. MCINNES / Tropical Cyclones and Coastal Inundation under Enhanced Greenhouse Conditions 159–169

A. ALI / Vulnerability of Bangladesh to Climate Change and Sea Level Rise through Tropical Cyclones and Storm Surges 171–179

WYSS W.-S. YIM / Vulnerability and Adaptation of Hong Kong to Hazards under Climatic Change Conditions 181–190

A.M. JOSE, L.M. SOSA, and N.A. CRUZ / Vulnerability Assessment of Angat Water Reservoir to Climate Change 191–201

G.A. MEEHL / Vulnerability of Freshwater Resources to Climate Change in the Tropical Pacific Region 203–213

R.A. WARRICK, G.J. KENNY, G.C. SIMS, N.J. ERICKSEN, Q.K. AHMAD, and M.Q. MIRZA / Integrated Model Systems for National Assessments of the Effects of Climate Change: Applications in New Zealand and Bangladesh 215–227

J.B. SMITH, S.E. RAGLAND, and G.J. PITTS / A Process for Evaluating Anticipatory Adaptation Measures for Climate Change 229–238

Workshop Participants 239–246

Author Index 247

Key Word Index 249

FOREWORD

The Philippines, an archipelago set in an ocean environment and a prevailing tropical climate, is no stranger to global climate change processes. Volcanoes and typhoons periodically alter the country's land forms, while nearly 7,000 islands are subject to the subtleties of sea level rise. It was against this backdrop that the Philippines offered to host the ***Regional Workshop on Climate Change Vulnerability and Adaptation in Asia and the Pacific,*** held in Manila on 15-19 January 1996. The workshop focused on climate vulnerability and adaptation studies ongoing in the Pacific Rim countries and oceanic island states. The articles prepared for this volume are representative of the many excellent papers presented at the workshop, which addressed climate change vulnerability and adaptation issues ranging from boreal forest changes in northern Asia to sea level rise in the Pacific Islands.

Island nations in Asia and the Pacific are intimately tied to their oceanic environment, geographically and culturally. The greatest vulnerability within the region might result from sea level rise — a subtle process accelerated by anthropogenic emissions of greenhouse gases into the atmosphere. Resulting changes to the climate regime may also alter the frequency and intensity of typhoons and monsoons, which so many Asian and Pacific economies depend on for water resources. The demand for surface water increases daily to support ever-growing populations, increased agricultural activities, and expanding modern urban infrastructure. The cumulative essence of these papers, and others of similar theme, is that we, as individual nations and as a region, are not immune to the effects of global warning. Sea level rise, fresh water supply, and food production — particularly rice — appear to be the predominant concerns for Asia and the Pacific region.

I take this opportunity to thank all the participants of the workshop for sharing with us work in their respective countries on vulnerability and adaptation to climate change. This work and the articles within represent a process that will continue in the years to come, as we try to fully comprehend the realities of what might lie ahead for this region. Finally, on behalf of the participating countries, we collectively thank the U.S. Country Studies Program for their support in our efforts to grapple with the long-term effects of climate change.

Leoncio A. Amadore, Director
Philippine Atmospheric, Geophysical
and Astronomical Services Administration (PAGASA)

PREFACE

The Regional Workshop on Climate Change Vulnerability and Adaptation in Asia and the Pacific was held in Manila, Philippines, on 15-19 January 1996. The meeting was attended by more than 100 technical and policy experts from 20 countries and territories. The workshop was cosponsored by the Philippine Atmospheric, Geophysical & Astronomical Services Administration (PAGASA), the U.S. Country Studies Program (U.S. CSP), and the Asian Development Bank (ADB).

The U.S. CSP is working with 55 developing countries and countries with economies in transition to help them fulfill their commitments as signatories to the U.N. Framework Convention on Climate Change. It is providing financial and technical assistance to aid these countries in developing inventories of greenhouse gas emissions, analyzing options for mitigating greenhouse gas emissions, and assessing their vulnerability to climate change and options for adapting to its impacts. A series of regional workshops are being held in Asia/Pacific, Africa, Eastern Europe, and Latin America for countries participating in the program and other experts in the field to present and discuss their results. In addition, as part of the U.S. Country Studies Program's efforts to build institutional and human capacity, the program is also publishing a vulnerability and adaptation synthesis report presenting results from countries throughout the world and is providing assistance to countries in development of their National Communications required under the U.N. Framework Convention on Climate Change.

At the Regional Workshop, parallel breakout sessions were held on agriculture, forests and biodiversity, coastal resources, and water resources and human settlements. In each session, international experts and researchers from Asian and Pacific countries presented papers on vulnerability and adaptation assessments. The participants then discussed relevant research and policy issues. The workshop convened in a final plenary session to discuss reports of the breakout sessions and to discuss and adopt the conference summary statement.

These proceedings present selected papers from the Workshop that address vulnerability and adaptation assessments and current research on agriculture, grasslands and livestock, sea level rise, extreme events and cyclones, water resources, integrated assessment methods, and adaptation. All of these papers address issues that are relevant to climate change, in which climate may be defined as a synthesis, using a variety of measures (i.e., mean, maximum, minimum), of weather data on decadal time scales. Many climate change researchers use weather data from a 30-year period to characterize a climate. Therefore, in addressing various climate change issues, some of these papers discuss weather events, which may be distinguished from climate in that weather consists of the short-term variations in the atmosphere. The 20 peer-reviewed papers are a significant contribution to the literature on climate change vulnerability and adaptation in Asia and the Pacific.

ACKNOWLEDGMENTS

The U.S. Country Studies Program (U.S. CSP) would like to acknowledge the 100 participants in the workshop, whose active participation in the discussions contributed to the development of the papers in this proceeding, and the authors, whose cooperation in responding to technical and editorial reviews was invaluable. The U.S. CSP would also like to acknowledge the assistance of Dr. Leoncio A. Amadore, Ms. Lourdes Tibig, and Ms. Imelda Valeroso of the Philippines Atmospheric, Geophysical and Astronomical Administration (PAGASA), Dr. Bindu Lohani and Ms. Annie Idanan of the Asian Development Bank, Dr. William C. Bolhofer of the U.S. National Oceanic and Atmospheric Administration, and Mr. Joel B. Smith, Ms. Betty Neal, and Ms. Stephanie Lenhart of Hagler Bailly Consulting, Inc. for their assistance in planning and running the workshop.

Introductory statements at the Regional Workshop on Climate Change Vulnerability and Adaptation in Asia and the Pacific were presented by Mr. Peter Sullivan, Vice President (East) of the Asian Development Bank; Ms. Lydia Tansinsin, Assistant Secretary, Philippines Department of Science and Technology; Dr. Roman Kintanar, Chairman of the Scientific and Technical Committee of the International Decade for Natural Disaster Reduction; and Dr. Robert Dixon, Director of U.S. Country Studies Program. The keynote address was delivered by Philippine Senator Heherson Alvarez.

Editorial assistance was provided by Ms. Christina Thomas and Ms. Stephanie Lupenski, and production assistance was provided by Ms. Diane Blagusz, Ms. Janice Pagel, Ms. Ashley Frazee, Ms. Sylvia Luedeking, Mr. Marty Mapes, Mr. Richard Fyfe, and Mr. Neil Paananen, all of Hagler Bailly Consulting, Inc.

CLIMATE CHANGE VULNERABILITY AND ADAPTATION IN ASIA AND THE PACIFIC: WORKSHOP SUMMARY

L. AMADORE,[1] W.C. BOLHOFER,[2] R.V. CRUZ[3]
R.B. FEIR,[4] C.A. FREYSINGER,[5] S. GUILL,[6] K.F. JALAL[7]
A. IGLESIAS,[8] A. JOSE,[1] S. LEATHERMAN,[9] S. LENHART[10]
S. MUKHERJEE,[7] J.B. SMITH,[10] and J. WISNIEWSKI[11]

[1] *PAGASA, Philippines*
[2] *U.S. National Oceanic and Atmospheric Administration, U.S.A.*
[3] *University of Los Baños, Philippines*
[4] *NAMIRA, Philippines*
[5] *Camber Corporation, U.S.A.*
[6] *U.S. Country Studies Program, U.S.A.*
[7] *Asian Development Bank, Philippines*
[8] *Instituto Nacional de Investigaciones Agrarias, Madrid, Spain*
[9] *University of Maryland, U.S.A.*
[10] *Hagler Bailly Consulting, Inc., U.S.A.*
[11] *Wisniewski and Associates, U.S.A.*

Abstract. The Regional Workshop on Climate Change Vulnerability and Adaptation Assessment in Asia and the Pacific met to present and discuss assessments of vulnerability and adaptation to climate change in agriculture, forests, coastal resources, and water resources. Discussions were held in breakout and plenary sessions about the state of the science for vulnerability and adaptation assessment, conclusions that can be drawn about the vulnerability of the region to climate change, and where future research efforts should be directed. The workshop concluded that sea level rise is of greatest concern to island and coastal nations in the region, climate change will have a significant effect on agriculture, water resources are sensitive to changes in average climate conditions and to tropical monsoons and cyclones, and forests could be significantly affected by climate change. The workshop recommended that efforts to improve general circulation models continue and that countries in the region cooperate on the analyses of vulnerability and addressing adaptation measures. The workshop also concluded that results of vulnerability and adaptation assessments should be presented to policy makers and the public and that assessments continue to be undertaken to improve our understanding of the issue.

Key words: Asia, Pacific island nations, adaptation, agriculture, forest, coastal resources, water resources, sea level rise, national action plans

1. Introduction

Ranging from the boreal expanse of Siberia to tropical river deltas subject to tropical storms to small islands whose very existence is threatened by sea level rise, the Asia and Pacific region includes a breadth of ecosystems and socioeconomic systems vulnerable to climate change. With over three-fifths of the world's population, natural resources in many parts of the vast continent, archipelagos, and islands are already under stress.

At the Regional Workshop on Climate Change Vulnerability and Adaptation in Asia and the Pacific, recent studies on climate change vulnerability and adaptation were presented and reviewed. Research needs, further technical assistance on vulnerability and adaptation, and preparation of national communications that are required for all signatories of the U.N. Framework Convention on Climate Change were also discussed. How the region can address adaptation and develop national action plans on climate change was also discussed.

Water, Air, and Soil Pollution **92:** 1–12.
© 1996 *Kluwer Academic Publishers. Printed in the Netherlands.*

2. Agriculture

2.1 CONCLUSIONS ABOUT VULNERABILITY

Agriculture is a key economic sector in the region and accounts for a high portion of the national GDPs. For example, 20% of Thailand's GDP is in agriculture. Substantial foreign exchange earnings are derived from exports of agricultural products, (e.g., 70% in the Philippines) and agriculture employs over 50% of the labor force in most countries (60% in Thailand). The region faces increasing population, spread of urbanization, lack of adequate water resources, and environmental pollution, which may hinder growth of the region's future agricultural productivity.

South and Southeast Asia are vulnerable to many environmental hazards, including frequent floods, droughts, cyclones, and storm surges that damage life, property, and agricultural production (e.g., Bangladesh is especially vulnerable). El Niño-Southern Oscillation (ENSO) events play a key role in determining yearly agricultural production across the entire region of South and Southeast Asia.

Many climate change impact studies have been conducted in various regions of South, Southeast, and East Asia. However, the impact of climate change on crop production remains uncertain not only because of uncertainties in climate projections, but also because of the lack of understanding of key processes in crop production, such as the direct effects of CO_2 and the complex interactions with water resources. Despite these significant uncertainties, and the limitation of the modeling studies outlined above, several conclusions can be made:

- Crop yields and productivity changes will vary considerable across regions. Thus, the pattern of agricultural production is likely to change across the region. Based on crop impact studies using $2\times CO_2$ equilibrium GCM scenarios, lower latitude countries have been shown to be more negatively affected. Nevertheless, crop yield simulation results vary widely (e.g., ±20% changes in yield) for specific countries sites across studies, and GCM scenarios.
- Vulnerability to climate change depends not only on physical and biological response but also on socioeconomic characteristics. Low income populations depending on isolated agricultural systems, especially dryland systems in semi-arid and arid regions, are particularly vulnerable. Many of these at-risk populations are found in South and Southeast Asia.
- Although global studies suggest that agricultural production appears to be sustainable under climate change as expressed by GCMs under doubled CO_2, the regions of South and Southeast Asia appear to be among the most vulnerable, and East Asia appears to be relatively less vulnerable. Furthermore, global studies have shown that incremental additional costs of agricultural production and additional risk of hunger under climate change, which could create a serious burden for some developing countries in the region.

Because of the key role that ENSO events play in determining yearly agricultural production in South and Southeast Asia, changes in ENSO frequency and severity would be likely to affect the agriculture of these regions.

2.2 SUITABILITY OF ADAPTATION MEASURES

Adaptation to climate change is likely in areas that are currently less climatically stressed; the extent depends on the affordability of adaptive measures, access to technology, and biophysical constraints such as land and water resource availability, soil characteristics, genetic diversity for crop breeding (e.g., crucial development of heat-resistant rice cultivars), and topography.

Many existing agricultural and resource policies are likely to discourage effective adaptation, and are a source of current land degradation and resource misuse.

2.3 FUTURE RESEARCH NEEDS

The assessment of the direct effects of CO_2 on crop production remains a crucial research question. Although many experiments have confirmed the beneficial effect of CO_2 on the mean response of crops (+30% for C_3 crops, including rice), variation in responsiveness between plant species and ecosystems persists. Response depends on available nutrients, species, cultivar, temperature, and other stressors, as well as differences in the experimental technique.

Development and broad application of integrated agricultural modeling efforts (those that consider interactions of biophysical and socioeconomic factors), and modeling approaches particularly applicable at the regional scale, warrant increased attention. Inclusion of multiple joint effects may change significantly the current "mean" estimate of impacts.

3. Forest Resources

3.1 ASSESSMENT METHODS

The suitability of current assessment methods to address vulnerability and adaptation to climate change in forests was discussed. The uncertainty in climate change scenarios should be carried through into these vulnerability assessments for the forest sector. For example, results may be presented as a range of potential impacts.

The current assessment methodologies generally do not consider how changes in water use efficiency resulting from increased atmospheric concentrations of CO_2 may influence the distribution of vegetation, particularly in the case of decreasing soil moisture, nor do they incorporate factors such as pests, fire regimes, or seasonality. Further refinement of the methodologies is needed to incorporate these factors. More sophisticated models are available and could be used to refine forest vulnerability assessments.

Modeling techniques that use climatic indices to predict the distributions of vegetation zones do not address dynamic or transient responses. Paleoclimate research could be one of the most promising methods for estimating how quickly vegetation could migrate. Paleoreconstruction research from Russia indicates that an increase in precipitation may allow for the expansion of forests on the plains and in the mountains, and that larch forests have migrated relatively quickly and can withstand a wide range of temperature fluctuations.

Additionally, none of the studies presented discussed nonclimate factors such as deforestation despite the significance of such influences on forest vulnerability. Potential changes in land use and socioeconomic conditions are difficult to estimate for the relevant time period. For example, there is significant uncertainty about future rates of deforestation and habitat fragmentation.

3.2 CONCLUSIONS ABOUT VULNERABILITY

Although the research presented is too limited to draw many general conclusions about the vulnerability of forests in the Asia Pacific region, a number of findings can be highlighted:

- Adding to the research on the vulnerability of lower latitude countries, the results of research from Thailand indicate that the potential impact on tropical forests may be significant. Under climate change scenarios generated by the United Kingdom Meteorological Office and the

Goddard Institute of Space Science general circulation models, the area of subtropical forests is estimated to decline from 50% of forest cover to either 20% or 12% depending on the model, and the area of tropical forests is estimated to increase from 45% of forest cover to 80%. This analysis suggests that global climate change will have a profound effect on the future distribution, productivity, and health of Thailand's forests.

- Research in the Russian Federation focused on larch forests (*Larix russica* and *Larix gmelinii*), which are the most widely distributed coniferous forests in the world. The major vulnerability factors for larch forest are fires and damage caused by insects. The frequency of these vulnerability factors is likely to increase with global warming; however, these forests may also benefit from soil nutrition improvements due to permafrost melting.
- Research in Mongolia estimated a shift to warmer and drier conditions, which would be consistent with other research predicting midcontinental drying. The Gobi Desert would change from warm temperate desert to warm temperate desert scrub and the cool temperate desert scrub would be replaced by warm temperate throne steppe. The area of cool temperate desert scrub would also expand to the Khangai mountains, supplanting forest areas.
- An equation relating Precipitation Effectiveness Indices (PEP) and net primary productivity was constructed for a study in the state of Kerala, India. Using this approach, it was estimated that due to the depletion of soil moisture estimated by GCMs [United Kingdom Meteorological Office, Commonwealth Scientific and Industrial Research Organisation (CSIRO), and Max Planck Institute], the productivity of teak plantations would decline from 5.40 m^3 ha^{-1} yr^{-1} to 5.07 m^3 ha^{-1} yr^{-1} and the productivity of moist deciduous forests would decline from 1.8 m^3 ha^{-1} yr^{-1} to 1.5 m^3 ha^{-1} yr^{-1}.

3.3 SUITABILITY OF ADAPTATION MEASURES

Adaptation assessments should proceed, despite the need for further research on vulnerability to climate change. At the same time, the difficulties in obtaining the data necessary to evaluate various forests management practices should be recognized.

Some of the research presented provides insight into the potential for forests to adapt to climate conditions. For example, research in the Russian Federation indicates that morphological and physiological features of *Larix gmelinii* (larch) allow this species to withstand a range of air temperature fluctuations of more than 100 °C and that one of the major ways of adaptation to unfavorable soil conditions is provided by a specific feature of root formation.

Work on vulnerability of forests to climate change could be integrated with vulnerability assessments from other sectors to identify socioeconomic activities or aspects of the biodiversity that are vulnerable to climate change. This type of research could provide information on the importance of various vulnerabilities that is necessary to begin developing policy strategies and options. For example, maps of changes in forest vegetation may be overlaid with maps of potential biodiversity to identify areas that are particularly vulnerable. These areas can then be the focus of an assessment of local management practices.

3.4 INTEGRATION OF VULNERABILITY AND ADAPTATION ASSESSMENTS INTO NATIONAL COMMUNICATIONS

The focuses of each country's national plan for the forest sector in the Asia Pacific region are extremely diverse, ranging from Malaysia, with large forests that are not yet managed, to

Thailand, with large areas of forest that are already being managed, to the Philippines, with limited forest resources remaining. Some of the countries (Philippines, Thailand) already have national forestry plans, which include the importance of influencing the forest planning process to include climate change. For example, climate change vulnerability research could be valuable to the current reforestation project in Thailand. Also, Russia, which has focused on forests as one of its most important sectors, has completed a comprehensive national communication.

3.5 REGIONAL COLLABORATION AND TECHNICAL ASSISTANCE

The vulnerability of the region is diverse and the most relevant options for adaptation will vary; however, there is a great benefit in collaborating in efforts to improve assessment methods. While it is valuable to use outside technical assistance, it is also important to do what is feasible within each country using existing capacity and resources.

In addition, acid deposition is a significant problem in the region, and it will require regional efforts to implement mitigation strategies and improve the overall situation/impacts.

3.6 FUTURE RESEARCH NEEDS

Future research needs include improved techniques for downscaling methods to a regional level, development of growth functions, collection of data on management practices, and incorporation of land use and socioeconomic conditions into the analysis.

4. Coastal Resources

4.1 ASSESSMENT METHODS

The Bruun rule, which estimates the amount of beach erosion for a given level of sea level rise, is being used worldwide to determine the amount of erosion in anticipating a response to a given sea level rise. There are real limitations in its application to many island states bounded by coral reef flats. The depth of closure (which measures the active part of the beach) varies by location. Without this critical parameter, the Bruun rule calculations are nearly meaningless. Also, there has been some attempt to use the Bruun rule along muddy shorelines such as Bangladesh. Here, the formulation loses all physical meaning because the equation is based on the equilibrium profile of a sandy beach. Clearly, more basic scientific research, including long-term beach profiling, is necessary to better model sedimentary dynamics in these environmental settings.

Beach erosion and inundation are controlled by relative sea level rise, which is the combination of eustatic (global) and local (uplift or land subsidence) components. Therefore, tide gauges are necessary throughout the Pacific Ocean because a gauge in one country may have little relationship to the actual sea level changes in another.

The seven-step common methodology of the IPCC is being used by most countries in their vulnerability analyses. This should be considered a very coarse, first-order approach as today's conditions are merely projected into the distant future when a 1 m sea level rise will be realized. This static approach is not very useful for planners who work on 5 to 10 year time horizons. What is needed is a dynamic approach where this big jump is broken down into smaller steps on a decadal basis. For example, beach nourishment can be used against sea level rise induced erosion, but typically such calculations are made for the baseline, 0.5 m, and 1.0 m scenarios.

In actuality, beach nourishment projects rarely last more than 20 years, and most require renourishment in 5 to 10 years.

4.2 CONCLUSIONS ABOUT VULNERABILITY

Accelerated sea level rise is one of the most certain outcomes of global warming. It is of great concern to Pacific island and coastal nations in Asia because most have long shorelines and major population centers located on low-lying areas susceptible to erosion, inundation, and storm surge flooding. Other coastal impacts include salt-water intrusion, which is a particular problem for coral reef atolls such as the Marshall Islands.

In addition to sea level rise is the potential for more frequent or severe tropical storms if the earth warms. Typhoons are a serious problem throughout most of Asia, causing massive destruction and loss of life where they make landfall. Especially impacted are Vietnam, South China, Hong Kong, Philippines, parts of Australia, India, and Bangladesh. Interestingly, only about 1% of global cyclones hit Bangladesh, yet almost 50% of global deaths from cyclones happen in Bangladesh. If climate change alters cyclonic size, frequency, track, etc., only slightly, then the impact could still be enormously important to Bangladesh because of its high sensitivity to these extreme events. The dual threat of sea level rise and storm surge flooding looms large in vulnerability and adaptation analyses for Bangladesh as well as for a number of other coastal nations.

4.3 SUITABILITY OF ADAPTATION MEASURES

Shoreline stabilization has been used widely to counter ongoing beach erosion, which is a worldwide problem. But many of these projects have been ill-conceived, often causing as much or more damage than proposed benefits. Hard shoreline engineering in the form of seawalls and rock groins/jetties have been oversold as a panacea to coastal erosion problems. The consequences are all too clear in places like Sri Lanka, where beautiful white sandy beaches have been squeezed out of existence. Seawalls also encourage unwise development practices by allowing hotels and infrastructure to be constructed too close to an eroding shore. Building setbacks must be instituted to provide the buffer size needed under existing conditions as well as to provide a measure of protection as the seas rise and beach erosion rates accelerate in the coming decades.

5. Water Resources

5.1 ASSESSMENT METHODS

Assessing the true vulnerability of water resources is difficult, especially given the uncertainties about regional climate change and the complex nature of modeling runoff. The scale of GCMs is far more coarse than the scale of processes that affect runoff. Although the resolution of the GCMs has been improved, they are still at too coarse a scale to resolve such important features as mountains. Limited area models (which model regional climate) are a promising development. In the workshop it was noted that some of the GCMs made available by the U.S. Country Studies Program (CSP), such as the Canadian Climate Centre Model and the Geophysical Fluid Dynamics Laboratory, show peak precipitation in July and August, similar to observed peaks. Runoff is a highly complex nonlinear process that is difficult to model.

Simple runoff models are relatively easy to use, but they leave out many features of water resource systems such as daily events. Estimates that assume that use of water per capita or per unit of GNP will stay the same ignore the potential effect of new technologies and management practices.

In spite of the limitations of methods to assess the affects of climate change on water resources, it is important to conduct vulnerability and adaptation assessments with the best tools available in order to inform policy makers about the potential risks from climate change. The best information on future climate change in Asia and the Pacific comes from GCMs, and the best information on how climate change will affect water supplies comes from the use of GCM-based scenarios with runoff models to assess the sensitivity of water supplies. Such studies should be conducted at the river basin level, which is most relevant to decision making on water resource management. Policy makers should be presented the results of assessments such as those discussed here, but should also be clearly made aware of the limitations of this type of analysis.

5.2 CONCLUSIONS ABOUT VULNERABILITY TO CLIMATE CHANGE

Water resources in many parts of Asia and the Pacific are quite vulnerable to climate change. Although there are many uncertainties concerning regional climate change and how it will affect water supplies and demand, it seems highly likely that climate change will result in significant changes in water supplies and demand. These changes could cause many problems, particularly in areas with high populations or limited water supplies. Changes in climate variability, particularly in storm tracks and ENSO events, could have an even greater effect on water supplies than mean climate changes.

The nature of many of the vulnerabilities of water resources in Asia and the Pacific is unique to the region. These unique vulnerabilities, particularly in South and Southeast Asia and the Pacific islands, stem from the combination of unique topography and current vulnerability to tropical cyclones. In Nepal, for example, runoff is strongly affected by the Himalayas and the presence of tropical monsoons and tropical cyclones. A recent and rare tropical cyclone in November dumped over 500 mm of precipitation in less than a day. Archipelagos such as the Philippines have no snow and therefore are highly dependent on rainfall. Changes in rainfall patterns can have a significant effect on water supplies. Atolls in the Pacific are very sensitive to precipitation patterns and changes in storm tracks; they capture rainwater in cisterns and rely on a below ground lens of fresh water that rides on salt water as a reserve source of water. Reductions in precipitation and sea level could diminish this fresh water lens.

Water resources in the region may be vulnerable to increases in demand from population growth and industrialization because they are already strained in many parts of the region. Asia has three-fifths of the world population and projections are for population to increase significantly over the next century. For example, the World Bank projects that the population of the Philippines will increase from 69 million in 1990 to 160 million in 2075. Industrialization could also result in increased demand for water. The Philippines government projects that demand for water will triple over the next half century. Not only could this strain water supplies, but water quality could also suffer. Climate change would be an additional stress on water resources in the region.

5.3 ADAPTATION MEASURES

The concern about potential climate change impacts on water resources is great enough to justify addressing adaptation in water resource management decisions. Climate change could have a very significant effect on the availability and quality of water supplies in Asia and the Pacific. In addition, the demand for water for uses such agriculture could also increase because of climate change. Policy makers should consider climate change when making decisions about construction of water resource infrastructure and developing water management systems. The difficulty in addressing climate change is that the direction of change is uncertain. For most if not all areas, we do not know whether climate change will increase or decrease water supplies. Nonetheless, it is likely that water supplies will be changed. Water supply systems can be made more robust if their storage capacity is increased or if adjacent basins are linked so there can be interbasin transfers. Use of market prices for water can reduce inefficient use of this precious resource and enable water users to quickly respond to changes in water supply conditions. Small islands and atolls may need to plan relief efforts or evacuations in case water supplies are reduced.

5.4 FUTURE RESEARCH NEEDS

Future analyses of climate change and water resources should address the likelihood of how well water resource development projects will cope with climate change. There are many plans to further develop water resource supplies in many parts of the region. These development projects should be assessed as to their capability to meet objectives under scenarios of climate change. The scenarios should incorporate changes in climate variability, particularly the timing of the monsoon and the location and timing of tropical cyclones. If the capability is compromised by climate change, analysis should be conducted on what reasonable and cost-effective modifications can be made to reduce vulnerability by increasing robustness and resiliency of water management systems.

6. National Action Plans

6.1 VULNERABILITY AND ADAPTATION SECTORS TO BE ADDRESSED

In the Philippines, the sectors that will be included in its national action plan are agriculture, water resources, coastal resources, forestry and biodiversity, health and human settlements, energy/transportation and industry, and adaptation technology assessment.

Micronesia's national action plan will address coastal resources (including marine biodiversity), fisheries (subsistence and offshore foreign fleets), fresh water resources, agriculture, forestry and biodiversity, and socioeconomic factors, including tourism and human settlement.

Indonesia is participating in the ADB sponsored study called Least-Cost Greenhouse Gas Abatement Strategy for Asia (ALGAS). At the time of the workshop, Indonesia was conducting its vulnerability and adaptation analysis, as well as identification of mitigation measures. When these phases are complete, work will begin on Indonesia's national action plan.

6.2 HOW ADAPTATION MEASURES MAY BE SELECTED

First and foremost, a country can select adaptation measures that make sense on their own regardless of climate change, which can be called "no regrets" measures. One example is using market-based systems to allocate water supplies, which results in less wasteful uses of water under current climate and enables water users to more efficiently adapt to climate change.

Policy makers tend to take a short-term view of issues affecting their countries. One response to this problem may be to include climate change considerations in the design of long-term infrastructure projects. Also, agricultural research, with 5- to 10-year time horizons, may be another appropriate arena to encourage policy makers' consideration of climate change.

6.3 GOVERNMENTAL ORGANIZATIONAL STRUCTURE FOR DEVELOPING NATIONAL ACTION PLANS

The approaches for the countries that are developing their plans are as follows:

- In the Philippines, the Inter-Agency Committee on Climate Change (IACCC) was created in 1992 to address the issues of climate change. It is co-chaired by the Department of Science and Technology and the Department of Environment and Natural Resources. Membership includes several agencies, and is being expanded to include Transportation as well as agencies with responsibility for cultural resources. The IACCC reports directly to the Philippine Council for Sustainable Development (PCSD), which comprises 14 agencies and seven NGOs. PCSD's main responsibility is to implement the Philippine Strategy for Sustainable Development (Agenda 21), and the council uses consensus-building in its operations. The PCSD reports directly to the President of the Philippines.
- In Micronesia, the U.S.-sponsored Country Study precipitated the formation of the Presidential Council on Environmental Management and Sustainable Development in 1995. The interdepartmental council, representing the four Micronesian states, meets monthly, and is chaired by the Vice President. In the absence of an environmental agency in Micronesia, the council takes on the role of coordinating across sectors, following up on UNFCCC commitments, and ensuring compliance with environmental provisions of other treaties that Micronesia has signed. Sustainable development councils are now being formed. It is at this level that NGOs, private businesses, and municipalities will be involved in climate change activities in Micronesia.
- Indonesia has established a National Committee on Climate Change, with government agencies, NGOs, and universities as members. The Indonesian government recognizes that the output of scientific research can help with informed policy decisions, and does not try to influence the group's deliberations. Each committee member has assigned responsibilities, which will be augmented upon the completion of the ALGAS project.

6.4 INTEGRATION WITH OTHER DEVELOPMENT PLANS

Several approaches for integrating national action plans with other development plans were discussed:

- In the Philippines, a number of ongoing and completed studies will need to be integrated with its national action plan, among them ALGAS, the Regional Study on Global Environmental Issues, and the EPA/IRRI Study on Global Impacts of Climate Change. With regard to mitigation measures involving alternative energy sources, coordination with the Philippine National Energy Plan will also be required.

- A problem cited by Micronesia is integration of the national action plan into other national development plans. National development plans, with their five-year time horizons, are already difficult for Micronesian policy makers to address. The time horizons of the vulnerability and adaptation analyses, 50 or 100 years, are too far in the future to be incorporated in such plans. Also, more immediate concerns push climate change further down the list of priorities.
- Indonesia views its national action plan in light of three positions. First, as a member of OPEC, Indonesia must ensure that its proposed measures do not contradict the wishes of fellow OPEC members. Second, Indonesia is a G-77 country, and so must also consider the aspirations of its fellow developing nations. Finally, Indonesia comprises 18,000 islands, and is a member of the Alliance of Small Island Nations. In this delicate position, Indonesia will be approaching its national action planning carefully. Nonetheless, Indonesia's national action planning is under way, and is being coordinated with U.S. CSP and USAID. It is being considered important to both sustainable development as well as continued economic growth.

6.5 RECOMMENDATIONS REGARDING CONTENT OF NATIONAL ACTION PLANS

In some of the panelists' opinions, the concept of "planning" still carries negative connotations in a region where market economies are dominant. Further, a national action plan carries the underlying assumption that the government can control the outcome of its policy actions.

A national action plan that looks broadly at vulnerability and adaptation should contain three sections:

- Vulnerability to climate change and potential adaptation measures; a plan based on several IPCC scenarios, with a focus on medium and long-term projects.
- GHG sources and sinks, a long-term GHG forecast under IPCC guidelines, and mitigation action plans.
- How the government will implement the action plan. This section would include recommendations such as legislative changes, emissions trading opportunities, or economic measures.

Climate change is one of a host of environmental problems affecting countries in the region. Each country will have to evaluate the tradeoffs between climate change actions and nearer-term needs (such as hunger, air and water pollution, energy demand).

Many countries still view climate change as a distant threat. For example, Indonesia's 200 million people, dealing with hunger and other pressing issues, may have difficulty incorporating climate change into their priorities; therefore, to be effective, a national action plan has to relate to the economic development plans across all sectors. Viewed in this manner, a national action plan is part of a country's total planning activity.

6.6 OPPORTUNITIES FOR REGIONAL COOPERATION

Regional cooperation is important, especially for countries that share water resources. Groups like the Association of Southeast Asian Nations (ASEAN) have environmental components, which national actions plans could build upon.

As countries move from technical studies to policies, more sharing of experience between the countries will be important to achieve results. In Africa, U.S. CSP countries started a newsletter to communicate and share experiences, and the countries in Asia and the Pacific may want to follow their approach.

7. Workshop Conclusions

There are significant vulnerabilities to climate change in the Asia and Pacific region. The major conclusions of the Regional Workshop on Climate Change Vulnerability and Adaptation in Asia and the Pacific are as follows:

- **Sea level rise is of great concern to residents of Pacific islands and coastal nations in Asia.** Asia and the Pacific countries are particularly vulnerable to sea level rise because of the large number of people living on small islands, archipelagoes, and in river deltas such as the Ganges-Brahmaputra and Mekong. These areas are already affected by storm surge flooding and other related impacts. In addition, many major Asian and Pacific Island cities such as Manila, Jakarta, Shanghai, and Hong Kong are at sea level. Coastal reefs and mangroves are also threatened by rising seas and higher temperatures. Many vulnerable cities and countries are already taking steps to cope with rising sea levels and the potential for accelerated sea level rise. Although these measures may enable populations to continue to live in low-lying coastal areas, they will do so at an increased risk of coastal flooding. Future assessments of sea level rise should include analysis of human impacts on coastal ecosystems, shoreline stabilization, and storm surge flooding from tropical cyclones.
- **Changes in average climate conditions and climate variability will have a significant effect on crop yields in many parts of Asia and the Pacific.** Crop production in South Asia, Southeast Asia, and China appears to be more vulnerable to climate change than in Northeast Asia. Low income populations dependent on isolated agricultural systems are particularly vulnerable. This includes many areas dependent on rice production, populations dependent on production of tropical crops such as tea and coconuts, and regions with limited access to agricultural markets. Changes in climate variability will affect the reliability of agricultural and livestock production in the region. Increasing population could also place stress on agricultural production in the region. The extent of adaptation depends on the affordability of adaptive measures, access to technology, land and water resource availability, and genetic diversity of crop breeding. Many agricultural and other natural resource management policies discourage effective adaptation and lead to misuse of natural resources. The region has a rich history of farmer adaptations to a variety of climatic conditions, and this broad experience can be used to help adapt Asian and Pacific agriculture to the effects of climate change.
- **Water resources are very sensitive to climate change, and there are sensitivities unique to the region.** Water resources in this region, particularly in South and Southeast Asia and the islands, are very sensitive not only to changes in temperature and precipitation, but also to changes in tropical monsoons and cyclones. For example, runoff in Nepal is affected by snowcover, the southwest monsoon, and southwest cyclones, all of which are likely to be affected by climate change. In contrast, atolls depend on a lens of fresh water below the surface of the ground as a reserve supply. This lens will probably be reduced if rainfall declines. In addition, countries such as the Philippines may experience a large increase in population and GNP, which could further strain water resources. Climate change will be an additional, but important, stress. Consideration of climate change should be incorporated in the design of water resource projects, water development plans, and other related national action plans.

- **Forests in many parts of Asia and the Pacific could be significantly affected by climate change.** A large fraction of the forests of Asia and the Pacific will undergo a significant change in ecological life zones under climate change. For example, the results of research from Thailand indicate that the potential impacts of climate change on tropical forests could be significant. Research from Siberia shows that boreal forests could be both positively and negatively affected by climate change. Warmer temperatures could result in improvement of soil conditions, but also exceed thermal tolerances of many species and result in forest fires and pests. Future assessments of climate change impacts on forests should account for CO_2 fertilization and nonclimate factors such as changes in land use.

The following recommendations resulted from the workshop:

- **Efforts to improve general circulation models should continue.** The general circulation models (GCMs) show large differences in the direction and magnitude of climate change at a regional scale. Such differences make it difficult to get policy makers to accept the results of vulnerability analyses and address adaptation; therefore, efforts to improve the resolution and accuracy of the GCMs and to resolve differences concerning predictions of regional climate change are needed.
- **Asian and Pacific countries should cooperate on the analysis of climate vulnerability and adaptation.** Cooperation among countries in the region is needed because vulnerabilities are not confined to national boundaries. River basins such as the Ganges-Brahmaputra and Mekong are vulnerable to climate change and traverse national borders. In addition, many nations in the region share similar problems such as vulnerability to tropical cyclones and storm surges.
- **Asian and Pacific countries should address adaptation to climate change.** Adaptation measures designed to anticipate the potential effects of climate change can help offset many of the negative effects. Many countries are addressing adaptation through the development of national action plans on climate change. These plans will be developed in cooperation with natural resource and other agencies across governments, and should also be developed in consultation with key stakeholders and the public. These plans will feed into the writing of the national communications.

8. Summary

Many uncertainties remain about the vulnerability of the Asia and Pacific region to climate change, and much research on vulnerability and adaptation is needed. The assessments presented at this workshop indicate that climate change is of great concern to the region. The studies present a sound basis for addressing the mitigation of greenhouse emissions and the inevitable need to adapt to climate change. The results of these assessments need to be presented to policy makers and the general public. Assessments of vulnerability should continue in order to improve our understanding of the vulnerabilities of Asia and the Pacific to climate change and how the region can adapt to these changes.

Acknowledgments

We acknowledge the 100 participants in the workshop, whose active participation in the discussions made it easy for us to write this paper.

CLIMATE CHANGE IN ASIA: A REVIEW OF THE VULNERABILITY AND ADAPTATION OF CROP PRODUCTION

A. IGLESIAS,[1] LIN ERDA,[2] and C. ROSENZWEIG[3]
[1]*INIA, Madrid, Spain*
[2]*Chinese Academy of Agricultural Sciences, Beijing 100081, China*
[3]*NASA/Goddard Institute for Space Studies, U.S.A.*

Abstract. A number of studies have provided quantitative assessments of the potential climate change impacts on crop production in Asia. Estimates take into account (a) uncertainty in the level of climate change expected, using a range of climate change scenarios; (b) physiological effects of carbon dioxide on the crops; and (c) different adaptive responses. In all cases, the effects of climate change induced by increased atmospheric carbon dioxide depended on the counteracting effects among higher daily evapotranspiration rates, shortening of crop growth duration, and changes in precipitation patterns, as well as the effects of carbon dioxide on crop growth and water-use efficiency. Although results varied depending on the geographical locations of the regions tested, the production of rice (the main food crop in the region) generally did not benefit from climate change. In South and Southeast Asia, there is concern about how climate change may affect El Niño/Southern Oscillation events, since these play a key role in determining agricultural production. Furthermore, problems arising from variability of water availability and soil degradation are currently major challenges to agriculture in the region. These problems may be exacerbated in the future if global climate change projections are realized. Many studies have considered strategies for improving agricultural management, based on the optimization of crop management decisions. Climate change analyses could be further strengthened by economic studies that integrate the potential use of natural resources across sectors.

Key words: Asia, agriculture, El Niño

1. Introduction

1.1 BACKGROUND AND AIMS OF THIS REVIEW

Scientists predict significant global warming in the coming decades due to increasing atmospheric carbon dioxide and other trace gases (IPCC, 1990a; 1992; 1996a). Substantial changes in hydrological regimes are also forecast (IPCC, 1996a). Understanding the potential effects of these changes on agriculture is critical because agriculture provides food for the world's population, now estimated at 5 billion and projected to rise to 8.5 billion by 2025 (International Bank for Reconstruction and Development/World Bank, 1990; United Nations, 1989). A large portion of this estimated population increase will take place in Asia.

Despite technological advances such as improved crop varieties and irrigation systems, weather and climate are still the key factors in agricultural productivity. For example, weak monsoon rains in 1987 caused large shortfalls in crop production in India, Bangladesh, and Pakistan, forcing these countries to import wheat. The close links between agriculture and climate, and the international nature of food trade and food security, highlight the need to consider the impacts of climate change in an interactive context.

In Asia, a substantial amount of research related to the impacts of climate change on crop production has focused on regional assessments using dynamic crop growth models, usually based on the simulation of the potential yield changes of a single crop. This review aims to increase understanding of potential simultaneous changes in production in the major

Water, Air, and Soil Pollution **92:** 13–27.
© 1996 *Kluwer Academic Publishers. Printed in the Netherlands.*

food-producing areas of the region by comparing and contrasting the available information, and thus analyzing the relative interregional competitiveness of crop production. The review also analyzes the potential for adaptation to climate change. Many of the selected studies centered on rice production because of the importance of rice to the food security of Asia.

1.2 AGRICULTURE IN THE REGION AND MAJOR CAUSES OF CONCERN

The enormous continent of Asia can be divided into the regions of South and Southeast Asia (Pakistan, India, Bangladesh, Thailand, Myanmar, Viet Nam, Cambodia, Malaysia, Indonesia, and the Philippines) and East Asia (China, Taiwan, North and South Korea, and Japan).

Agriculture is an important sector in South and Southeast Asia. For example, in Pakistan, agriculture accounts for about 30% of the GDP and employs over 50% of the labor force (Qureshi and Iglesias, 1994). Agriculture is a key contributor to the economy of Thailand: agricultural systems (arable crops, rangelands, forestry, and fisheries) employ over 60% of the labor force during the cropping season and account for about 20% of the national GDP; their products are the major export (Tongyai, 1994). Agriculture is considered to be the economic lifeline of the archipelago nation of the Philippines (Escaño and Buendia, 1994). More than 50% of the working population is engaged in agriculture, and more than 70% of foreign exchange earnings is derived from exports of agricultural products. Rice is the staple food crop of the Philippines, and is grown on 3.4 million ha of land. Crops such as tea, cocoa, coffee, and rubber are exported by some countries in the region.

Rice is by far the most important food crop across the humid areas of South and Southeast Asia. Most production is in paddy rice fields, farmed by smallholders, where relatively high yields are achieved because of new varieties and fertilizer. Upland and rainfed rice generally achieve lower yields. Improved rice varieties, developed primarily at the International Rice Research Institute (IRRI) of the Consultative Group for International Agriculture Research (CGIAR), and greatly increased fertilizer applications have raised rice yields in the region. In arid and semiarid areas, a wide variety of other crops are grown, including soybeans, maize, wheat, roots, tubers, fruits, and vegetables. Production is often highly dependent on irrigation; in fact, irrigation here is used on a higher proportion (about 70%) of the agricultural land than anywhere else in the world.

Agriculture in South and Southeast Asia is very vulnerable to weather-related hazards, especially tropical cyclones, storm surges, floods, and droughts, which damage life, property, and crop production. In January 1991, the Philippine weather bureau (PAGASA) reported that after the onset of an ENSO-related drought, rice and maize crops suffered estimated damages of $753 million (Escaño and Buendia, 1994). Seasonal monsoons are a dominant climate feature. The shallow waters off Malaysia and Indonesia are a key part of the ENSO processes. Changes in these dominant climate features would have important impacts on agriculture in this area. Climate change also raises the potential for increases in sea level, leading to flooding of low-lying coastal regions that support high populations and extensive agricultural production (e.g., in Bangladesh). A decrease in rice production due to climate effects and sea level rise, combined with rapidly increasing population, would threaten food security.

For East Asia, rice, wheat, and maize are the three leading food crops. The region accounts for 40% of the world rice area. More than 90% of the rice area is irrigated. Food grain production averages about 2 t ha^{-1} in the warm arid and semiarid subtropical agroecological zones, about 4 t ha^{-1} in the warm subhumid areas, over 6.5 t ha^{-1} in the humid areas, and about 2 t ha^{-1} in the cool subtropics (IRRI, 1993). China has only 7% of the world's cultivated land, but supports more than one-fifth of the world's population. China is the largest rice producer and consumer in the world. Increasing population, spread of urbanization, lack of water resources, and environmental pollution, however, are pressures that may hinder growth in China's agricultural productivity in the future. The important agroecological zones for East Asia are (1) warm arid and semiarid subtropical; (2) warm subhumid subtropical; (3) humid subtropical; and (4) cool subtropical with summer rain.

2. Methods

2.1 DESIGN OF THE IMPACT STUDIES

Most regional studies are designed with a suite of climate change scenarios, dynamic process crop models, and simulation experiments. Climate change scenarios comprise climate sensitivity tests and scenarios devised from general circulation models (GCMs). The crop models respond to climate changes and to the direct physiological effects of increased atmospheric CO_2 on crop growth and water use. They also allow for the simulation of both rainfed and irrigated agricultural systems, and potential farmer adaptations to climate change.

2.2 CLIMATE SCENARIOS

Climate scenarios are sets of climatic perturbations that are used with impact models to test the sensitivity of the system to the projected changes. The design of an impact study should include more than one scenario, so that a range of possible effects may be defined. It is also preferable for the climate change scenarios to be internally consistent (Wigley, 1987). This means that climate variables within the scenario should vary in a physically realistic way. For example, surface air temperature and solar radiation tend to be linearly correlated, and should be in the climate change scenario.

Scenarios are often devised by changing an original set of climatological data according to prescribed anomalies. These anomalies may be derived from historical climate or from GCMs. The studies included in this review consider either doubled CO_2 GCM climate change scenarios and/or sensitivity tests for the crop modeling analysis. Table I shows, as an example, the global and regional $2{\times}CO_2$ sensitivity of three GCMs widely used in agricultural impacts. In general, GCM projections agree in the temperature changes in a given region, but differ greatly in the precipitation changes. For example, in southern China, the average annual changes in precipitation projected by the Goddard Institute for Space Studies (GISS) scenario are +18%, but site differences range from -20 to +80%.

An alternative approach used to analyze the possible impacts of different climate on crop yield is to specify incremental changes to temperature and precipitation and to apply these changes uniformly to the baseline climate. This type of sensitivity study allows the consideration of the question: "What type, magnitude, and rate of climate change would seriously perturb the agricultural system in question?"

TABLE I
GCM climate change global and regional sensitivity. Annual temperature differences ($2\times CO_2$ GCM - $1\times CO_2$ GCM) and precipitation changes (% change of $2\times CO_2$ GCM from $1\times CO_2$ GCM).

Geographical Area[§]	Model[†]	Change in Average Temperature (°C)	Change in Average Precipitation (%)
Global	GISS	4.2	11
	GFDL	4.0	8
	UKMO	5.2	15
Pakistan	GISS	4.3	16
	GFDL	4.7	14
	UKMO	6.4	4
Bangladesh	GISS	3.4	10
	GFDL	2.8	33
	UKMO	4.3	41
Philippines	GISS	3.2	-10
	GFDL	2.5	5
	UKMO	3.0	-4
South China	GISS	4.6	18
	GFDL	4.2	23
	UKMO	5.6	15
Japan	GISS	3.3	4
	GFDL	4.2	15
	UKMO	5.3	-2

[§] Global changes include all gridpoints. Regional changes include the gridpoints in the agricultural regions.
[†] GISS (Hansen *et al.* (1983); resolution 7.83° lat. × 10.0° long.); GFDL (Manabe and Wetherald (1987); resolution 4.4° lat × 7.5° long.); UKMO (Wilson and Mitchell (1987); resolution 5.0° lat. × 7.5° long.).

2.3 CROP MODELS

Many of the results summarized in this review are based on the simulation of crop responses to climate using process-based crop models. The process-based models employ simplified functions to express the interactions between the growth of crops and the major environmental factors that affect it (i.e., climate, soils, and management). Many of the crop models developed during the last decades have been applied in climate impact assessment. Recently, efforts have been made to provide descriptions of the available models for crop production (i.e., by IGPB Programmes) and to organize model intercomparison exercises. The International Benchmark Sites Network for Agrotechnology Transfer (IBSNAT, 1995) has standardized model structure, and provides models in a single package that includes a decision support system for the evaluation of management alternatives.

The IBSNAT crop models have been used in a wide range of climate impact assessments (e.g., Rosenzweig and Iglesias, 1994). Models are available for wheat (Godwin *et al.*, 1990a), maize (Jones and Kiniry, 1986), paddy and upland rice (Godwin *et al.*, 1990b; Singh *et al.*, 1993), soybean (Jones *et al.*, 1989), and other crops. The models have been validated over a wide range of environments (e.g., Otter-Nacke *et al.*, 1986) and are not specific to any particular location or soil type. Thus they are suitable for use in wide-geographical area studies in which crop-growing conditions differ greatly. Furthermore, because management practices such as cultivar, planting date, plant population, row spacing, and sowing depth may be varied in the models, they permit experiments that simulate management adjustments by farmers to climate change.

Modeled processes in the IBSNAT models include phenological development, i.e., duration of growth stages, growth of vegetative and reproductive plant parts, extension growth of leaves and stems, senescence (aging) of leaves, biomass production and partitioning among plant parts, and root system dynamics. Some crop models also have the capability to simulate the effects of nitrogen deficiency and soil-water deficit on photosynthesis and pathways of carbohydrate movement in the plant. The models require daily values for solar radiation, maximum and minimum temperature, and precipitation. Soil data needed are values for the functions of drainage, runoff, evaporation and radiation reflection, soil water-holding capacities and rooting- preference coefficients for each soil layer, and initial soil water content.

2.4 EFFECTS OF CO_2 ON CROPS

Most plants growing in atmospheric CO_2 higher than ambient exhibit increased rates of net photosynthesis (i.e., total photosynthesis minus respiration) (see Pickering, 1995 for a review). High CO_2 also reduces the stomatal openings of some crop plants. By so doing, CO_2 reduces transpiration per unit leaf area while enhancing photosynthesis. Thus, it often improves water-use efficiency (the ratio of crop biomass accumulation or yield and the amount of water used in evapotranspiration). Experimental effects of CO_2 on crops have been reviewed by Acock and Allen (1985) and Cure (1985). In a compilation of greenhouse and other experimental studies, Kimball (1983) estimated a mean crop yield increase of 33 ±6% for a doubling of CO_2 concentration from 300 to 600 ppm.

To project the impact of increasing CO_2 on agricultural production, these beneficial direct effects should be considered along with the climatic effects of the radiatively active trace gases. The assessment of the relative contributions of the direct effects of CO_2 and the predicted climate changes to agricultural crop responses remains a crucial research question.

3. Limitations

This section outlines the main limitations and sources of uncertainty of the studies included in the review. Among them the most important are those derived from the climate change scenarios, the lack of consideration of climatic variability in the studies, the limitations of the crop models, and the uncertainties of the effects of CO_2 and other environmental factors on the crops. Finally, a major limitation of the studies is derived from the lack of integration of the crop studies with other sectors.

3.1 GCM SCENARIOS, CLIMATE VARIABILITY, AND ENSO EVENTS

Current general circulation models employ grids of ~4° latitude by ~5° longitude. At this resolution, many smaller scale elements of climate are not properly represented, such as warm and cold fronts and hurricanes, as well as the diversities of ecosystems and land use.

Accurate modeling of hydrological processes is particularly crucial for determining climate change impacts on agriculture, but GCM simulation of infiltration, runoff, evaporation, and other hydrological processes is highly simplified. Precipitation, in particular, is poorly represented in the GCM results (Grotch, 1988). High or low pressure systems may be misplaced or even absent, and monsoonal circulations are often poorly estimated.

There is also uncertainty in the prediction of rate and magnitude of climate change. Ocean heat transport is a key, though not well understood, process that affects how fast the climate may warm. Most of the climate change simulations to date have assumed an abrupt doubling of the CO_2 concentration in the atmosphere and have then allowed the simulated climate to come to a new equilibrium. The more recent time-dependent scenarios derived from simulations of transient trace gas and aerosol emissions will enhance the vulnerability assessments, as methodologies are developed to simulate transient crop responses.

None of the studies considered in this review include the evaluation of impacts on crop production with altered climatic variability. Such changes could affect plant functioning (Panturat and Eddy, 1989; Kuiper, 1993; Mearns *et al.,* 1992, 1996). ENSO events, which tend to recur every two to nine years, are related oceanic and atmospheric phenomena, characterized by increases in sea-surface temperatures of the tropical Pacific Ocean, suppression of upwelling nutrient-rich water along the coast of South America, and disruption of the trade winds. The cycle has long been known to be a large component of natural interannual climate variability in the Asian tropics and subtropics, and to a lesser extent, in the midlatitude regions. Changes in ENSO event frequency and severity would affect the agriculture of the region.

3.2 OTHER ENVIRONMENTAL FACTORS

Other environmental changes will interact with changes in climatic variables and elevated CO_2 to affect crop yields. Among these are exposure to O_3 (ozone), surface concentrations of which have doubled in the last 100 years in the Northern Hemisphere to a level that is estimated to reduce yields in the range of 1% to 30% (e.g., Ashmore, 1988; Bosac *et al.,* 1993), and exposure to UV-B radiation that is expected to increase due to stratospheric ozone depletion. Tropospheric aerosols (SO_2) will also affect crop growth and yield.

3.3 CROP GROWTH MODELS AND DIRECT CO_2 EFFECTS

The crop models contain many simple, empirically derived relationships that do not completely mimic actual plant processes. These relationships may or may not hold under differing climatic conditions, particularly the higher temperatures predicted for global warming. Other simplifications of the crop models are that weeds, diseases, and insect pests are controlled; there are no problem soil conditions such as high salinity or acidity; and response to catastrophic weather events such as heavy storms is not included. The crop models simulate the current range of agricultural technologies available around the world; they do not include potential improvements in such technology, but may be used to test the effects of some potential improvements, such as improved varieties and irrigation schedules.

The results of a large number of experiments have confirmed a beneficial effect of elevated CO_2 on crops (IPCC, 1996b). Although the mean response (+30% for C3 crops) has been confirmed, variations in crop responsiveness among plant species exist (from -10 to +80%). Response depends on availability of plant nutrients, temperature, and water stress, as well as differences in experimental technique (IPCC, 1996b). Furthermore, field responses may differ from experimental responses (Korner, 1990).

3.4 INTEGRATED MODELING

Development and broad application of integrated agricultural modeling that considers interactions of biophysical and socioeconomic factors (particularly applicable at the regional scale) warrant increased attention. Validation, testing, and comparison of alternative approaches are also needed (IPCC, 1994). Climate effects on soils and plant pests, other environmental changes, adaptation options, and economic responses should be an integrated part of the studies rather than treated on an *ad hoc* basis or as separate modeling exercises (IPCC, 1994). Inclusion of these multiple, joint effects may significantly change our estimates of "mean" impacts.

4. Results And Discussion

4.1 VULNERABILITY OF CROP PRODUCTION TO CLIMATE CHANGE

This section analyses the sensitivity of different components of the agricultural systems in Asia to changes in climate conditions. In general, the published studies focus on crop production changes under climate change scenarios. Only a limited number of studies have included farm-level adaptations (e.g., changes in planting dates) to climate change. Furthermore, very few studies have evaluated the economic or social effects of changes in crop production. The recent IPCC Impacts Assessment (1996b) includes most of the published studies in the region (Tables II and III).

4.1.1. South and Southeast Asia

Matthews *et al.* (1995) estimated the impacts on rice yields for many countries in the region. GCM equilibrium climate change scenarios predicted temperature increases and, for 60% of the area of South, Southeast, and East Asia, precipitation increases. The results show substantial variation in impact across the entire Asian region among the GCMs (+30 to -38% of rice yields). Decreased rice yields were projected for low latitude areas in south and southeast Asia, and increased yields were projected for higher latitudes; such results indicate a possible shift in the rice-growing regions away from the equatorial regions to higher latitudes. Temperature effects alone were generally found to reduce yields, but CO_2 fertilization was a significant positive effect.

Qureshi and Iglesias (1994) used GCMs and dynamic crop models to estimate the potential agricultural effects of climate change in Pakistan. Under present climate conditions, wheat is currently under stress due to high temperatures and arid conditions. Projected climate change caused simulated wheat yields to decrease dramatically in the major areas of agricultural production, even under fully irrigated conditions. Decreases in modeled grain yields were caused primarily by temperature increases that shortened the duration of the life cycle of the crop, particularly the grain-filling period. These decreases were somewhat counteracted by the beneficial physiological CO_2 effects on crop growth. Adaptation

TABLE II
Summary of some recent impact studies in South and Southeast Asia

Study	Scenario	Geographic Scope	Crops	Yield Impact (%)	Comments
Rosenzweig and Iglesias, 1994[§]	GCMs	Pakistan India Bangladesh Thailand Philippines	Wheat Wheat Rice Rice Rice	-61 to +67 -50 to +30 -6 to +8 -17 to +6 -21 to +12	UKMO, GFDL, GISS, and +2, +4°C and +/- 20% precip. Range is over sites and GCM scenarios with direct CO_2 effects; scenarios w/o CO_2 and w/ adaptation also were considered. CO_2 effects important in offsetting losses of climate-only effects; adaptation unable to mitigate losses.
Parry *et al.*, 1992	GISS	Indonesia	Rice Soybean Maize	apprx. -4 +10 to -10 -25 to -65	Indonesia: Low estimates consider adaptation; also estimated overall loss of farmer income ranging from $10 to $130 annually.
		Malaysia	Rice Maize Oil Palm Rubber	-12 to -22 -10 to -20 increase -15	Malaysia: Maize yield affected by reduced radiation (increased clouds); variation in yield increase; range is across seasons.
		Thailand	Rice	5 to 8	
Matthews *et al.*, 1995	3 GCMs	India Bangladesh Indonesia Malaysia Myanmar Philippines Thailand	Rice	-12 to +23 -12 to -2 -6 to +22 +21 to +26 -9 to +30 -2 to +12 -20 to -34	GISS, GFDL, UKMO scenarios; included the direct CO_2 effects; range is across GCMs; varietal adaptation was shown to be capable of ameliorating detrimental effects of a temperature increase in currently high temperature environments.

[§] Country studies were by Qureshi and Iglesias, 1994; Rao and Sinha, 1994; Karim *et al.*, 1994; Tongyai, 1994; Escaño and Buendia, 1994, for Pakistan, India, Bangladesh, Thailand, and the Philippines, respectively. As reported in IPCC, 1996b.

strategies that included shifting cultivars and delaying planting mitigated the simulated yield losses partially, but not totally.

Studies of climate change impacts on wheat and sorghum production in India (Rao and Sinha, 1994; Rao *et al.,* 1995) also found that yields would generally decrease, although responses varied by crop and season. Wheat yield decreases could have a serious impact on food security of the country, in view of the increasing population and its demand for grains. Most of the wheat production in India comes from the northern plains, where it is almost impossible to increase the present area of wheat under irrigation. Sorghum is relatively insensitive to temperature, and simulated yields were not affected as much by climate change scenarios as wheat.

In Bangladesh, climate change raises the potential for increases in sea level leading to flooding of low-lying coastal regions, which support high populations and extensive agricultural production. A decrease in rice production due to climate effects and sea level rise, combined with rapidly increasing population, would threaten the food security (Karim *et al.,* 1994). Brammer *et al.,* (1993) also found negative effects on rice production in the country.

The U.N. Environmental Programme sponsored a project that considered the potential socioeconomic effects of climate change in Southeast Asia (Parry *et al.,* 1992). A range of yield impacts (-65 to +10%) across Thailand, Indonesia, and Malaysia and across seasons

TABLE III
Summary of recent current impact studies in East Asia

Study	Scenario	Geographic Scope	Crops	Yield Impact (%)	Comments
Zhan Tao, 1992	2×CO_2; +1°C	China	Wheat Rice Cotton, Fruits, Oil Crops, Potatoes, Corn	-8 -6 +1 to -4	Agricultural productivity loss, >5%; included direct effects; positive effects in NE and NW; negative in most of the country; no change in SW
Zhang Houxuan, 1993	+1.5°C	South of China	Rice	-7 to -11	Double crop; included CO_2 effect
Jin Zhiquin *et al.*, 1994	GCMs	South of China	Irrigated Rice	-37 to +15	Range across GISS, GFDL, UKMO scenarios; no consideration of enrichment effects of CO_2
Sugihara, 1991	2×CO_2; +3°C	Japan	Rice	+10	
Suyama, 1988	+2°C	Japan	Temperate Grass	-10 to +10	Average +5.6% in productivity for grass; included direct CO_2 effects
Yoshino, 1991		Japan	Sugar Cane	-8	Rainfall was reduced 25% to 30% May to October
Seino, 1994	GISS, GFDL, UKMO	Japan	Rice Maize Wheat	-11 to +12 -31 to +51 -41 to +8	Impacts vary by GCM scenarios and area; included direct CO_2 effects; generally positive in north, negative in south
Horie, 1993	GCMs	Hiroshima and Akita, Japan	Rice	-45 to +30	Range based on different crop models, GCMs, and across sites; included direct CO_2 effects
Matthews *et al.*, 1995	GCMs	S. Korea M. China Taiwan Japan	Rice	-1 to +23 -24 to +25 +7 to +23 -14 to +12	GISS, GFDL, UKMO scenarios; included direct CO_2 effects; range is across GCMs; varietal adaptation was shown to ameliorate the negative effects

As reported in IPCC, 1996b.

culminated in an overall loss to farmer income of US$10 to US$130 per year. Coastal inundation was also estimated to be a threat to rice and to fish, prawn, and shrimp ponds. A 1 m sea level rise could cause landward retreat of 2.5 km in Malaysia and would threaten 4200 ha of productive agricultural land, an area equal to about 0.63% of Malaysia's paddy rice area (0.61% of total cereal area). Under GISS 2×CO_2 climate, erosion rates in Malaysian river basins increased from 14 to 40% across three watersheds considered, and average soil fertility declined by 2 to 8%.

In Thailand, Tongyai (1994) tested GCM climate change scenarios on simulated upland and paddy rice and found that yields decreased by -2 to -17% in two of the scenarios (with direct CO_2 effects included); the third GCM scenario implied little change (-1 to +6%). Escaño and Buendia (1994) found that simulated rice yields in the Philippines declined under GCM scenarios with direct CO_2 effects taken into account in at least one important agricultural region of the Philippines.

4.1.2. East Asia

Possible climatic impacts in East Asia span a wide range depending on climate scenario and location. For China, results show generally negative yield effects, but these range from less than -10% (Zhang, 1993) to over -30% (Jin *et al.*, 1994). The Matthews *et al.* (1995) study found that rice yields in China varied across region and across GCM $2\times CO_2$ scenario. While finding large changes for all of China, Hulme *et al.* (1992) conclude that warming would be beneficial, with increasing yield due to a diversification of cropping systems. Studies for Japan also show a wide variation, but the range of results is centered near zero or toward yield increases.

Impact studies in Asia also indicate that warming would cause a general northward movement of agroclimatic regions, with certain exceptions in the south in some areas where the moisture deficit may increase more than in the north. The general possibility of increased summer dryness in the continental midlatitudes suggests the following six areas of China as most likely to be negatively affected by climate change (Lin Erda, 1994): the area around the Great Wall lying southeast of the transition belt between crop agriculture and animal husbandry; the Huang-Hai Plains, where dryland crops such as wheat, cotton, corn, and fruit trees are grown; the area north of the Huai river, including eastern Shandong, which lies along the south edge of the temperate zone; the central and southern areas of the Yunnan Plateau; the middle and lower reaches of the Yangtze River; and the Loess Plateau.

In general, these areas would be at heightened risk of drought and would suffer potential increases in soil erosion. The Yunnan Plateau, with generally abundant rainfall, is subject to alternating droughts and waterlogging; production here would be sensitive to changes that would increase variability of climate. Indices of vulnerability based on physical productivity and socioeconomic capability to adapt show that among China's 30 provinces, Shanxi, Inner Mongolia, Gansu, Hebei, Qinhai, and Ningxia are particularly vulnerable and less able to adapt to climate change. These seven provinces produced 12% of China's total agricultural output value in 1990 (Editorial Board, 1991). Changes in climate will occur against a steadily increasing demand for food in China over the next 55 years (Lu and Liu, 1991a; 1991b). More intensive use of inputs is likely to exacerbate land degradation and other environmental problems.

For Japan, several climate change impact studies have examined both major production areas and vulnerable regions. The studies indicate that the positive effects of CO_2 on yields would generally more than offset the negative climatic effects in the central and northern parts of Japan, leading to yield gains. In the southwestern part of Japan, particularly in Kyushu, the yield effects were, on balance, estimated to be negative. Uchijima (1987), Horie (1987), and Uchijima and Seino (1988) have shown that warming could expand the total area suitable for agricultural production and increase potential rice yields.

4.2. ADAPTATION OF CROP PRODUCTION TO CLIMATE CHANGE

Only a limited number of studies in Asia have evaluated adaptation in a systematic way. The dynamic reactions of farmers to changes in environmental conditions (i.e., autonomous adaptation, without government intervention) should be part of all crop vulnerability assessments. The system's potential for adaptation should also include the evaluation of changes in technologies and policies that can be taken to prepare for or respond to climate change.

The adaptation strategies tested involved changes in current management practices (e.g., planting date, fertilizer, and irrigation) and changes in crop variety to either existing varieties or hypothetical new varieties. In addition, some studies have considered the expansion of crops (e.g., rice in China) to areas that are temperature-limited under current conditions.

Changes in planting date. The most likely response of farmers to warmer temperatures would be to plant earlier to utilize the cooler early season and to avoid high temperatures during the grain-filling period. A relatively small change in planting date, perhaps up to four weeks, should be easily supportable, but with longer shifts in the sowing date, soil moisture may be either too wet or too dry, and amounts of solar radiation in the region may be lower at planting. In some cases, such as in the Philippines (Escaño and Buendia, 1994) and China (Jin *et al.,* 1994), the sensitivity tests on planting date suggest large changes in seasonal agricultural activities, implying major changes in the agricultural systems. For example, farmers in Batac, the Philippines, report that planting one to two months earlier may not be appropriate for rice because strong winds later may coincide with the grain-filling stage (Escaño and Buendia, 1994).

Irrigation. The crop model simulations in India and Pakistan suggested that climate change may bring significant increases in the need for irrigation to compensate for moisture losses caused by increased evapotranspiration. The amount of irrigation water used for crops already under irrigation would increase, and there may be a switch from rainfed to irrigated production, if economic and water resources are available. The potential problems associated with increased irrigation as an adaptive strategy are the questionable availability of water resources, the associated costs, and the environmental problems of soil salinization and water pollution.

Changes in fertilizer. An increase in the amount of fertilizer applied can compensate in some cases for yield losses caused by climate change. This strategy may have significant implications for future fertilizer use under climate change conditions.

Changes in crop variety. Since most areas are predicted to experience substantially higher growing season temperatures, substituting varieties with higher thermal requirements for currently grown varieties could be an important adaptation measure. Some researchers have also considered hypothetical new varieties genetically adapted to climate change conditions. While this is a useful evaluation for establishing new breeding objectives, the option has to be carefully evaluated by plant breeders, considering the genetic pool available. For rice, spikelet sterility emerged as a major factor determining the differential predictions of yield; however, genetic variability suggests that it may be relatively easy to adapt varieties to new climate conditions (Matthews *et al.*, 1995).

Expansion of cropping area. Jin (1993) estimates that climate change would bring significant shifts in the rice-cropping pattern in China, based on the extension of the growing period and increased thermal regime during the rice-growing season (Gao and Jin, 1987). An increase in temperature will expand the northern limits for single-rice, double-rice, and triple-rice cropping systems by 5° to 10° of latitude, depending on the scenario (Jin *et al.*, 1995). There would be increased sowing of *indica* rice now grown in southern China, replacing the current *japonica* types.

5. Conclusions

A large number of climate change impact studies have been conducted in many regions of South, Southeast, and East Asia. However, the impact of climate change on crop production remains uncertain, due to uncertainties not only in climate projections but also in the understanding of key processes in crop production, such as the direct effects of CO_2 and the complex interactions with water resources. Despite these significant uncertainties, and the limitations of the modeling studies outlined above, the following conclusions can be drawn:

1. Crop yield and productivity changes will vary considerably across the region. Thus, the pattern of agricultural production is likely to change. Based on the crop impact studies using $2\times CO_2$ equilibrium GCM scenarios, lower latitude countries have been shown to be more negatively affected than higher latitude countries. Nevertheless, crop yield simulation results vary widely, e.g., ±20% changes in yield, for specific sites, countries, and GCM scenarios.

2. Adaptation strategies that consider changes in crop varieties or in the timing of agricultural activities imply low costs and, if readily undertaken, can compensate for some of the yield loss simulated with the climate change scenarios. Adaptation to climate change is likely in areas that are currently less climatically stressed; the extent depends on the affordability of adaptive measures, access to technology, and biophysical constraints such as water resource availability, soil characteristics, and genetic diversity for crop breeding (especially crucial is the development of heat-resistant rice cultivars).

3. Because of the key role of ENSO events in determining variability of agricultural production in South and Southeast Asia, changes in ENSO event frequency and severity would affect the agriculture of the region.

While global agricultural production appears to be sustainable under climate change as expressed by GCMs under doubled CO_2 (Rosenzweig and Parry, 1994), the studies reviewed here suggest that the regions of South and Southeast Asia appear to be among the more vulnerable; East Asia appears to be relatively less vulnerable. Furthermore, global studies have shown incremental costs of agricultural production and additional risk of hunger under climate change that could create a serious burden for some developing countries in the region. The resiliency of the agricultural sector depends on improved use of systems for monitoring weather, soil moisture, nutrient requirements, and pest infestations, and on adequate agricultural research capacity. Finally, strong links among the agricultural research, production, and policy sectors are essential to build appropriate preparedness for climate change in Asia, as elsewhere.

Acknowledgment

We thank the U.S. Country Studies Program for their support of our research on agricultural vulnerability and adaptation to climate change.

References

Acock B. and Allen, Jr. L.H.: 1985, Crop responses to elevated carbon dioxide concentrations, in: *Direct Effects of Increasing Carbon Dioxide on Vegetation*, Strain B.R. and Cure J.D. (eds), DOE/ER-0238, U.S. Department of Energy, Washington, DC, pp. 53-97.

Ashmore M.S.: 1988, *Atmos. Environ.* **22**, 2060.

Bosac C., Black V.J., Black C.R., Roberts C.R., and Lockwood F.: 1993, *New Phytol.* **124**, 439.

Brammer H., Asaduzzaman M., and Sultana P.: 1993, Effects of Climate and Sea-Level Changes on the Natural Resources of Bangladesh, Briefing Document No. 3, Unnayan Parishad, Dhaka, Bangladesh.

Cure J.D.: 1985, Carbon dioxide doubling responses: a crop survey, in: *Direct Effects of Increasing Carbon Dioxide on Vegetation,* Strain B.R. and Cure J.D. (eds), DOE/ER-0238, U.S. Department of Energy, Washington, DC, pp. 33-97.

Editorial Board: *1991, China Agriculture Yearbook,* Beijing, China.

Escaño C.R. and Buendia L.V.: 1994, Climate impact assessment for agriculture in the Philippines: simulation of rice yield under climate change scenarios, in: *Implications of Climate Change for International Agriculture: Crop modeling Study*, Rosenzweig C. and Iglesias A. (eds), U.S. Environmental Protection Agency, Washington, DC.

Gao L., Li L., and Jin Z.Q.: 1987, *Agricultural Forest Meteorology* **39**, 55.

Godwin D., Ritchie J., Singh U., and Hunt, L.: 1990a, *A User's Guide to CERES-Wheat, Version 2.10*, International Fertilizer Development Center, Muscle Shoals, Alabama, 94 pp.

Godwin D.C., Ritchie J.T., Singh U., and Hunt L.: 1990b, *A User's Guide to CERES-Rice, Version 2.10*, International Fertilize Development Center, Muscle Shoals, Alabama, 86 pp.

Grotch S.L.: 1988, *Regional Intercomparisons of General Circulation Model Predictions and Historical Climate Data*, DOE/NBB-0084, U.S. Department of Energy, Washington, DC.

Hansen J., Russell G., Rind D., Stone P., Lacis A., Lebedeff S., Ruedy R., and Travis L.: 1983, Efficient three-dimensional global models for climate studies: models I and II, *Monthly Weather Review* **111**, 609.

Horie T.: 1987, The effects on rice yields in Hokkaido, in: *The Impact of Climatic Variations on Agriculture. Vol. 1. Assessment in Cool Temperate and Cold Regions,* Parry *et al.* (eds), Kluwer Academic Publishers, Dordrecht, The Netherlands, pp. 809-825.

Horie T.: 1993, *Journal of Agricultural Meteorology* (Tokyo) **48**, 567.

Hulme M., Wigley T., Jian T., Zhao Z., Wang F., Ding Y., Leemans R., and Markham A.: 1992, Climate Change due to the Greenhouse Effect and its Implications for China, CRU/WWF/SMA, World Wide Fund for Nature, Gland, Switzerland.

IBSNAT (International Benchmark Sites Network for Agrotechnology Transfer Project): 1995, *Decision Support System for Agrotechnology Transfer Version 3.0 (DSSAT V3.0)*, Hawaii Dept. Agronomy and Soil Sci., College of Tropical Agr. and Human Resources, University of Hawaii, Honolulu.

International Bank for Reconstruction and Development/World Bank: 1990, *World Population Projections*, Johns Hopkins University Press, Baltimore, Maryland.

IPCC (Intergovernmental Panel on Climate Change): 1990a, *Climate Change: The IPCC Impacts Assessment,* W.J. McG. Tegart, Sheldon G.W., and Griffiths D.C. (eds), World Meteorological Organization and United Nations Environmental Programme, Bracknell, United Kingdom.

IPCC: 1990b, *Climate Change: The IPCC Impacts Assessment*, Tegart, W.J. McG., Sheldon G.W., and Griffiths D.C. (eds), Australian Government Publishing Service, Canberra, Australia.

IPCC: 1992, *Climate Change 1992*, Houghton, J.T., Callander B.A., and Varney S.K. (eds), The Supplementary Report to the IPCC Scientific Assessment, Cambridge University Press, Cambridge, England.

IPCC: 1994, *IPCC Technical Guidelines for Assessing Climate Change Impacts and Adaptations*, Carter T.R., Parry M.L., Harasawa H., and Nishioka S., WMO and UNEP, Center for Global Environmental Research, University College of London, London.

IPCC: 1996a, *Climate Change 1995: Impacts, Adaptations, and Mitigation of Climate Change: Scientific-Technical Analyses*, Watson R.T., Zinyowera M.C., and Moss R.H. (eds), Contribution of Working Group II to the Second Assessment Report of the Intergovernmental Panel on Climate Change, University Press, Cambridge, 890 pp.

IPCC: 1996b, *Climate Change 1995: the IPCC Second Assessment Report, Volume 2: Scientific-Technical Analyses of Impacts, Adaptations, and Mitigation of Climate Change,* Watson R.T., Zinyowera M.C., and Moss R.H. (eds), Cambridge University Press, Cambridge and New York.

IRRI: 1993, *IRRI Rice Almanac 1993-1995*, International Rice Research Institute, Los Baños, Philippines.

Jin Zhiqing: 1993, Impacts of climate change on rice production and strategies for adaptation in the southern China, in: *Climate Change, Natural Disasters and Agricultural Strategies,* China Meteorological Press, Beijing,

Jin Z., Ge D., Chen J., and Fang J.: 1994, Effects of climate change on rice production and strategies for adaptation in southern China, in: *Implications of Climate Change for International Agriculture: Crop Modeling Study,* Rosenzweig C. and Iglesias A. (eds), U.S. Environmental Protection Agency, Washington, DC.

Jin Z., Ge D., Chen H., and Fang J.: 1995, Effects of climate change on rice production and strategies for adaptation in southern China, in: *Climate Change and Agriculture: Analysis of Potential Impacts,* Rosenzweig C., Allen L.H., Harper L.A., Hollinger S.E., and Jones J.W. (eds), American Society of Agronomy No. 59, Madison, Wisconsin.

Jones C.A. and Kinir J.R., (eds): 1986, *CERES-Maize: A Simulation Model of Maize Growth and Development,* Texas A&M University Press, College Station.

Jones J.W., Boote K.J., Jagtap S.S., Hoogenboom G., and Wilkerson G.G.: 1988, *SOYGRO V5.41: Soybean Crop Growth Simulation Model. User's Guide.* Florida Agr. Exp. Sta. Journal No. 8304, IFAS, University of Florida, Gainesville.

Karim Z., Ahmed M., Hussain S.G., and Rashid Kh.B.: 1994, Impact of climate change on the production of modern rice in Bangladesh, in: *Implications of Climate Change for International Agriculture: Crop Modeling Study,* Rosenzweig C. and Iglesias A. (eds), U.S. Environmental Protection Agency, Washington, DC.

Kimball B.A.: 1983, *Agronomy Journal* **75**, 779.

Korner C.: 1990, CO_2 fertilization: the great uncertainty in future vegetation development, in: *Global Vegetation* Change, S.A. and Reidel D. (eds), D. Reidel and Co., Hingman, Massachusetts.

Kuiper P.J.C.: 1993, Diverse influences of small temperature increases on crop performance, in: *International Crop Science, Vol. I,* Crop Science Society of America, Madison, Wisconsin, pp. 309-313.

Lin Erda: 1994, *Rural Eco-Environment* **10**, 1.

Lu Liangshu and Liu Zhicheng: 1991a, *Studies on the Medium and Long-Term Strategy of Food Development in China,* Agricultural Publishing House, Beijing, China.

Lu Liangshu and Liu Zhicheng: 1991b, *Productive Structure and Developmental Prospects of Planting Industry,* Agricultural Publishing House, Beijing, China.

Manabe S. and Wetherald R.: 1987, *Journal of Atmospheric Science* **44**, 1601.

Matthews R.B., Kropff M.J., Bachelett D., and Van Laer H.H.: 1995, Modelling the Impacts of Climate Change on Rice Production in Asia, CAB International, Oxford, and International Rice Research Institute, Manila, Philippines.

Mearns L.O., Rosenzweig C., and Goldberg R.: 1992, *Sensitivity Analysis of the CERES-Wheat Model to Changes in Interannual Variability of Climate,* U.S. Environmental Protection Agency, Washington, DC.

Mearns L.O., Rosenzweig C., and Goldberg, R.: (in press), *Journal of Agric. For. Meteorol.*

Otter-Nacke S., Godwin D.C., and Ritchie J.T.:1986, *Testing and Validating the CERES-Wheat Model in Diverse Environments,* AgGRISTARS YM-15-00407, Johnson Space Center, Houston, Texas.

Panturat S. and Eddy A.: 1989, *Clim. Bull.* **24**, 16.

Parry M.L., Blantran de Rosari, Chong A.L., and Panich S. (eds): 1992, *Socioeconomic Impacts of Climate Change in Southeast Asia,* U.N. Environment Programme, Nairobi, Kenya.

Pickering N.B., Jones J.W., and Boote K.: 1995, Adapting SOYGRO V5.42 for prediction under climate change conditions, in: *Climate Change and Agriculture: Analysis of Potential Impacts,* Rosenzweig C., Allen L.H., Harper L.A., Hollinger S.E., and Jones J.W. (eds), American Society of Agronomy No. 59, Madison, Wisconsin.

Qureshi A. and Iglesias A.: 1994, Implications of global climate change for agriculture in Pakistan: Impacts on simulated wheat production, in: *Implications of Climate Change for International Agriculture: Crop Modeling Study,* Rosenzweig, C. and Iglesias A. (eds), U.S. Environmental Protection Agency, Washington, DC.

Rao D.G. and Sinha S.K.: 1994, Impact of climate change on simulated wheat production in India, in: *Implications of Climate Change for International Agriculture: Crop Modeling Study,* Rosenzweig C. and Iglesias A. (eds), U.S. Environmental Protection Agency, Washington, DC.

Rao D.G., Katyal J.C., Sinha S.K., and Srinivas K.: 1995, Impacts of climate change on sorghum productivity in India: Simulation study, in: *Climate Change and Agriculture: Analysis of Potential Impacts,* Rosenzweig C., Allen L.H., Harper L.A., Hollinger S.E., and Jones J.W. (eds), American Society of Agronomy No. 59, Madison, Wisconsin.

Rosenzweig C. and Iglesias A. (eds): 1994, *Implications of Climate Change for International Agriculture: Crop Modeling Study,* U.S. Environmental Protection Agency, Washington, DC.

Rosenzweig C. and Parry M.L.: 1994, *Nature* **367**, 133.

Seino H.: 1994, Implications of climate change for Japanese agriculture: Evaluation by simulation of rice, wheat, and maize growth, in: *Implications of Climate Change for International Agriculture: Crop Modeling Study,* Rosenzweig C. and Iglesias A. (eds), U.S. Environmental Protection Agency, Washington, DC.

Singh U., Ritchie J.T., and Godwin D.C.: 1993, *A User's Guide to CERES-Rice Version 2.10,* International Fertilizer Development Center, Muscle Shoals, Alabama.

Sugihara S.: 1991, *Kikogaku-Kishogaku Kenkyu Kokoku* **16**, 32.

Suyama T.: 1988, *Kisho-Kenya Note.* **162**, 123.

Tongyai C.:1994, Impact of climate change on simulated rice production in Thailand, in: *Implications of Climate Change for International Agriculture: Crop Modeling Study,* Rosenzweig C. and Iglesias A. (eds), U.S. Environmental Protection Agency, Washington, DC.

Uchijima T.: 1987, The effects of latitudinal shift of rice yield and cultivable area in northern Japan, in: *The Impact of Climatic Variations on Agriculture. Vol. 1. Assessment in Cool Temperate and Cold Regions,* Parry *et al.* (eds), Kluwer Academic Publishers, Dordrecht, The Netherlands, pp. 797-808.

Uchijima T. and Seino H.: 1988, *Bull. Natl. Inst. Agro-Environ. Sci.* **4**, 67.

United Nations: 1989, *World Population Prospects 1988*, United Nations, New York.

Wigley T.M.L.: 1987, Climate Scenarios, Prepared for the European Workshop in Interrelated Bioclimate and Land-Use Changes, NCAR 3142-86/3, National Center for Atmospheric Research, Boulder, Colorado.

Wilson C.A. and Mitchell J.F.B.: 1987, *Journal of Geophysical Research* **92**, 13315.

Yoshino M.: 1991, Global Climate Change in the Coming Half Century and an Estimate of its Impact on Agriculture, Forestry, Fishery, and Human Environment in Japan, Report on Climatology and Meteorology, 16, University of Tsukuba, Tsukuba, Japan.

Zhan Tao: 1993, Influences of global climate change on agriculture in China, in: *Climate Biosphere Interactions,* John Wiley & Sons, New York.

Zhang Houxuan: 1993, The impact of greenhouse effects on double rice in China, in: *Climate Change and Its Impact,* Meteorology Press, Beijing, China.

EFFECTS OF INTERANNUAL CLIMATE VARIABILITY AND CLIMATE CHANGE ON RICE YIELD IN JAVA, INDONESIA

I. AMIEN, P. REJEKININGRUM, A. PRAMUDIA, and E. SUSANTI

Center for Soil and Agroclimate, Jalan M.H. Juanda 98, Bogor, 16123 Indonesia

Abstract. About 60% of the nearly 40 × 10^6 t of rice produced in Indonesia are from the island of Java. However, the rice self-sufficiency that has been attained and maintained since 1984 could be threatened by changing climate, and has been affected by the climate variability effects of the El Niño/Southern Oscillation phenomenon. To aid policy makers and planners in formulating strategic policy options, the effects of recurring droughts and possible climate change on rice yields were studied using climate and crop models. Three models were used to simulate climate change: those of the Goddard Institute for Space Studies, Geophysical Fluid Dynamics Laboratory, and the United Kingdom Meteorological Office. Several climate scenarios were generated for Ngawi, in East Java, and Sukamandi, in West Java. These models indicate that doubling greenhouse gases would increase solar radiation by 1.2-2.1%, minimum and maximum temperatures by 7.6-16.8°C, and precipitation by 20.5-91.7%. The Goddard Institute for Space Studies transient climate change scenarios indicate that maximum and minimum temperatures would increase by 3.5 and 4.9%, respectively, in 2010, 6.9 and 9.8%, respectively, in 2030, and 11.1 and 15.7%, respectively, in 2050. The rainfall increase varies from 7.0% for West Java in 2010 to 8.7% for East Java in 2050. The Decision Support System for Agrotechnology Transfer crop model slightly under-predicts lowland rice yields of several experimental plots in three sites in Java and one site in Sumatra, but the results are almost equal to or a little higher than farm level yields. Nevertheless, the simulation outputs and experimental plots yields are closely related with a coefficient of determination value of 87%. Changes in climate in the decades of 2010, 2030, and 2050 could drastically reduce rice yield: the rice yield is estimated to decrease by about 1% annually in East Java and less in West Java. Currently, the rice yields in dry years are about one half those of normal years.

Key words: Indonesia, Java, agriculture, rice, drought, El Niño, GFDL, GISS, UKMO

1. Introduction

As a staple food in Indonesia, rice is strategically important in the agricultural development and economy of the country. About 60% of the rice produced in the country comes from the fertile, volcanic ash soils of the island of Java. The introduction of high yield rice varieties and improvements in traditional techniques over the last several decades have boosted rice production. Except during recurring El Niño/Southern Oscillation episodes, which cause droughts in Indonesia, rice production has steadily increased since 1969 from 12.8 × 10^6 t to almost 31 × 10^6 t in 1993 (Binus Pangan, 1994). Rice self-sufficiency was attained in 1984, when Indonesia produced 25.9 × 10^6 t of rice and exported 11,000 t. However, population increase, albeit at a reduced growth rate from 2.3% in the 1970s to 1.7% in the early 1990s (BPS, 1994), the recent prolonged droughts such as those of 1991 and 1994, and the decreasing availability of land for rice cultivation could threaten the food supply in the country. The effects of El Niño in 1991 and 1994 caused widespread crop failures that compelled Indonesia to import about 6 × 10^5 t and over 1 × 10^6 t of rice, respectively. It is estimated that about 2 × 10^4 ha of irrigated paddy fields on Java are converted to residential and industrial uses every year.

Rice production in Indonesia is also threatened by a decrease in yield because of potential climate change. Long-term records indicate that atmospheric concentrations of CO_2 and the global temperature have increased during the past 100 years. Air temperature and air CO_2 concentration directly affect plant growth through photosynthesis and respiration.

Water, Air, and Soil Pollution **92:** 29–39.
© 1996 *Kluwer Academic Publishers. Printed in the Netherlands.*

Higher CO_2 concentrations in the atmosphere do, to a certain extent, enhance photosynthesis. But plant growth is notoriously sensitive to temperature. Often a difference of only a few degrees leads to a noticeable change in growth rates. Higher air temperatures increase the respiration rate of rice plants, reduce net photosynthesis, and hence eventually reduce plant yield.

Science and technology have generated a wealth of information, but the functional knowledge necessary to support decision making is not accessible to those who need it most. To help remedy this situation, accelerating the process of knowledge dissemination to provide decision makers with alternative scenarios of potential outcomes is a necessity. Systems scientists use models as a means to capture, condense and organize knowledge. Models are the means by which knowledge about systems and their performance is made portable and accessible to users whose livelihood and welfare depend on this performance.

To provide policy makers and planners with information to formulate strategies to cope with the effects El Niño events and climate change, a clear picture of what is going to happen in the future is necessary. For that purpose, climate change scenarios are simulated here using three general circulation models (GCMs). The responses of food crops, including rice, to environmental effects now can be predicted using crop simulation models (IBSNAT, 1991). Rice growth models simulate crop responses to changes in climate, soil, cultivar, and management. Thus, the climate scenarios are used as input to the rice crop model to predict the effect of changing climate on rice yield. This paper discusses the climate models used and their limitations, the rice simulation model and its validation, the results of the analysis, and possible adaptation options.

2. Methods

2.1 CLIMATE MODELS

Because it is not possible to investigate new climatic phenomena through laboratory research, computer models of the climate system are the most useful tool to investigate future climate. To incorporate all the important physical interactions, especially with atmospheric circulation, and to calculate the location of future climate change, fully three-dimensional GCMs are necessary. These sophisticated models solve simultaneous equations for all the important climate variables in three dimensions. The world is broken up into a discrete grid of boxes placed side by side and stacked to cover the globe. The equations used to represent the physical and chemical processes involved are also simplifications of real-world processes (Robock, 1989).

When discussing climate change, it is sometimes convenient to refer to an equivalent doubling of CO_2, which means the effect of all the greenhouse gases together that would have the same effect as doubling CO_2.

Several climate modeling groups have conducted GCM experiments to calculate the equilibrium climate response to doubled CO_2. These include researchers at NOAA's Geophysical Fluid Dynamics Laboratory (GFDL), NASA's Goddard Institute for Space Studies (GISS), and the United Kingdom Meteorological Office (UKMO). The results from the different experiments depend on the assumptions made, especially on the treatment of clouds and of oceans. The models predicted global temperature increases of 2.8 to 5.2°C and global precipitation increases of 7 to 16% (Robock, 1989).

2.2 STUDY AREAS

Two main rice production areas in Java with different geographic conditions, Ngawi in East Java and Sukamandi in West Java, were selected for this study. Sukamandi is on lowlands in the northern coast of West Java at an elevation of 12 m above sea level, and Ngawi is farther inland, on the flood plain of the Bengawan Solo River. At an elevation of 75 m above sea level, Ngawi is flanked by two volcanoes, Lawu and Wilis. In the last five years, nearly 2×10^6 ha were cultivated in West Java, or 21% of the country rice cultivation, and more than 1.5×10^6 ha were cultivated in East Java, or over 16% of the country total (BPS, 1994).

To simulate future weather of the areas, daily climate data, consisting of rainfall, minimum and maximum temperature, and solar radiation or day length, from 1973 to 1992 are used as the baseline. At Sukamandi, the mean annual rainfall during that period was 1633 mm, maximum and minimum temperatures were 30.3 and 22.8°C, respectively, and the average monthly solar radiation was 17.16 MJ m^{-2}. At Ngawi, the mean annual rainfall during that period was 1,875 mm, maximum and minimum temperatures were 31.3 and 22.0°C, respectively, and monthly solar radiation was 12.32 MJ m^{-2}.

Although the weather in the tropics is warm all year round, rainy seasons alternate with dry seasons and each has it own distinct pattern of prevailing winds. The rainfall pattern is also affected by the El Niño phenomenon, which is primarily centered in the Indo-Pacific basin, but has wider near-global impacts and implications. At least four dry years (i.e., 1976, 1982, 1987, and 1991) in the standard climate data from 1973 to 1992 were probably related to El Niño. The average annual rainfall during those years in Ngawi was 1,262 mm, or only 67.3% of the 20 year average. In Sukamandi, it was 1,112 mm, or 68.1% of the 20 year average.

The future climate in both areas was simulated based on doubling the CO_2 level from pre-industrial times to 555 ppm using the GFDL, GISS, and UKMO climate models. The GISS transient model was used to simulate the climate in the decades of 2010, 2030, and 2050.

2.3 CROP GROWTH SIMULATION MODEL

The cereal, or CERES (Crop-Environment Resource Synthesis), family of crop models is used in the Decision Support System for Agrotechnology Transfer (DSSAT) v.3 crop model to predict the performance of six grain crops (IBSNAT, 1993). All six of these models are designed to use a minimum set of soil, weather, genetic, and management information. The models are daily incrementing and require daily weather data of maximum and minimum temperature, solar radiation, and rainfall. They calculate crop phasic and morphological development using temperature, daylength, genetic characteristics, and vernalization where appropriate. Leaf expansion, growth, and plant population provide information for determining the amount of light intercepted, which is assumed to be proportional to biomass production. The biomass is partitioned into various growing organs in the plant using a priority system. A water and nitrogen balance submodel provides feedback that influences the development of growth processes.

The CERES-Rice model has the same features and characteristics as those described for the other CERES models. It differs from them, however, in the following ways: it can simulate the establishment of a rice crop from dry sowing (pregerminated seeding to transplanting) and provision has been made within the model to calculate the effect of transplant shock on crop duration. Also, water balance modeling simulates crop water uptake under intermittent flooding and drying, and fully upland conditions where the soil is never flooded. In addition, the nitrogen (N) submodel of the CERES-Rice model simulates transformations of N in the plant in both upland and lowland conditions. The model simulates the effects of N deficiency on photosynthesis, leaf area development, tillering, senescence, and remobilization of N during grain filling.

In simulating rice growth, rainfed conditions were applied by using weather data from both normal and dry years. The baseline climate data as well as the climate scenarios from the GISS transient model were utilized in the DSSAT rice growth model. The rice cultivars used in running the model were IR 36 for Ngawi and IR 42 for Sukamandi. A standard management practice in the rice intensification program in Java, i.e., transplanting rice seedling 18 days after germination, planting two seedlings per hill at 20 cm by 20 cm spacing, using two applications of urea fertilizer at 58 kg N per hectare before planting and 30 days after planting at 5 cm depth, was selected as well. The simulation began at transplanting. Full irrigation with no water stress was applied to simulate years with normal rainfall by using the weather data of the rainy season in normal years. (The rice simulation model has several water management options, including fully irrigated with no water stress, and rainfed, which fully depends on rainfall. By using the weather data from normal years during the rainy season, the condition of no water stress can be assured.)

The baseline climate data for 1976, 1982, 1987, and 1991 and the corresponding years in the transient climate scenario, which are years when El Niño events occurred, show very little rainfall. The weather data for those years were utilized to simulate the rice performance for dry years in current and future climate. In dry years, water in many of the upstream reservoirs is far less than normal. In competition with human and animal uses, agriculture, and particularly rice cultivation, is given the last opportunity in utilizing the limited water. In dry years, many farmers opt for secondary crops such as corn, peanut, or cassava that require less water.

3. Limitations

GCMs do a reasonable job of simulating global values of surface air temperature and precipitation, but do poorly at the regional scale (Grotch, 1988). Computer speed and size constraints limit the size of the grid boxes to very large areas horizontally, but only 1 to 5 km vertically. The details of coastlines and topography are not resolved. For a country that consists of more than 13,000 islands in the tropics, the GCMs are not yet suitable to simulate future climate. Attempts to increase the resolution of the CSIRO9 model have only been able to simulate data at points 200 km apart (Whetton, 1994). In Java, this size of grid covers an area from coast to coast that has several distinct rainfall patterns.

Global sensitivity of climate to greenhouse gases is expressed as the change in the global temperature when the equivalent CO_2 concentration in the atmosphere is doubled and reaches a new equilibrium climate, which is estimated by IPCC (1996) at a range between 1.5 and 4.5°C. That is a factor of three; the difference between the last glacial period and the present is only 5°C.

Another other source of uncertainty is the spatial pattern of changes in temperature, precipitation, and solar radiation. Pittock (1994) found a greater measure of agreement in the more recent, improved GCMs, and there is some confidence that at the subcontinent scale the spatial patterns are correct, but there is still little agreement in Java, particularly in precipitation. Also, the effects of natural variability due to other causes such as volcanic eruptions, changes in deep ocean circulation, and solar variability are still not understood. The selection of El Niño years in the future climate for crop growth simulation probably is not quite correct, because the events were not always similar as in the past. El Niño events seem to be more frequent in the last decade.

Although the GCMs do not exactly reproduce current climate, the differences between their simulations of current and future climates may be used as an estimate of potential future changes. In addition, the GCMs produce a data set of all the variables needed for impact assessment that are physically consistent within the physics of the model. The actual model projections are useful in providing scenarios for impact assessment. As model projections of current climate become more accurate, the scenarios they generate will become more accurate.

4. Results and Discussion

4.1 CLIMATE CHANGE SCENARIOS

The changes from current climate (1973-1992) as simulated by the three GCMs are presented in Table I. Using the GISS transient climate model, the climate scenarios in the decades of 2010, 2030, and 2050 were also simulated, and are presented in Table II.

Except for solar radiation in the decade of 2010, all the climate variables increase from current conditions. Higher increases in solar radiation and temperature were produced by the GISS model, and higher increases in rainfall were produced by the UKMO model.

In both locations, minimum and maximum temperatures increase by more than 1 °C in the decade of 2010 and by more than 3 °C in the decade of 2050 under the GISS transient scenario. This increase is more than twice the scenario produced by CSIRO9 model, and far higher than past observed data, which probably implies that the rice yield reduction will not be as high as predicted.

TABLE I
Increases in rainfall, temperature, and solar radiation in Sukamandi and Ngawi at CO_2 concentration of 555 ppm as simulated by the GCMs

	Sukamandi	Ngawi
GISS Scenario		
Monthly Rainfall, mm (% change)	19.6 (20.5%)	24.1 (16.9%)
Maximum Temperature, °C (% change)	3.8 (12.6%)	3.8 (12.2%)
Minimum Temperature, °C (% change)	3.8 (16.8%)	3.8 (17.3%)
Solar Radiation, MJ m^{-2} (% change)	0.3 (1.9%)	0.2 (1.9%)
GFDL Scenario		
Monthly Rainfall, mm (% change)	20.7 (21.8%)	24.4 (17.4%)
Maximum Temperature, °C (% change)	2.3 (7.6%)	2.3 (7.3%)
Minimum Temperature, °C (% change)	2.3 (10.1%)	2.3 (10.4%)
Solar Radiation, MJ m^{-2} (% change)	0.3 (2.1%)	0.1 (0.9%)
UKMO Scenario		
Monthly Rainfall, mm (% change)	59.1 (91.7%)	76.5 (61.2%)
Maximum Temperature, °C (% change)	2.9 (9.6%)	2.9 (9.3%)
Minimum Temperature, °C (% change)	2.9 (12.8%)	2.9 (13.1%)
Solar Radiation, MJ m^{-2} (% change)	0.2 (1.2%)	0.1 (1.1%)

TABLE II
Change in climate in the 2010, 2030, and 2050 decades as simulated by GISS transient model in Sukamandi and Ngawi

	Sukamandi	Ngawi
The 2010 Decade		
Monthly Rainfall, mm (% change)	6.1 (3.4%)	10.8 (7.0%)
Maximum Temperature, °C (% change)	1.1 (3.6%)	1.1 (3.5%)
Minimum Temperature, °C (% change)	1.1 (4.8%)	1.1 (4.9%)
Solar Radiation, MJ m^{-2} (% change)	0.0 (-0.01%)	0.0 (-0.1%)
The 2030 Decade		
Monthly Rainfall, mm (% change)	12.2 (10.3%)	10.3 (6.2%)
Maximum Temperature, °C (% change)	2.2 (7.2%)	2.2 (6.9%)
Minimum Temperature, °C (% change)	2.2 (9.6%)	2.2 (9.8%)
Solar Radiation, MJ m^{-2} (% change)	0.2 (1.1%)	0.1 (1.1%)
The 2050 Decade		
Monthly Rainfall, mm (% change)	11.8 (8.9%)	13.5 (8.7%)
Maximum Temperature, °C (% change)	3.5 (11.4%)	3.5 (11.1%)
Minimum Temperature, °C (% change)	3.5 (15.3%)	3.5 (15.7%)
Solar Radiation, MJ m^{-2} (% change)	0.2 (1.2%)	0.1 (1.1%)

For example, the warming was thought to be least in island and coastal areas and greater inland (Whetton, 1994). Data from several places in Java indicated only a small rise in minimum temperature, but no clear indication of an increase in maximum temperature in the last 100 years (Karyoto, *personal communication*). Average summer daily maximum temperature by 2275 in Jakarta, about 100 km west of Sukamandi, is estimated to increase to 31.6°C, or 2.6°C higher than the current 29°C (Cline, 1992). Under the GISS transient model, that temperature would be reached in the mid or late 2030s.

Higher temperatures may decrease solar radiation because of increased clouds, particularly in a maritime archipelago like Indonesia. As these clouds reach higher altitudes and lower barometric pressure, they expand and cool adiabatically, and then condense on airborne particles.

4.2 CROP MODEL VALIDATION

For validation purposes, the crop model outputs were compared to the results of field experiments in various sites in Indonesia. The rice field experiments were conducted at Sitiung in West Sumatra, Cianjur and Sukamandi in West Java, and Ngawi in East Java during 1981-1982 and 1991-1992. The experiment in Sumatra was testing the response of several rice cultivars to soil management practices on acidic soil. The experiments in Java were testing the source, rate, and methods of nitrogen fertilizer application on IR 36 and IR 42 rice cultivars. The two cultivars were the main rice cultivated in Java because of their relatively high yield and their resistance to brown plant hopper. As these new hopper resistant rices were introduced, new populations of brown plant hopper developed to feed on them. When IR 42 became susceptible to brown plant hopper in late 1980s, a new variety of rice, Cisadane, was introduced, particularly in West Java. At present six rice cultivars — IR 36, 42, 46, 64, Cisadane, and Krueng Aceh — account for 70-80% of Indonesia's rice production (Fox, 1994).

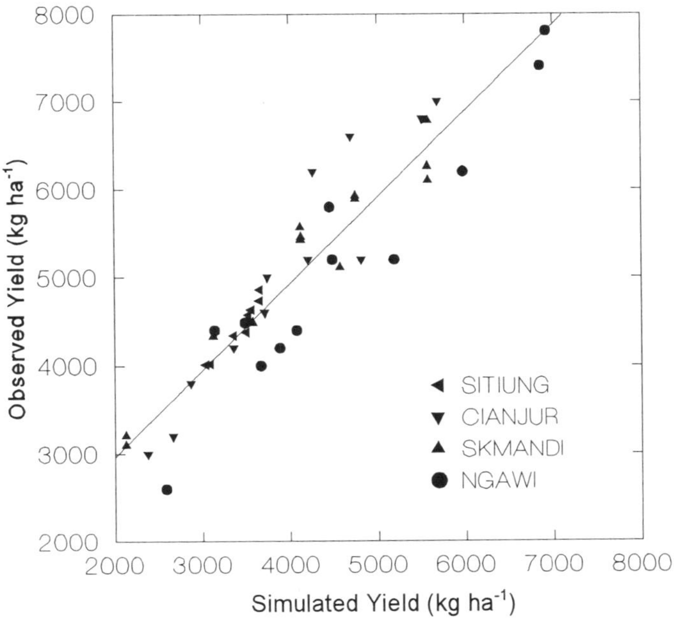

Fig. 1. Observed and simulated rice yields in three locations in Java and in one location in Sumatra.

The DSSAT rice crop model slightly under-predicts lowland rice yields compared to the yields of the experimental plots, but is almost equal to or a little higher than farmers' level yields. It seems that in all locations except Sukamandi the model is under-predicting the grain weight from the experimental plots by almost 1 mg. Nevertheless, the simulation outputs and experimental plots yields are closely related, with a coefficient of determination of 87% (Figure 1). Intensive rice cultivation at the farmer level with 250 kg of urea fertilizer per hectare produced 5,115 kg of rice in the East Java area compared to 5,789 kg of rice in the simulation. In West Java, the farmer produced 4,101 kg compared to 4,426 kg as simulated at Sukamandi and 3,974 kg compared to 4,683 kg as simulated at Cianjur (BPS, 1981). The small experiment plots are certainly better managed than the larger fields of the farmers. Experimental plots were used in validating the rice simulation model because the experiment is designed to investigate the particular effect of treatments on crop performance. The experimental plots are properly controlled according to the designed treatments, and precise measurements are carried out in observing the yield components.

4.3 SIMULATED RICE YIELDS

The rice yield at Ngawi as simulated by the DSSAT rice growth model was 6,925 kg per hectare during a normal year but only 2,939 kg during a dry year. In Sukamandi, the rice yields were 5,578 kg and 2,637 kg per hectare, respectively for the normal and dry years. The biomass production in both location was significantly reduced as well during the dry years.

Cell enlargement is extremely sensitive to desiccation and can be affected by the normal changes in plant water status that follow changes in evaporative demand during the day. Although photosynthetic activity of the leaves is considered less sensitive to drought, any inhibition during grain filling has a critical effect on agronomic yield. The flowering process requires a specialized set of events that may be controlled to an extent by photosynthate supply and cell enlargement. Floral development, flowering, pollination, and fertilization represent the stages of crop development most sensitive to drought inhibition (Boyer and McPherson, 1976).

Rice, more than any other food crop, has a critical requirement for high and regular water availability. The water required is lost through evapotranspiration, seepage, and percolation since very little is actually retained by the plant (Lawson and Alluri, 1980). In lowland rice systems, the seepage and percolation losses are minimized because of the bunds (small dikes that keep the water level in flooded rice system) and development of the plough pan (a hard, impermeable layer in the soil developed by the accumulation of manganese and iron oxides) at about 50 cm depth. High temperatures and solar radiation increase evapotranspiration and water requirements, but also favor growth and yield.

Large decreases in both grain and biomass yields were obtained under the GISS transient climate scenarios for the 2010, 2030, and 2050 decades (Table III). Within the optimal temperature range, from 15-18°C to 30-33°C, the higher temperatures in general favor vegetative growth and development of rice plants. However, root growth has an optimal temperature of about 25°C, and lower temperatures are better for raising strong seedlings (Nishiyama, 1976). In the climate scenarios of the 2010 decade , the temperature increases a modest 1°C, well within the optimal temperature for rice plants. However, the higher temperature probably is unfavorable for reproductive growth. Panicle differentiation occurs favorably under a temperature range of approximately 18-30°C. Above 30°C, it is delayed or inhibited under some conditions (Adachi and Inouye, 1972, in Nishiyama, 1976). The critical temperature for seed ripening is in the range from 12 to 18°C (Sinitsyna and Chan, 1972). Higher air temperatures averaging approximately 30°C are not favorable for ripening (Osada *et al.*, 1973, in Nishiyama, 1976).

The number of spikelets plays an important role in determining the grain yield of rice plants through its effect on the total sink size. Spikelets are the future rice grains, and all the photosynthesis product during the generative stage of rice growth will go to the spikelets. In analyzing rice yields from 200 farmer fields ranging from cool to warm regions in Japan, Munakata (1976) concluded that spikelets production have a clear optimum at 20-23°C and a rise again at 27-28°C. The complex temperature response may be explained by analyzing the competing biological reaction in plants. Tsunoda (1964, in Munakata, 1976) observed that the number of tillers (the number of rice plants within a hill) had an inverse proportion to stem elongation (plant height), and the optimum temperature for stem elongation was about 25-30°C. These results suggest that inhibition of stem elongation at low and high temperatures brought about the increase of tillers, resulting in the temperature curve with two optima for the number of panicles and spikelets. It seem likely that the temperature curve with two optima is attributed to the increase in ammonification of soil by high temperature, as well as by the inhibition of stem elongation at low and high temperatures (Munakata, 1976). High temperatures also increased sterility as a result of smaller pollens and nondehiscence of the anthers (not splitting open to discharge their contents) (Sato *et al.*, 1973).

In the simulation results, both vegetative and reproductive yield decreased in the higher temperature projected for the 2010, 2030, and 2050 decades. Although the yield reduction is higher in grain than in biomass in the decade of 2010, in the decade of 2050 the yield reduction in biomass is higher. The annual yield decrease of over 1% in Ngawi is almost twice that in Sukamandi in the climate scenario of the 2010 decade. In the scenarios of the 2030 and 2050

TABLE III

Simulated rice yield under climate change and interannual climate variability using the GISS transient model

	1990	2010	2030	2050
Ngawi				
Grain Yield (kg ha^{-1})				
Normal Years	6,925	3,569	2,956	2,423
Percent Change	—	-48%	-17%	-18%
El Niño Years	2,969	1,721	1,529	1,370
Percent Change	—	-42%	-11%	-10%
Biomass (kg ha^{-1})				
Normal Years	10,644	6,748	5,595	4,584
Percent Change	—	-37%	-17%	-18%
El Niño Years	5,975	2,977	2,931	2,597
Percent Change	—	-50%	-2%	-11%
Sukamandi				
Grain Yield (kg ha^{-1})				
Normal Years	5,578	4,220	3,473	2,890
Percent Change	—	-24%	-18%	-17%
El Niño Years	2,637	1,988	1,985	1,969
Percent Change	—	-25%	0	-1%
Biomass (kg ha^{-1})				
Normal Years	9,777	7,607	6,569	5,366
Percent Change	—	-22%	-14%	-18%
El Niño Years	5,461	2,717	2,603	2,551
Percent Change	—	-50%	-4%	-2%

decades, the annual yield reductions are similar, at less than 1%. Part of the higher yield reduction in 2010 is probably due to the decrease in solar radiation in the climate scenario of that decade, and the lower yield reduction in the later decades is probably due to the estimated increase of CO_2. Higher CO_2 concentration will increase photosynthesis: in a CO_2 enrichment experiment, Yoshida (1976) found that the optimum CO_2 concentration for growth and yield of rice lies between 1,500 and 2,000 ppm. These values are far higher than any model estimates of future CO_2 concentrations in the atmosphere. However, the yield increase brought about by increased CO_2 concentrations may not be able to compensate for the yield reduction due to increasing temperature.

Caution has to be taken because the yield reduction due to climate change as simulated by the climate and crop models is probably a little exaggerated. At least two main reasons need to be considered. First, the temperature increase seems to be higher than predicted by other models and the changes that have been observed in the past. Second, the rice growth model under-predicts the experimental yields.

The yield decline in El Niño years is far greater than the threat of yield decrease by possible climate change. El Niño events also have become more frequent in recent years. During the El Niño episode of 1991, about 3.64×10^5 ha of rice field were affected by drought, and crops failed on about 1.18×10^5 ha, mostly in Java (Dirjen, 1995). Simulating the rice growth in the dry years using rainfed conditions does not really represent actual conditions in the field. Under rainfed conditions all the rainfall either percolates through the soil profile or is lost through runoff. However, in rice cultivation on bunded fields there is always standing water after the rain for a certain time, although it probably is lost through evaporation later. Farmers, learning from experience, usually plant rice after sufficient rainfall. The time of planting rice in the areas experiencing a prolonged dry season is not always in November, as in normal years. Delaying planting time until sufficient water is available, or supplementing irrigation water, will greatly affect the rice yield.

Greater amounts of rainfall and its variability could enable farmers to grow more upland rice, but the already intensive land utilization in Java and the low yield of upland rice, as well as competition with other crops that have higher market value, would hinder more development in the uplands. Also, higher rainfall in the overutilized upstream areas of Java would induce more erosion, increase siltation of dams and harbors, raise river beds, and increase flooding during the rainy season. Currently, intensive upland rice cultivation is promoted outside of Java, where it is intercropped on tree plantations before the establishment of the full canopy of the trees.

To address the problem of securing the food supply of the country, several actions are being taken, including producing better near real time analyses of climate to improve preparedness; using early rice varieties; using *gogo rancah,* that is, planting early at the end of dry season in the uplands and flooding the rice when sufficient rains come; and more efficient water usage such as intermittent flooding and improved rural irrigation systems. Improvement of the cultural techniques used in many parts of the outer islands could contribute substantially to the country's rice production. Currently the yield levels in these places are about 50%-60% of those in Java. In the long run, expansion of rice cultivation, particularly to the under-utilized swampy areas of the outer islands, as well as crop diversification seems to be inevitable.

The simulation results indicate that rice yield is adversely affected, particularly by increasing temperature. The analysis is based on the assumption that varietal characteristics are the same in the future as now. It is highly likely that, give the time scale involved, plant breeding programs will more closely adapt to the gradually changing conditions, thereby mitigating the negative and enhancing the positive effects of climate change. In the tropics, using rice varieties that are more tolerant to high temperature probably can minimize the effect of climate change. The microclimate in rice fields can be improved to support better rice growth through plant and water management. In irrigated rice, the existence of water in the rice field reduced the temperature by absorbing heat. Planting trees parallel to the rice field will induce wind that can reduce temperature.

5. Conclusions

Climate models do not yet predict climate change at the regional scale particularly well, especially for an island system like Indonesia. The GISS, GFDL, and UKMO climate models predict higher temperature increases at both study sites compared to what was predicted by the CSIRO9 model. Data from observations in the archipelago of Indonesia indicate only slight increases in minimum temperature, and no significant increases in maximum temperature.

The DSSAT crop model predicted lowland rice yields quite well for different management options, with a coefficient of determination value of 87%. Limited experimental data with standard management practices for sufficient spans of time restrain the model validation for a changing climate.

The rice crop model predicted yield reductions of about 1% annually in East Java and less in West Java because of possible climate change. Development of more tolerant varieties probably can compensate the yield losses due to climate change in the future. Higher yield losses are predicted because of interannual climate variability. Because the dry spell threat is more imminent and frequent, a short-term climate prediction for the tropical region is urgently needed to improve preparedness.

Acknowledgments

The authors owe a debt of gratitude to the U.S. Environmental Protection Agency that provides funding for this research through the Institute of Space Study, Columbia University, New York, and Indonesia's State Ministry for the Environment.

References

Adachi K. and Inouye J.: 1972, *Proc. Crop Sci. Soc. Japan* **41**, 78.

Binus Pangan: 1994, Vademecum Pemasaran (Marketing Vademecum), Direktorat Bina Sarana Usaha Tani Tanamn Pangan dan Hortikultura (Directorate of Farm Inputs for Food and Horticultural Crops), Jakarta.

Biro Pusat Statistik: 1981, *Statistik Indonesia*, BPS, Jakarta.

Biro Pusat Statistik: 1994, *Statistik Indonesia*, BPS, Jakarta.

Boyer J. and McPherson H.G.: 1976, Physiology of water deficits in cereal grains, in: *Proc. Symposium on Rice and Climate*, IRRI, Los Baños, The Philippines, pp. 321-343.

Cline W.R.: 1992, *The Economics of Global Warming*, Institute for International Economics, Washington, DC.

Dirjen T.P.H.: 1995, Kebijakan penanggulanngan kekeringan (Government policies to overcome the drought problem), Keynote Speech in Panel Discussion on Anticipation and Action to Overcome Drought, Proceeding of a seminar held at Sukamandi, West Java on August 26-27, 1994, Perhimpi and Peragi, pp. xii-xxii.

Fox J.: 1994, Ecology of agricultural production: The Indonesian case, in: *Climate Impact Assessment Methods for Asia and the Pacific*, Jackman A.J. and Pittock A.B. (eds), AIDAB, Canberra, Australia, pp. 123-127.

Grotch S.L.: 1988, Regional Intercomparisons of General Circulation Models Predictions and Historical Climate Data, (DOE/NBB-0084), U.S. Department of Energy, Washington, DC.

IBSNAT: 1993, *The IBSNAT Decade: Ten Years of Endeavor at the Frontier of Science and Technology*, IBSNAT, University of Hawaii, Honolulu, Hawaii.

IPCC (Intergovernmental Panel on Climate Change): 1996, *Climate Change 1995: The IPCC Second Assessment Report, Volume 2: Scientific-Technical Analyses of Impacts, Adaptations, and Mitigation of Climate Change*, Watson R.T., Zinyowera M.C., and Moss R.H. (eds), Cambridge University Press, Cambridge and New York.

Lawson T.L. and Alluri K.: 1980, Maximum water requirement of upland rice variety OD6 in the humid/subhumid zone of West Africa, in: *Proc. Symposium Agrometeorology of Rice Crop*, WMO-IRRI, Los Baños, The Philippines, pp. 181-185.

Munakata K.L.: 1976, Effects of temperature and light on the reproductive growth and ripening of rice, in: *Proc. Symposium on Rice and Climate*, IRRI, Los Baños, The Philippines, pp. 187-210.

Nishiyama I.: 1976, Effect of temperature on the vegetative growth of rice plants, in: *Proc. Symposium on Rice and Climate*, IRRI, Los Baños, The Philippines, pp. 159-185.

Osada A., Nara M., Chakrabandhu H., Rahong M., and Gesprasert M.: 1973, *Proc. Crop Sci. Soc. Japan* **42**, 351.

Pittock B.: 1994, Constructing climate scenarios: The philosophy, in: *Climate Impact Assessment Methods for Asia and the Pacific*, Jackman A.J. and Pittock A.B. (eds), AIDAB, Canberra, Australia, pp. 15-20.

Robock A.: 1989, Global climate change, in: *The Potential of Global Climate Change on the United States*, Smith J.B. and Tirpak D. (eds), U.S. Environmental Protection Agency, Washington, DC, pp. 9-28.

Sato K., Inaba K., and Tozawa M.: 1973, *Proc. Crop. Sci. Soc. Japan* **42**, 207.

Schlesinger M.E. and Mitchell J.F.B.: 1985, Model projections of the equilibrium climatic response to increase carbon dioxide, in: *Projecting the Climatic Effects of Increasing Carbon Dioxide*, MacCracken M.C. and Luther F.M. (eds), (DOE/ER-0237), U.S. Department of Energy, Washington, DC, pp. 81-147.

Sinitsyna N.I. and Chan D.: 1972, *Meteorol. Klimatol. Gidrol. Mezhved. Nauchn. Sb.* **8**, 83: also published in *Field Crop Abstr.* **26**, 432, 1973.

Tsunoda K.: 1964, *Bull. Natl. Inst. Agric. Sci. A* **11**, 75.

Whetton P.: 1994, Constructing climate scenarios: The practice, in: *Climate Impact Assessment Methods for Asia and the Pacific*, Jackman A.J. and Pittock A.B. (eds), AIDAB, Canberra, Australia, pp. 21-27.

Yoshida S.: 1976, Carbon dioxide and yield of rice, in: *Proc. Symposium on Rice and Climate*, IRRI, Los Baños, The Philippines, pp. 211-221.

VULNERABILITY OF RICE AND CORN TO CLIMATE CHANGE IN THE PHILIPPINES

R.D. BUAN,[1] A.R. MAGLINAO[2]
P.P. EVANGELISTA,[3] and B.G. PAJUELAS[1]

[1] *Philippine Atmospheric, Geophysical and Astronomical Services Administration (PAGASA)*
1424 Quezon Avenue, Quezon City, Philippines

[2] *Philippine Council for Agriculture, Forestry and Natural Resources Research and Development (PCARRD)*
Los Baños, Laguna, Philippines

[3] *Bureau of Soils and Water Management (BSWM)*
Elliptical Road, Quezon City, Philippines

Abstract. The potential impact of climate change on rice and corn crops in the Philippines was assessed using preliminary results from four general circulation models and the Decision Support System for Agrotechnology Transfer. The results showed both increases and decreases in the yield of two rice varieties. The results for corn showed a consistent decrease in yield. The decrease in crop yield may be due to a shorter maturity period and an increase in potential evapotranspiration as a result of increased daytime and nighttime temperatures, which are critical to dry matter production. Rainfall in tropical areas such as the Philippines is generally high, and so a decrease in rainfall of only 10% may not affect the water supply significantly, but an increase of the same magnitude may affect crop production tremendously because of frequent occurrences of floods. One potential negative impact not included in this analysis is the effects of strong winds due to typhoons.

Key words: Philippines, agriculture, rice, corn, CCCM, CERES-Corn, CERES-Rice, GFDL, GISS, UKMO

1. Introduction

1.1 BACKGROUND

Agriculture is the largest contributor to the Philippine's economy (DENR, 1991). In 1988, agriculture generated 36% of the country's export earnings and 28% of the gross national product. Likewise, the gross value added by agriculture to the economy ranged from 29% to 27% between 1970 and 1990 (NSCB, 1992).

The agricultural land in the country is estimated at 9.7 million ha, and the cultivated area is still expanding to the more fragile and marginal uplands. However, because of the problem of soil erosion in these sloping areas, and the continued conversion of prime agricultural lands to nonagricultural uses, agricultural production is very much threatened. This is further aggravated by the increasing population and its food demands. Because of the rapid increase in population, the per capita land share decreased from 2,870 m^2 per person in 1960 to 1,600 m^2 per person in 1990 (NSCB, 1992).

The Philippines, like other tropical countries, is considered to be extremely vulnerable to extreme meteorological events such as typhoons, storm surges, floods, and droughts. It is also affected by volcanic eruptions, earthquakes, and tsunamis, which have caused significant changes in the land's configuration and coastlines.

Property damages caused by disastrous typhoons alone are close to about 30% of annual GNP. For instance, damages caused by typhoons, floods, and droughts (droughts generally caused by El Niño events) to rice and corn alone is estimated to be 2% to 3% of annual production (ADB, 1994). Because of the economic impact of these extreme events, the Philippines may find it difficult to sustain its economy under current climatic and geologic

Water, Air, and Soil Pollution **92:** 41–51.
© 1996 *Kluwer Academic Publishers. Printed in the Netherlands.*

conditions. It is therefore important to assess the vulnerability of agricultural production in the Philippines to climate change. This vulnerability assessment will be useful in the planning, development, and implementation of adaptation strategies to address potential impact of climate change on agriculture.

1.2 OBJECTIVES OF THE STUDY

The general objective of this study is to conduct an assessment of the vulnerability and adaptation of rice and corn production to climate change. Specifically, the study aims to:

- identify GCMs for developing climate change scenarios applicable under Philippine conditions
- conduct an assessment of the vulnerability of rice and corn production to climate change using the Decision Support System for Agrotechnology Transfer version 3 (DSSAT 3) model (Benioff *et al.*, in press)
- identify, develop, and validate adaptation options to address the impact of climate change on rice and corn production.

This paper presents research conducted to identify general circulation models (GCMs) and to assess the vulnerability of rice and corn production to climate change.

1.3 AGRICULTURAL PRODUCTION IN THE PHILIPPINES

Agricultural production in the Philippines is traditionally concentrated on a few main crops. Rice and corn are the major food crops. Coconut and sugarcane are the major commercial crops that constitute important export commodities (NSCB, 1992). Of the 9.7×10^6 ha of agricultural land in the country, about 3.7×10^6 ha are planted with rice (about 1.5×10^6 ha are irrigated; irrigation of other crops is negligible), 2.0×10^6 ha with corn, and 2.8×10^6 ha with coconut. The remaining 1.2×10^6 ha are planted with tobacco, sugarcane, citrus, vegetables, abaca, and others (NSCB, 1992).

Rice dominates the agricultural food crop sector. About one-third of the country's farmers are engaged in rice production, mostly on a subsistence basis. Until the 1960s, rice production had increased primarily because the area under cultivation was expanded. After the 1970s, increases were achieved by improved productivity, mainly through the use of high yielding varieties.

Rice production grew at an average of 2.8% per annum from 1970 to 1992. From 1981 to 1990, a mean total production of 8.57×10^6 t yr^{-1} had been achieved. From 1981 to 1992, rainfed farms produced an average annual yield of 2.83×10^6 t of rice. This is only about half of the total yearly production of 5.88×10^6 t in irrigated areas, which, incidentally, accounts for only about 56% of the total area devoted to rice production.

Corn is a major substitute crop for rice, particularly in the central part of the Philippines, and is the main ingredient for livestock feeds, food products, and is important in industrial uses.

Before the launching of the MAISAGANA program in 1981 and the advent of high-yielding hybrid corn varieties, the average yield of corn was only about 0.95 t ha^{-1}. In 1980, the total production was about 3.1×10^6 t from an area of 3.3×10^6 ha. In 1987, the average yield increased up to 1.2 t ha^{-1} for a total production of 4.2×10^6 t from an area of 3.7×10^6 ha (PCARRD, 1990).

To achieve self-sufficiency in food and feed requirements, corn production will also be enhanced through the Grains Production Enhancement Program of the Department of Agriculture. An annual average growth rate of 6.2% is targeted for corn production.

1.4 EFFECTS OF CLIMATE VARIABILITY ON CROP PRODUCTION

Rice and corn crops are highly vulnerable to climate variability. Records show that 48% of the losses in rice and corn production between 1968 and 1990 were due to tropical cyclones and floods. Some 33% of the production losses for the same period were due to droughts, and 18% of the losses were attributed to weather related incidences of pests and diseases. Agricultural losses attributed to salt water intrusion in the Bulacan province alone were estimated at P 10.7 million (1985) per year.

The present day vulnerability of agricultural areas in the Philippines to climatic variability is due to the high frequency of tropical cyclones and floods, the seasonal occurrence of droughts, especially during El Niño episodes, and the incidence of saline water intrusion in coastal areas due to coastal inundation caused by storm surges.

Studies on the effects of higher temperature and CO_2 concentration have shown varied results. The effect of higher rainfall coupled with an increase in temperature and CO_2 may have positive effects on rice production. Another study conducted by the Philippine Council for Agriculture, Forestry and Natural Resources Research and Development (PCARRD) using output from GCMs to run the DSSAT 2 model noted a decrease in rice yield in two locations (Batac, Ilocos Norte, and Los Baños, Laguna) (Escaño and Buendia, 1994).

The effects of climate variability and frequency of extreme events may be more important than the combined effects of increased CO_2 and temperature on crop yield. The variability of crop yield is likely to increase proportionally to the frequency of extreme events, which could threaten some of the economically marginal farming areas. Risk analysis based on multiyear simulations of locally calibrated crop models such as the DSSAT 3 model is needed to assess the implications of climatic changes on crop production.

2. Methods

2.1 STUDY SITES

Six sites were identified and used in this crop yield simulation study for rice and corn. These sites represent the major rice and corn growing areas in the country. The site used for corn was the Isabela State University (ISU) in Echague, Isabela; the three sites used for rice were the Central Luzon State University (CLSU) in Munoz, Nueva Ecija; the University of the Philippines (UPLB) in Los Baños, Laguna; and the Camarines Sur State Agricultural College (CSSAC) in Pili, Camarines Sur. The sites used for both rice and corn were the Central Mindanao University (CMU) in Musuan, Bukidnon; and the University of Southern Mindanao (USM) in Kabacan, North Cotabato (Table I and Figure 1). The site elevations range from 22 m to 302 m above sea level. The soils of the study sites belong to the order Vertisol, Alfisol, and Ultisol. The soil taxonomy classification for each study site is presented in Table I. The base climate data used in the study are daily maximum and minimum temperatures, rainfall, and solar radiation for each study site. The period of records for these data ranges from 13 to 22 years (Table I).

TABLE I
Statistics on the sites used in the simulations

Location	Coordinates	Elevation (m)	Record of Weather Data (years)	Soil Taxonomy	Crop for Assessment
ISU, Isabela State University Echague, Isabela	16.70°N, 121.67°E	83.24	1977-1990	Typic Haplustalf	Corn
CLSU, Central Luzon State University Munoz, Nueva Ecija	15.72°N, 120.90°E	76.00	1974-1990	Entic Pellustert	Rice
UPLB, University of the Philippines Los Baños, Laguna	14.70°N, 121.25°E	21.70	1977-1990	Typic Pellustert	Rice
CSSAC, Camarines Sur State Agricultural College Pili, Camarines Sur	13.57°N, 123.27°E	35.70	1975-1990	Typic Pelludert	Rice
CMU, Central Mindanao University Musuan, Bukidnon	7.93°N, 125.07°E	302.00	1978-1990	Typic Paleudult	Rice and Corn
USM ,University of Southern Mindanao Kabacan, North Cotabato	7.12°N, 124.83°E	52.00	1969-1990	Typic Haplustalf	Rice and Corn

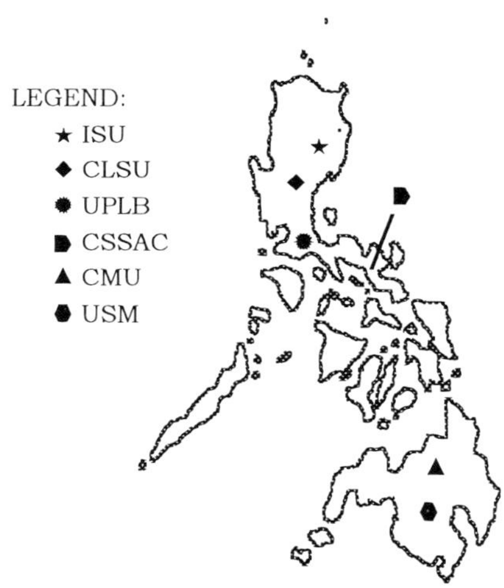

Fig. 1. Study sites for vulnerability assessment for rice and corn crops in the Philippines.

2.2 ASSUMPTIONS REGARDING CULTURAL PRACTICES

Rice varieties IR 64 and IR 72 and sweet corn P3228 were used in the simulation. The maturity dates ranged from 115 to 125 days for rice, and from 75 to 114 days for corn. The standard cultural practices followed in the National Cooperative Testing (PCARRD, 1990) for rice and corn were employed throughout the growing period. These practices include transplanting 10-day to 14-day old rice seedlings at three to four seedlings per hill, and application of fertilizer in the amount of 60-30-0 kg (N, P_2O_5, K_2O). This practice includes application of all of the phosphorus and 40 kg N at the time of transplanting, and the application of the remaining

20 kg N as topdressing one week before panicle initiation. The standard cultural practices for corn include planting two seeds per hill at a spacing of 75 cm between furrows and 20 cm between hills, and application of fertilizer in the amount of 120-60-60 or 100-40-40 kg per hectare (N, P_2O_5, K_2O), depending on soil fertility requirements.

2.3 DEVELOPMENT OF CLIMATE CHANGE SCENARIOS

Climate change scenarios were created from the outputs of four GCMs at $2\times CO_2$ (Benioff *et al.*, in press). The GCMs used in the comparative analyses were the Canadian Climate Centre Model (CCCM) (Boer *et al.*, 1992), the Geophysical Fluid Dynamics Laboratory (GFDL) model (Mitchell *et al.*, 1990), the Goddard Institute for Space Studies (GISS) model (Hansen *et al.*, 1983), and the United Kingdom Meteorological Office (UKMO) model (Wilson and Mitchell, 1987). The seasonal base climate and $2\times CO_2$ GCM climate scenarios, including the change from base climate (adjustment statistics for development of climate change scenarios), are presented in Table II.

2.4 AGRICULTURAL CROP IMPACT ANALYSIS

The CERES-Rice and -Corn models from DSSAT 3 were used for rice and corn simulations. The DSSAT 3 model now includes sensitivity analysis to simulate increase in CO_2 (Rosenzweig *et al.*, 1995).

To capture the impact of climate changes, three planting dates were simulated for all sites for the first and second cropping seasons. For rice, the planting dates for the first cropping season were the 15th days of June, July, and August; for the second cropping season, the 15th days of September, October, and November were selected. The planting dates used for corn were the 15th days of April, May, and June for the first cropping season, and the 15th days of October, November, and December for the second cropping season.

The simulations started with the validation of crop models using base climate and yield data from each site with the current CO_2 concentration. Then the equilibrium ($2\times CO_2$) output of the GCM models was used to develop climate change scenarios for the analysis, and the atmospheric CO_2 concentration was set to 555 ppm. This procedure was repeated using all GCM scenarios, crops, sites, and planting dates. All the simulations assume a rainfed condition.

3. Limitations of the Study

The scope of the study is limited to the simulation of rice and corn yields using GCM equilibrium ($2\times CO_2$) temperatures, rainfall, and solar radiation data output. Other important crops that are expected to be studied are cassova, soybean, and sugarcane. Another limitation of the study is that the effect of extreme climatic events is not represented in the scenarios. Also, it assumed that the soil conditions during the simulation are the same after the doubling of carbon dioxide, which may not be true. Finally, the effect of strong winds due to tropical cyclones was not considered.

TABLE II
Development of adjustment statistics for six study sites

Study Sites	Climatic Elements	Cropping Season	Mean Values: Base	Mean Values: CCCM	Mean Values: GFDL	Mean Values: GISS	Mean Values: UKMO	Change from Base Climate§: CCCM	Change from Base Climate§: GFDL	Change from Base Climate§: GISS	Change from Base Climate§: UKMO
ISU	Total Rainfall	First	894.9	902.8	1115.2	648.1	1047.9	0.9	24.6	-27.6	17.1
		Second	789.0	688.1	855.1	845.9	717.7	-12.8	8.4	7.2	-9.0
	Solar Radiation	First	14.5	14.2	14.8	14.9	15.0	-2.1	2.1	2.8	3.5
		Second	11.0	10.9	12.2	11.0	11.7	-0.9	10.9	0.0	6.4
	Min. Temp.	First	23.5	25.6	25.9	26.5	26.2	2.1	2.4	3.0	2.7
		Second	20.9	23.0	23.7	24.0	24.1	2.1	2.8	3.1	3.2
	Max. Temp.	First	33.7	35.7	35.9	36.5	36.2	2.0	2.2	2.8	2.5
		Second	29.6	31.7	32.3	32.5	32.7	2.1	2.7	2.9	3.1
CLSU	Total Rainfall	First	1556.4	1585.9	1862.7	1133.9	1818.2	1.9	19.7	-27.2	16.8
		Second	353.6	313.4	366.5	415.0	300.6	-11.4	3.7	17.4	-15.0
	Solar Radiation	First	15.7	15.4	16.2	16.2	16.2	-1.9	3.2	3.2	3.2
		Second	16.4	16.2	18.0	16.3	17.4	-1.2	9.8	-0.6	6.1
	Min.Temp.	First	20.4	25.5	25.7	26.4	26.0	5.1	5.3	6.0	5.6
		Second	21.4	23.5	24.2	24.5	24.7	2.1	2.8	3.1	3.3
	Max. Temp.	First	32.1	34.2	34.4	35.2	34.7	2.1	2.3	3.1	2.6
		Second	30.9	33.0	33.6	34.0	34.1	2.1	2.7	3.1	3.2
UPLB	Total Rainfall	First	1206.9	1197.5	1384.7	860.3	1453.3	-0.8	14.7	-28.7	20.4
		Second	765.0	683.9	817.9	879.4	616.2	-10.6	6.9	15.0	-19.5
	Solar Radiation	First	17.4	17.1	18.0	18.0	17.8	-1.7	3.5	3.5	2.3
		Second	15.6	15.4	17.1	15.6	16.4	-1.3	9.6	0.0	5.1
	Min. Temp.	First	23.7	25.7	25.9	26.7	26.3	2.0	2.2	3.0	2.6
		Second	22.0	24.0	24.7	25.2	25.7	2.0	2.7	3.2	3.7
	Max. Temp.	First	33.1	35.1	35.3	36.2	35.7	2.0	2.2	3.1	2.6
		Second	30.4	32.5	33.1	33.6	34.2	2.1	2.7	3.2	3.8

§ Change in temperature is in °C; change in total rainfall and solar radiation is in percent.

TABLE II (cont.)
Development of adjustment statistics for six study sites

Study Sites	Climatic Elements	Cropping Season	Mean Values					Change from Base Climate[§]			
			Base	CCCM	GFDL	GISS	UKMO	CCCM	GFDL	GISS	UKMO
CSSAC	Total Rainfall	First	1143.7	1104.8	1286.0	807.9	1393.8	-3.4	12.4	-29.4	21.8x
		Second	946.3	835.3	1124.5	1066.9	841.1	-11.7	18.8	12.7	-11.1
	Solar Radiation	First	14.4	14.2	14.8	14.9	14.5	-1.4	2.8	3.5	0.7
		Second	13.1	12.9	14.4	13.1	13.9	-1.5	9.9	0.0	6.1
	Min. Temp.	First	23.8	25.9	26.0	26.9	26.4	2.1	2.2	3.1	2.6
		Second	21.8	23.8	24.4	25.1	25.4	2.0	2.6	3.3	3.6
	Max. Temp.	First	32.5	34.6	34.7	35.6	35.1	2.1	2.2	3.1	2.6
		Second	30.5	32.4	33.0	33.7	34.0	1.9	2.5	3.2	3.5
CMU	Total Rainfall	First	1485.2	1476.8	1847.1	1150.9	1433.9	-0.6	24.4	-22.5	-3.5
		Second	676.9	658.1	1023.7	760.8	706.5	-2.8	51.2	12.4	4.4
	Solar Radiation	First	18.5	17.9	19.0	18.9	19.2	-3.2	2.7	2.2	3.8
		Second	17.1	16.5	17.7	17.1	17.6	-3.5	3.5	0.0	2.9
	Min. Temp.	First	21.2	23.2	23.8	24.5	24.5	2.0	2.6	3.3	3.3
		Second	20.9	22.9	23.3	24.3	23.4	2.0	2.4	3.4	2.5
	Max. Temp.	First	31.8	33.9	34.1	35.0	35.1	2.1	2.3	3.2	3.3
		Second	31.4	33.5	33.9	34.9	33.9	2.1	2.5	3.5	2.5
USM	Total Rainfall	First	1000.5	1016.8	1261.1	787.5	1123.0	1.6	26.1	-21.3	12.2
		Second	670.0	636.6	941.5	767.2	695.7	-5.0	40.5	14.5	3.8
	Solar Radiation	First	18.0	17.3	18.3	18.5	18.4	-4.0	1.7	2.8	2.2
		Second	17.4	16.6	17.9	17.4	17.9	-4.5	2.8	0.0	2.9
	Min. Temp.	First	22.7	24.6	25.0	25.9	25.6	1.9	2.3	3.2	2.9
		Second	22.0	24.1	24.5	25.4	24.4	2.1	2.5	3.4	2.4
	Max. Temp.	First	33.2	35.2	35.4	36.4	36.0	2.0	2.2	3.2	2.8
		Second	33.0	35.0	35.4	36.3	35.4	2.0	2.4	3.3	2.4

[§] Change in temperature is in °C; change in total rainfall and solar radiation is in percent.

4. Results and Discussion

4.1 CHANGE FROM BASE CLIMATE

The percentage changes in mean total rainfall and solar radiation and the difference in temperatures from base climate are shown in Table II. All the GCMs used in simulating a doubled atmospheric CO_2 level resulted in higher minimum and maximum temperatures at all sites and in both cropping seasons.

Although all the GCMs indicated the same trend in temperature, they did not estimate similar trends for rainfall and solar radiation. The CCCM estimated a decrease in solar radiation, while the other three models estimated an increase. There was more variation in the estimated change in total rainfall among sites, seasons, and models used as shown in Table II.

4.2 IMPACT OF CLIMATE CHANGE ON CROP YIELDS

The simulated impact of $2\times CO_2$ conditions on crop production is generally varied. For rice, the results of the four GCMs were very inconsistent (Table III). However, for corn, the results showed consistent decreases in yield (Table IV). At Central Mindanao University and the University of Southern Mindanao, where both rice and corn yields were simulated, the results showed that for both crops, the yields were less negatively affected (or positively affected) by climate change in the second cropping season.

4.2.1 Effects on Rice Yields

Model simulation using both the CCCM and the GFDL climate scenarios resulted in an increase in rice yield with the doubling of atmospheric CO_2, except at the University of Southern Mindanao site and the Central Luzon State University site.

More varied results were obtained using the climate scenarios generated by the GISS and UKMO scenarios. The GISS scenario simulated a decreased yield at the Camarines Sur State Agricultural College and University of Southern Mindanao sites, but an increased yield at the Central Mindanao University site for both cropping seasons and rice varieties. At the University of the Philippines, Los Baños, both rice varieties had an estimated decrease in yield during the second cropping, but an increase in yield during the first cropping. Mean total rainfall during the second cropping is much less than the mean total rainfall during the first cropping. At Central Luzon State University, IR 64 yield increased in both seasons, but IR 72 had a decreased yield only in the first cropping.

For both seasons and for the two rice varieties, the UKMO climate scenario resulted in an estimated decreased yield at the University of the Philippines, Los Baños, but an increased yield at the University of Southern Mindanao. At Central Luzon State University, estimated rice yields increased during the first cropping but decreased during the second cropping. The opposite trend was observed at Central Mindanao University.

All GCM climate scenarios simulated a decrease in the maturity period of the two rice varieties in both cropping seasons. The decreases ranged from 2.56 to 13.79%. This corresponds to about 3 to 16 day decreases in the maturity period. While a decrease in yield was not shown under all climate scenarios, it may have been affected by this decrease in maturity period. It is to be noted that the decrease in the maturity period will also decrease dry matter accumulation.

TABLE III

Change in yield and maturity period of two rice varieties for selected climate change scenarios for five study sites

Study Site	Rice Varieties	Cropping Season	Yield Base (t ha^{-1})	Change in Yield (%)			
				CCCM	GFDL	GISS	UKMO
CLSU	IR 64	First	4.76	3.78	6.09	2.73	4.20
		Second	3.02	7.28	1.99	7.28	- 5.63
	IR 72	First	3.86	1.55	3.63	- 6.73	0.26
		Second	2.82	0.00	- 2.84	4.26	-10.28
UPLB	IR 64	First	4.71	7.64	11.46	9.13	-12.10
		Second	4.42	3.85	8.60	-13.12	- 6.56
	IR 72	First	3.51	6.27	11.40	9.67	-21.65
		Second	4.61	1.95	6.72	-13.67	- 7.38
CSSAC	IR 64	First	5.14	3.89	8.37	-22.37	4.47
		Second	4.23	4.49	8.51	-13.95	- 9.46
	IR 72	First	3.46	1.73	7.80	-27.17	- 0.58
		Second	4.30	3.02	6.51	-14.19	- 9.77
CMU	IR 64	First	5.22	1.92	1.53	1.92	- 6.51
		Second	3.81	11.29	18.11	7.61	5.25
	IR 72	First	4.50	7.56	6.89	2.44	- 4.44
		Second	3.63	12.95	16.53	6.34	3.03
USM	IR 64	First	4.00	- 1.50	10.50	-84.00	10.25
		Second	5.20	0.00	23.27	-12.31	16.73
	IR 72	First	3.28	- 7.01	14.02	-85.06	10.37
		Second	4.97	- 2.01	20.72	-16.50	15.29

TABLE IV

Change in yield and maturity period of two corn varieties for selected climate change scenarios for three study sites

Study Sites	Corn Varieties	Cropping Season	Yield Base (t ha^{-1})	Change in Yield (%)			
				CCCM	GFDL	GISS	UKMO
ISU	P 3228	First	6.73	-11.14	-8.62	-13.08	-5.79
		Second	5.71	-3.68	0.35	-11.21	-2.98
	SWEET	First	5.63	-10.48	-8.88	-11.37	-6.75
		Second	4.91	-7.94	-4.68	-17.72	-9.16
CMU	P 3228	First	9.54	-11.43	-8.39	-13.10	-12.68
		Second	7.96	-1.38	4.65	-7.16	2.76
	SWEET	First	8.49	-11.66	-8.95	-12.25	-12.37
		Second	7.42	-8.36	-1.75	-15.09	-2.02
USM	P 3228	First	7.14	-15.27	-17.93	-17.79	-14.57
		Second	6.94	-16.14	-16.14	-17.29	-8.07
	SWEET	First	5.99	-15.86	-22.87	-18.36	-15.86
		Second	5.61	-18.18	-18.72	-18.18	-6.60

4.2.2 Effects on Corn Yields

Unlike the results for rice yield, simulations using all the GCM climate scenarios estimated a decrease in corn yield under 2×CO_2 conditions (Table IV). Based on the mean values of the outputs of the four climate scenarios, the decrease was lower during the second cropping except for sweet corn at Isabela State University. This, however, did not follow the general trend observed of a decrease in the length of the maturity period. The maturity period was estimated to decrease by between 2.67 and 12.79%, depending on the site, crop variety, and cropping season, and was more negatively affected during the second cropping season (except for the sweet crop variety at the University of Southern Mindanao site).

5. Discussion and Conclusions

The results of crop simulations conducted using four GCM climate change scenarios showed that the impact of climate change on two rice crop varieties varied widely. The CCCM, GFDL, and UKMO scenarios used as inputs to the DSSAT 3 model to simulate the impact of climate change on crop yields showed varied but generally positive results; however, the GISS scenarios showed generally negative results for rice crops. The results for sweet and P3228 corn crops showed a decrease in yield under all climate change scenarios. This decrease may be due to the decrease in maturity period as a result of increased CO_2 concentration; CO_2 concentrations would increase the day and night temperatures, which are very important in dry matter accumulation. The rainfall in tropical areas like the Philippines is generally high, therefore a decrease in rainfall by only 10% may not affect the crop water requirements significantly, but an increase by the same magnitude may affect the production tremendously because of more flooding.

It is possible that rice, which is a C3 plant, would respond to the direct effects of CO_2 both in photosynthetic carbon assimilation and in regulating the gaseous exchange of CO_2 and water vapor through stomata, thereby contributing to the estimate of mixed impacts on rice yields. It is also possible that corn, which is a C4 plant, would not respond positively to the direct effects of CO_2. Corn would be more affected by the increase in air temperature, enhancing respiration and shortening the maturity period, thereby decreasing the yield (WMO Technical Note No. 196).

References

ADB: 1994, Climate Change in Asia: Philippines Country Report on the Regional Study on Global Environmental Issues, Asian Development Bank, Manila.

Benioff R., Guill S., and Lee J. (eds): In press, *Vulnerability and Adaptation Assessments: An International Guidebook*, Kluwer Academic Publishers, Dordrecht, The Netherlands.

Boer G.J., McFarlane N.A., and Lazare M.: 1992, *Bulletin of the American Meteorological Society* **5**, 1045.

DENR: 1991, A Report on Philippine Environment and Development — Issues and Strategies to the UNCED at Rio de Janeiro, Brazil, EMB, DENR, December 1991.

Escaño C.R. and Buendia L.V.: 1994, Climate impact assessment for agriculture in the Philippines: simulation of rice yield under climate change scenarios, in: *Implications of Climate Change for International Agriculture: Crop Modeling Study*, Rosenzweig C. and Iglesias A. (eds), U.S. Environmental Protection Agency, Washington, DC.

Hansen J., Russell G., Rind D., Stone P., Lacis A., Lebedeff S., Ruedy R., and Travis L.: 1983, *Monthly Weather Review* **3**, 609.

Mitchell J.F.B., Manabe S., Tokioka T., and Meleshko V.: 1990, Equilibrium change, in: *Climate Change: The IPCC Scientific Assessment*, Houghton J.T., Jenkins G.J., and Ephramus J.J. (eds), Cambridge University Press, New York.

NSCB: 1992, *Philippine Statistical Yearbook*, National Statistical Coordination Board, Manila.

PCARRD (Philippine Council for Agriculture, Forestry and Natural Resources Research and Development) National Post Harvest Institute for Research and Extension and the International Development Research Center: 1990, The Philippines Recommendations for Corn Post-Production Operations, Los Baños, Laguna, Technical Bull. Ser. No. 71.

Rosenzweig C., Jones J., Kiker G., Adams R., and Tubiello F.: 1995, *Handbook on Climate Change Crop Modeling Techniques*, U.S. Country Studies Program, Vulnerability and Adaptation Assessments, Honolulu, Hawaii.

Wilson C.A. and Mitchell J.F.B.: 1987, *Journal of Geophysical Research* **92**, 13315.

WMO: 1994, Climate Variability, Agriculture and Forestry, Report of the CAgM-1X Working Group on the Study of Climate Effects on Agriculture, including Forests, and of the Effects of Agriculture and Forest on Climate, WMO Technical Note #196, World Meteorological Organization, Geneva, Switzerland.

ASSESSING IMPACTS OF CLIMATIC VARIATIONS ON FOODGRAIN PRODUCTION IN BANGLADESH

Z. KARIM, S.G. HUSSAIN, and M. AHMED
Bangladesh Agricultural Research Council, Farm Gate, New Airport Road
Dhaka 1215, Bangladesh

Abstract. A simulation study was conducted to assess the vulnerability of foodgrain production in Bangladesh to potential climate change. Simulation runs were made for high yield varieties of rice for Aus (March-August), Aman (July-November), and Boro (February-July), the traditional growing seasons, using the CERES-Rice model. Simulation runs were made for wheat, which is grown from November through March, using the CERES-Wheat model. Three scenarios (baseline, Canadian Climate Centre Model, and Geophysical Fluid Dynamics Laboratory) and sensitivity analyses for temperature increases of 2 and 4°C at three levels of CO_2 (330, 580, and 660 ppm) were used. In the simulation, increased CO_2 levels increased rice yields over baseline, and considerable spatial and temporal variations were noted. Higher temperatures reduced the yields in almost all study locations and in all seasons, and it was particularly pronounced with a 4°C increase. The detrimental effect of temperature rise was observed even with increased CO_2 levels. Wheat yields increased with increased CO_2 level in all three locations. The adverse effects of increased temperature were more pronounced for wheat than for rice at all levels of increased CO_2. In the scenarios of the Canadian Climate Centre Model and Geophysical Fluid Dynamics Laboratory, both rice and wheat yields were decreased (35% and 31%, respectively) over baseline in all seasons, especially in the Aus season, and in all locations. The adverse effects of the climate under the Geophysical Fluid Dynamics Laboratory scenario were more pronounced for wheat than for rice. The greatest reductions in aggregated production for both crops were noted at 330 ppm CO_2 with a 4°C temperature rise. The greatest increases in aggregated production for rice and wheat were observed at a 660 ppm CO_2 with no temperature increase.

Key words: Bangladesh, agriculture, rice, wheat, CCCM, CERES-Rice, CERES-Wheat, GFDL

1. Introduction

Global warming due to increasing concentrations of greenhouse gases poses a threat to human society by changing the living and working environment to which society has adapted over many generations (Jodha, 1989). Agricultural impacts of climate change could have profound consequences in poor and developing countries.

Bangladesh, a developing country in South Asia, is primarily a deltaic floodplain, and elevations in most of the country do not exceed 10 m. The country has a humid tropical climate. Rains start in March-April and end in October, and average 1,500 mm per year in drier regions to 5,000 mm per year in wetter regions. Winter in Bangladesh is from November to February, when the average minimum and maximum daily temperatures are about 9.7 and 26.6°C, respectively. Summer is the hottest period of the year, with temperatures generally above 32.2°C but rarely exceeding 40.0°C. Maximum solar radiation occurs during March, April, and May and ranges from 18 to 16 MJ m^{-2} per day; minimum solar radiation occurs during December and ranges from 14 to 12 MJ m^{-2} day (BARC, 1991).

Bangladesh is highly vulnerable to severe weather events; floods and droughts are frequent and damaging (Karim, 1995). The crop most affected by floods and droughts is rice, which contributes 93.96% of the total cereal production in the country, followed by wheat (5.81%) and other crops (0.23%) (computed from BBS, 1993).

This study considers the possible impacts of climate change on the production of high yield varieties of rice and on wheat production. The study looks at the effects of two GCMs and eight

Water, Air, and Soil Pollution **92:** 53–62.
© 1996 *Kluwer Academic Publishers. Printed in the Netherlands.*

combinations of temperature and CO_2 increases in six locations for rice, and in three locations for wheat. These locations represent the major rice and wheat growing regions of the country.

2. Methods

Six locations, representing the drier (Jessore and Rajshahi), wetter (Sylhet and Mymensingh), and coastal regions (Chittagong and Barisal) of the country, were chosen for the rice study, and three (Jessore, Rajshahi, and Mymensingh) were chosen for the wheat study (Figure 1). Table I presents 1991-1992 statistics of the major cereal crops in Bangladesh. A simulation study was conducted for three high yield variety (HYV) rice crops, grown during three seasons: Aus (March-August), Aman (July-November), and Boro (December-May). The selected varieties were BR14, BR11, and BR3 for Aus, Aman, and Boro, respectively. In the simulation experiments, Aus rice was planted on 1 April, Aman rice on 15 July, and Boro rice on 1 February. The wheat variety chosen was Akbar; it was planted on 5 November in Jessore and Rajshahi and on 15 November in Mymensingh. Two planting dates for wheat were used because the median planting dates varied with location.

2.1 CROP MODELS

The simulation runs for rice and wheat were made by using CERES-Rice and CERES-Wheat models of Decision Support System for Agrotechnology Transfer version 3.0 (Tsuji *et al.*, 1994). Eighteen experiments (three crops at six locations) for rice and three experiments for wheat (one crop at three locations) were performed. Each experiment was composed of 11 simulations.

There are several input requirements of these crop models. These inputs and how they were generated are described in this section.

Genetic coefficients. Both CERES models use as inputs coefficients that account for differences in growth and development among cultivars. These coefficients are referred to as genetic coefficients. Genetic coefficients allow the model to simulate performance of different varieties under diverse weather and management conditions. Therefore, to obtain reasonable outputs

TABLE I
Crop statistics of major cereals for fiscal year 1991-1992

Crop	Area (ha)	National Average Yield ($t\ ha^{-1}$)[§]	Current Production (t)
HYV Aus	412,805	2.42	765,990
HYV Aman	1,763,756	2.96	4,016,720
HYV Boro	2,290,053	3.56	6,264,020
Other Rice	5,778,114	1.63	7,204,910
Rice Total	10,244,727	2.29	18,251,640
Wheat	574,662	1.85	1,065,050
Major Cereal Total	10,819,389	—	19,316,690

[§] Rice average yields are expressed as rough rice.
Source: BBS, 1993.

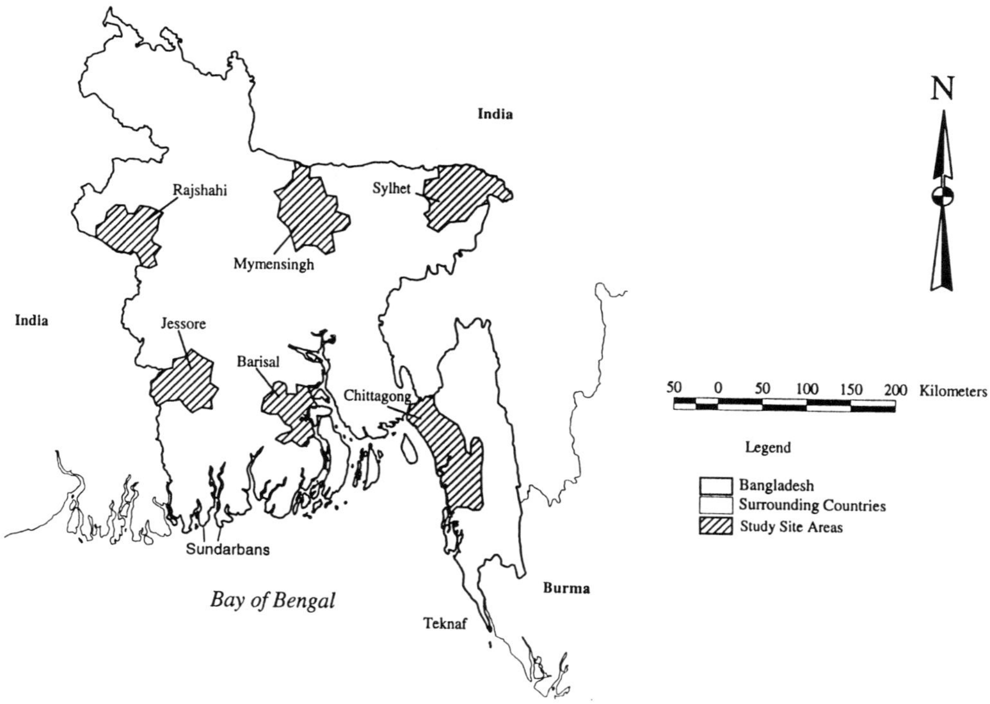

Fig. 1. Administrative map of Bangladesh. Regions selected for this study are marked with a dot.

from the crop simulation models, it is necessary to have the appropriate genetic coefficients for the selected cultivars. Since the genetic coefficients for the varieties considered in this study were not available, they were estimated by using GENCALC of DSSAT v3.0., a computer software program that facilitates estimation of crop genetic coefficients (Hunt *et al.*, 1993).

Soils. Soil information is contained in a soil data file. The file contains information collected for the soil profile at a specified site along with supplemental information extracted from soil survey reports for a soil with similar taxonomic classification. The file is created by maintaining the format required by the DSSAT v3.0 crop models. These data are used by the soil water, nitrogen, and root growth submodels of the CERES crop models. Soil profiles in DSSAT v3.0 input file format were created from information compiled from various sources (FAO/UNDP, 1971, 1988; Joshua and Rahman, 1983) and were added to the SOIL. SOL file of DSSAT v3.0.

Crop data. The required crop data on development and growth characteristics and cultural practices for the selected varieties were collected from BRRI (1981, 1984, 1985a, 1985b, 1985c, 1985d, 1991a, 1991b) and BARI (1987, 1988, 1989) publications. The development and growth characteristics information was used to estimate genetic coefficients of the selected varieties. The cultural practices information was used for creating different simulation experiments.

Weather data. Ten years of weather data (daily) for the six sites (see Figure 1) was obtained from the Bangladesh Meteorological Department and stored in the BARC Computer Center as

ASCII files. The weather data include daily values for the following variables: rainfall (mm), maximum and minimum temperatures (°C), and bright sunshine hours (h).

The weather data files were converted to DSSAT v3.0 weather file format by a FORTRAN program and the bright sunshine hour data were converted to solar radiation (MJ m^{-2} per day) using the WFORM.EXE (EPA/USAID, 1990) program. Ten years of bright sunshine hour (daily) data were used to calculate the daily solar radiation for each station.

2.2 CLIMATE CHANGE SCENARIOS

Three scenarios were used: one baseline, and two climate change scenarios. The climate change scenarios were generated by two general circulation models (GCMs) — the Canadian Climate Centre Model (CCCM) (Boer *et al.*, 1992) and the Geophysical Fluid Dynamics Laboratory (GFDL) model (Mitchell *et al.*, 1990).

For the sensitivity analyses, temperature increases of 0, 2, and 4°C at three levels of CO_2 (330, 580, and 660 ppm) were used. Different combinations of these variables can show the independent effects of temperature and CO_2 increases, and how they interact.

3. Results

3.1 IMPACT OF CLIMATE CHANGE SCENARIOS ON CROP YIELDS

3.1.1 Rice

Under the baseline scenario, Aus rice yields varied from a minimum of 5.10 t ha^{-1} in Chittagong to a maximum of 5.66 t ha^{-1} in Mymensingh. Similar minimum (5.31 t ha^{-1}) and maximum (6.35 t ha^{-1}) of Aman rice yields were obtained for Mymensingh and Chittagong. Maximum (8.61 t ha^{-1}) and minimum (8.02 t ha^{-1}) Boro rice yields were recorded in Sylhet and Chittagong, respectively.

Yield reductions for all rice varieties in all locations were noted for both CCCM and GFDL scenarios. The yield of Aus rice was reduced 19-35% under the CCCM scenario, and 22-31% under the GFDL scenario at different locations. For Aman rice, yield reductions varied from 3 to 18% and from 8 to 19% for the CCCM and GFDL scenarios, respectively. For Boro rice, yields under the CCCM and GFDL scenarios were also reduced, but the yield reductions ranged from 0 to 11%.

3.1.2 Wheat

Under the baseline scenario, a maximum (3.32 t ha^{-1}) wheat yield was recorded at Mymensingh and a minimum (2.90 t ha^{-1}) was recorded at Jessore. The yields under the CCCM and GFDL scenarios were reduced. Under the CCCM scenario yield reductions varied from 17 to 22%, and under the GFDL scenario yield reductions were as high as 58-63%.

3.2 SENSITIVITY ANALYSES

3.2.1 Rice

In general, increased levels of CO_2 increased all rice yields at all locations. The maximum yield increase of 44% was noted for a CO_2 level of 660 ppm. With increased temperatures, yields of Aus, Aman, and Boro rice decreased for CO_2 levels of 330 ppm. However, Boro yields generally increased for a 2°C warming with CO_2 levels of 580 and 660 ppm. Even at 4°C temperature rise, yield increases were generally found for Aman and Boro rice at 580 and 660 ppm CO_2 levels. The percentages of yield difference as compared to baseline are presented for Aus, Aman, and Boro rice at the six study sites in Table II.

3.2.2 *Wheat*

Increased wheat yields were noted for increased levels of CO_2 with no rise in temperature (Table III). Yields were drastically reduced by increased temperature, even at higher levels of CO_2. Yield reductions for a 2°C temperature increase at 330 ppm CO_2 ranged from 29% in Mymensingh to 44% in Jessore; for a 4°C increase at 330 ppm CO_2, reductions varied between 59 and 73% for the same locations. Significant yield reductions were also observed for the 4°C increase, even at 580 and 660 ppm CO_2 levels. At 580 ppm CO_2, yields were reduced between 29 and 44%, and at 660 ppm CO_2, the reductions ranged from 21 to 35%.

3.3 EFFECT OF CLIMATE CHANGE ON AGGREGATED RICE AND WHEAT PRODUCTION

Production of rice and wheat for the entire country was aggregated by crop and by season (Table IV). The observations are discussed in the following sections.

3.3.1 *Aus Rice*

The average aggregated production of Aus rice in Bangladesh was 7.7×10^5 t during 1991-1992. Under both CCCM and GFDL climatic scenarios, production was reduced by 27%. When the CO_2 levels were increased to 580 and 660 ppm without any increase in temperature, production was increased by 31 and 40%, respectively.

A 2°C increase at 330 ppm CO_2 reduced aggregated production by 19%. A 2°C increase at 580 and 660 ppm CO_2 increased aggregated production by 13 and 22%, respectively. An increase of 4°C at 330 and 580 ppm CO_2 reduced production by 38 and 6%, respectively, and increased production by only 4% at 660 ppm $_{CO_2}$.

3.3.2 *Aman Rice*

The aggregated production of Aman rice for 1991-1992 was 4.02×10^6 t. Simulation results show that under the CCCM scenario, production was lowered by 7%. Under the GFDL scenario, it was reduced by 13%. At 330 ppm CO_2, the production reductions were 13 and 25% for temperature increases of 2 and 4°C, respectively. At 580 and 660 ppm CO_2, the effect of temperature was also detrimental, but production was higher than baseline.

3.3.3 *Boro Rice*

The baseline production of Boro rice for 1991-1992 was 6.26×10^6 t. Under the CCCM and GFDL climate scenarios, production decreased by 3 and 7%, respectively. The production decreases were similar for temperature increases of 2°C (4%) and 4°C (7%), respectively. The increased production over baseline varied between 16 and 30% for combinations of increased temperature and CO_2. In general, though, the 4°C increase was more detrimental than the 2°C increase, despite of the level of CO_2.

3.3.4 *Wheat*

The baseline production of wheat for 1991-1992 was 1.07×10^6 t. Under the CCCM and GFDL scenarios, production was reduced by 20 and 61%, respectively. At 330 ppm CO_2, temperature

TABLE II
Rice yields under different climate change scenarios

	Chittagong			Barisal			Mymensingh		
Simulation	Aus	Aman	Boro	Aus	Aman	Boro	Aus	Aman	Boro
Baseline (t ha^{-1})	5,101	5,311	8,026	5,571	6,178	8,328	5,661	6,031	8,235
	Percent Change over Baseline								
CCCM	-25	-18	-6	-28	-10	-3	-24	-7	0
GFDL	-24	-19	-6	-25	-11	-4	-27	-14	-10
330 ppm CO_2 + 2°C	-19	-24	-8	-18	-12	-4	-17	-12	-1
330 ppm CO_2 + 4°C	-37	-42	-10	-40	-23	-6	-30	-20	-5
580 ppm CO_2 + 0°C	32	31	23	30	23	24	29	24	23
580 ppm CO_2 + 2°C	13	7	15	13	10	20	13	11	22
580 ppm CO_2 + 4°C	-5	-10	13	-8	1	18	0	5	17
660 ppm CO_2 + 0°C	41	40	29	39	31	31	37	31	30
660 ppm CO_2 + 2°C	22	16	22	22	17	28	21	18	29
660 ppm CO_2 + 4°C	5	-1	20	2	8	25	9	12	24

	Sylhet			Rajshahi			Jessore		
Simulation	Aus	Aman	Boro	Aus	Aman	Boro	Aus	Aman	Boro
Baseline (t ha^{-1})	5,528	6,355	8,613	5,487	6,118	8,291	5,067	5,717	8,260
	Percent Change over Baseline								
CCCM	-19	-12	-6	-31	-3	-3	-35	-4	-4
GFDL	-22	-12	-11	-31	-12	-6	-30	-8	-4
330 ppm CO_2 + 2°C	-16	-13	-9	-18	-11	-3	-22	-10	-2
330 ppm CO_2 + 4°C	-34	-22	-13	-41	-22	-6	-47	-22	-9
580 ppm CO_2 + 0°C	29	24	22	30	24	24	34	24	24
580 ppm CO_2 + 2°C	14	9	14	13	12	21	13	14	22
580 ppm CO_2 + 4°C	-3	2	8	-9	2	18	-14	5	14
660 ppm CO_2 + 0°C	37	31	29	39	31	32	44	32	32
660 ppm CO_2 + 2°C	22	15	20	23	19	28	24	21	30
660 ppm CO_2 + 4°C	6	9	14	1	10	25	-4	13	22

TABLE III
Wheat yields under different climate change scenarios

Simulation	Mymensingh	Rajshahi	Jessore
Baseline (t ha^{-1})	3,317	3,265	2,901
	Percent Change over Baseline		
CCCM	-17	-21	-22
GFDL	-58	-62	-63
330 ppm CO_2 + 2°C	-29	-39	-44
330 ppm CO_2 + 4°C	-59	-71	-73
580 ppm CO_2 + 0°C	32	39	44
580 ppm CO_2 + 2°C	4	-2	-4
580 ppm CO_2 + 4°C	-29	-43	-44
660 ppm CO_2+ 0°C	41	49	56
660 ppm CO_2 + 2°C	12	9	7
660 ppm CO_2 + 4°C	-21	-34	-35

TABLE IV
Aggregated HYV rice and wheat production of Bangladesh under different climate change scenarios

Simulation	HYV Aus (10^6 t)	Percent Change	HYV Aman (10^6 t)	Percent Change	HYV Boro (10^6 t)	Percent Change	Rice Total (10^6 t)	Percent Change	Wheat (10^6 t)	Percent Change
Baseline	0.77	—	4.02	—	6.26	—	11.05	—	1.07	—
CCCM	0.56	-27	3.75	-7	6.09	-3	9.93	-10	0.85	-20
GFDL	0.56	-27	3.51	-13	5.81	-7	9.18	-17	0.42	-61
330 ppm CO_2 + 2°C	0.63	-19	3.48	-13	6.04	-4	9.46	-14	0.67	-37
330 ppm CO_2 + 4°C	0.48	-38	3.00	-25	5.81	-7	7.99	-28	0.34	-68
580 ppm CO_2 + 0°C	1.01	31	5.02	25	7.73	23	15.01	36	1.47	38
580 ppm CO_2 + 2°C	0.87	13	4.47	11	7.49	20	13.39	21	1.06	-1
580 ppm CO_2 + 4°C	0.72	-6	4.06	1	7.24	16	12.03	9	0.64	-40
660 ppm CO_2 + 0°C	1.08	40	5.33	33	8.17	30	16.19	47	1.58	48
660 ppm CO_2 + 2°C	0.94	22	4.76	19	7.95	27	14.56	32	1.16	9
660 ppm CO_2 + 4°C	0.80	4	4.37	9	7.69	23	13.24	20	0.74	-31

increases of 2 and 4°C reduced production by 37 and 68%, and at 580 ppm CO_2, the production loss associated with these temperature increases was 1 and 40%. At 660 ppm CO_2, the production increased by 9% under a temperature increase of 2°C, but was reduced by 31% with a 4°C increase. In general, higher levels of CO_2 with no temperature change increased production by 38 to 48%.

3.4 EFFECT OF CLIMATE CHANGE ON CROP GROWING SEASON

In general, increases in temperature decreased the crop growing season by 2 to 12 days compared to baseline temperatures. Under CCCM and GFDL, the reductions were 7 to 10 days. Season length under the CCCM scenario was shorter than under the GFDL scenario.

4. Discussion

The growth and yield of crops are directly related to the rate of photosynthesis and phenology, and their response to temperature and radiation. Optimum temperatures for maximum photosynthesis range from 25 to 30°C for rice and from 15 to 20°C for wheat under the climatic conditions of Bangladesh. In general, the length of the cool winter periods in different zones of the country determines the rate of growth and development of wheat and Boro rice.

The results show that the effects of the CCCM and GFDL scenarios on the Aus and Aman rice crops were detrimental. Under the CCCM scenario the average growing season temperatures for Aus, Aman, and Boro rice were 31.2, 29.9, and 28.7°C, respectively. Under the GFDL scenario, the average growing season temperatures for Aus, Aman, and Boro rice were 31.0, 29.2, and 29.8°C, respectively. Increased temperatures during the growing season cause grain sterility. At flowering, the rice plant is most susceptible to high temperature (>35°C), which causes high spikelet sterility (Sakate and Yoshida, 1978). In Bangladesh, this problem is encountered in the dry-season crop grown in the drier region (Jessore and Rajshahi). The effect of increased temperatures for CCCM and GFDL scenarios on Boro rice was mild because the increase was within the tolerance level of the crop.

Wheat was severely affected by the increased temperatures, even at higher levels of CO_2. For optimum growth of wheat, 60 days with minimum temperatures of less than 15°C is considered the minimum requirement for adequate tillering and panicle development (BARI, 1986). During the wheat growing period, a 4°C increase considerably reduces time under cool temperatures. A result of the rise in temperature during panicle development, which is a very vulnerable stage of growth for wheat, is grain sterility.

Sensitivity analyses show that the crop yields as well as production increased with the increase in CO_2 levels. This indicates that the crop models are sensitive to CO_2 levels and the crops are capable of assimilating higher levels of CO_2. Up to a certain limit, the detrimental effects of increased temperatures are either tolerated or balanced at higher levels of CO_2. However, in general, rice crops were less susceptible than wheat to the detrimental effects of increased temperature. Rice and wheat have maximum rates of photosynthesis in the range of 15-30 mg CO_2 $dm^{-2}h^{-1}$ with light saturation at 0.2-0.6 cal cm^{-2} min^{-1}. Wheat is adapted to grow under conditions of moderately cool temperatures (10-20°C), but rice is adapted to grow under conditions of moderately warm temperatures (25-30°C) (BARC, 1991). Mahtab and Karim (1992) reported that with the current technology it would be possible to produce 25.52 and 31.95×10^6 t of foodgrain by 2000 and 2010, respectively. But this level of increased production would not be sufficient in view of estimated population increases from 115 million today to 145, 177, and 232 million by 2000, 2010, and 2030, respectively (World Bank, 1990). In Bangladesh, the population pressure is already affecting the sustainability of agricultural development.

To meet the challenges of food security and growing demand, agricultural production must increase at a faster rate. The growth of food production must take place in the existing land base by increasing cropping intensity, developing crops and cultivars more resistant to stresses due to foreseeable climate changes, and adopting technologies more appropriate for conditions of increased CO_2 and radiation.

Possible response strategies to impacts of climate change may be the following: limiting greenhouse gas emissions, substituting low yielding local varieties with HYVs, reducing dependency on rice by increasing fruits and animal protein in daily diet, and substantially reducing the population growth rate. These are all possible, if the people of the country are properly motivated. But, if climate change has a negative impact on other sectors then these measures may not work. It is anticipated that if sea level rises about 16% of Bangladesh could be covered by the sea to a depth of 1 m (Mahtab, 1987). This will reduce the area of cultivated as well as homestead lands, and as a result the crop production will be reduced. The surface and groundwater quality and availability for irrigation may also change.

5. Conclusions

The impact of climate change on foodgrain production may not always be a problem. The rice crop cultivars available today could produce more grain under the higher temperatures associated with the speculated CO_2 rise. If the CO_2 level increases without increasing the temperatures, especially in the tropics, it would be a blessing. A doubling of CO_2, as estimated by different GCMs, would produce a temperature increase in the range of >1 °C to 4.8 °C. The rice varieties now in use may not be able to tolerate the projected temperature increase.

If the population grows to 232 million by the year 2030, then the problem would be more acute. Currently, about 44% of the total rice area in Bangladesh is under HYV cultivation. It is expected that by the year 2050 the total rice area would be under HYV cultivation. However, even then the rice production is not projected to be adequate to meet demand. Therefore, development of cultivars that would be able to withstand a 4 °C to 5 °C temperature rise and have greater photosynthesis efficiency may mitigate the production problems projected for the next 50 years. Public awareness of the impact of climate change on the agricultural production systems deserves priority consideration, and mitigating technologies must be developed, which will require increased public and private investment.

Acknowledgments

The authors acknowledge the U.S. Country Study on Climate Change and the UNEP/FAO-BARC Climate Change and Global Agricultural Potentials projects for providing financial support to carry out this study. The cooperation rendered by the Bangladesh Centre for Advanced Studies in letting us use their E-mail facilities is also appreciated.

References

BARC: 1991, *Agroecological Database*, BARC Computer Centre, Bangladesh Agricultural Research Council, Dhaka.

BARI: 1986, *Wheat Research Centre Annual Report 1986*, Bangladesh Agricultural Research Institute, Joydebpur, Gazipur.

BARI: 1987, *BARI Annual Report 1987-88*, Bangladesh Agricultural Research Institute, Joydebpur, Gazipur, pp. 1-34.

BARI: 1988, *BARI Annual Report 1988-89*, Bangladesh Agricultural Research Institute, Joydebpur, Gazipur, pp. 3-31.

BARI: 1989, *BARI Annual Report 1989-90*, Bangladesh Agricultural Research Institute, Joydebpur, Gazipur, pp. 1-36.

BBS: 1993, *Yearbook of Agricultural Statistics of Bangladesh 1992*, Bangladesh Bureau of Statistics, Statistics Division, Ministry of Planning.

Boer G.J., McFarlane N.A., and Lazare M.: 1992, *Bulletin of the American Meteorological Society* **5**, 1045.

BRRI: 1981, BRRI Annual Report for 1978-79, Pub. No. 50: 6-1, Bangladesh Rice Research Institute, Gazipur, Bangladesh.

BRRI: 1984, BRRI Annual Report for 1980, Pub. No. 71: 14-15, Bangladesh Rice Research Institute, Gazipur, Bangladesh.

BRRI: 1985a, BRRI Annual Report for 1981, Pub No. 71: 7-9, Bangladesh Rice Research Institute, Gazipur, Bangladesh.

BRRI: 1985b, BRRI Annual Report for 1982, Pub. No. 79: 4-5, Bangladesh Rice Research Institute, Gazipur, Bangladesh.

BRRI: 1985c, BRRI Annual Report for 1983, Pub. No. 80: 6-9, Bangladesh Rice Research Institute, Gazipur, Bangladesh.

BRRI: 1985d, BRRI Annual Report for 1984, Pub. No. 81: 7-2, Bangladesh Rice Research Institute, Gazipur, Bangladesh.

BRRI: 1991a, BRRI Annual Report for 1988, Pub. No. 98: 5-10, Bangladesh Rice Research Institute, Gazipur, Bangladesh.

BRRI: 1991b, *Adhukin Dhaner Chash* (Bengla), Bangladesh Rice Research Institute Publication No. 5, Sixth Edition, Gazipur, Bangladesh.

CSE (Centre for Science & Environment): 1995, Science of Global Warming and Climate Change, working paper presented at the CSE-Organized South Asian Regional Workshop on Science and Climate Change, 21-23 November, 1995, New Delhi, India, pp. 1-41.

EPA/USAID (Environmental Protection Agency/U.S. Agency for International Development): 1990, *Climate Change Crop Modeling Study Handbook*, Rosenzweig C. and Jones J.W. (eds), International Climate Change and Crop Modeling Workshop, 28 January-2 February 1990, Washington, DC, pp. 20-21.

Hunt L.A., Pararajasingham S., Jones J.W., Hoogenboom G., Imamura D.T., and Ogoshi R.M.: 1993, *Agron. J.* **85**, 1090.

Jodha N.S.: 1989, Potential strategies for adapting to greenhouse warming: Perspectives from the developing world, in: *Greenhouse Warming: Abatement and Adaptation*, Rosenberge N.J., Eaterling III W.E., Crosson P.R., and Darmstadter J. (eds), RFF Proceedings, Climate Resources Program, Energy and Natural Resources Division, pp. 147-158.

Joshua W.D. and Rahman M.: 1983, Physical Properties of Soils of the Tista River Floodplain and Barind Tract of Bangladesh, Strengthening of Soil Resources and Development Institute FAO/UNDP Project BGD/81/023, SRDI, Government of the People's Republic of Bangladesh, Dhaka.

Karim Z.: 1995, Agricultural vulnerability and poverty alleviation in Bangladesh, in: *Climate Change and World Food Security*, Springer, pp. 307-346.

Mahtab F.U.: 1987, *Effects of Climate Change and Sea Level Rise on Bangladesh*, Expert Group on Climate Change and Sea Level Rise, Commonwealth Secretariat, London, England.

Mahtab F.U. and Karim Z.: 1992, *Ambio* **21**, 50.

Mitchell J.F.B., Manabe S., Tokioda T., and Meleshko V.: 1990, Equilibrium change, in: *Climate Change: The IPCC Scientific Assessment*, Houghton J.T., Jenkins G.J, and Ephramus J.J. (eds), Cambridge University Press, New York.

Sakate T. and Yoshida S.: 1978, *J. Crop Sci.* **47**, 6.

Soil Survey Staff: 1995, *Soil Taxonomy: A Basic System of Soil Classification for Making and Interpreting Soil Surveys*, Soil Conservation Service, U.S. Dept. of Agriculture Handbook 436, U.S. Government Printing Office, Washington, DC.

Tsuji G.Y., Uehara G., and Balas S. (eds): 1994, DSSAT Version 3.0. IBSNAT, Department of Agronomy and Soils, University of Hawaii, Honolulu.

UNDP/FAO: 1971, Soil Survey Project: Bangladesh Soil Resources, AGL:SF/PAK 6, Tech, Report 3, Rome, Italy.

World Bank: 1990, Bangladesh Managing the Adjustment Process — An Appraisal, Statistical Appendix.

AGRICULTURAL VULNERABILITY AND ADAPTATION TO GLOBAL WARMING IN CHINA

LIN ERDA

Agrometeorology Institute, Chinese Academy of Agricultural Sciences, Beijing 100081, China

Abstract. This paper discusses the vulnerability and adaptation of the agricultural sector of China to global warming. Based on a summarization of Chinese agricultural and general circulation model trends, adverse impacts on China's agriculture caused by a warming and drying climate were identified. Because of limited irrigation potential, the sustainable development of Chinese agriculture will be difficult. Six sensitive agricultural areas located on the edges of different agroecological zones, and seven provinces with high vulnerability to the impacts on agriculture, were identified. On the basis of an estimation of the potential supply of agricultural products and demand for food, the annual incremental costs for adaptation to climate change would be US$0.8-3.48 billion; without adaptation, the annual agricultural loss due to global warming would be US$1.37-79.98 billion from 2000 to 2050. Adaptive measures discussed include intensive management and the possibility of a tripartite structure of planting that would entail coordinated development of grain crops, feed crops, and cash crops.

Key words: China, agriculture, maize, rice, wheat, GFDL, MPI, UKMO

1. Introduction

Climate change would affect agriculture through effects on crops; soils; insects, weeds, and diseases; and livestock. Over the past nearly 40 years, most of China, especially in the north, has experienced a general warming trend. There has been a substantial decrease in precipitation over eastern China's farming regions, except for part of the northeast and the middle and lower Yangtze River. Irrigation, already important, has become even more vital, with two thirds of the country's grain, (60% of its economic crops), and 80% of its vegetables produced on the 45×10^6 ha (49.6% of farmland in 1990) equipped with effective irrigation facilities. If these climate trends continue, much of Chinese agriculture is likely to face shorter growing periods and increased water deficits, requiring even more irrigation. Yet the maximum irrigable potential of China's farmland under "normal" conditions is estimated at only 64×10^6 ha, or 60% of the total cultivated lands. Because China is already approaching the maximum irrigation of farmland, potential future water shortages may threaten the sustainability of China's agriculture development.

Agriculture is a very important sector for China's development. The overall objective of this paper is to forecast changes in crop production potential, and resulting economic impacts, for wheat and maize throughout the country and rice in some areas, given alternative general circulation model (GCM) scenarios. This paper also identifies sensitive agriculture areas and regions with high vulnerability. This paper describes the methodologies used for crop simulation, identification of sensitive regions, and identification of regions with high vulnerability as well as the limitations of these methodologies. The results and discussion section of this paper includes the output of crop simulations, analysis of sensitive areas, and a description of relative vulnerability. The final sections of the paper discuss potential agricultural supply and demand and potential adaptation to climate change.

Water, Air, and Soil Pollution **92:** 63–73.
© 1996 *Kluwer Academic Publishers. Printed in the Netherlands.*

2. Methodology

2.1 CROP SIMULATION METHOD

In China's Country Study analyses, the results of three equilibrium GCMs are being used as the scenarios of climate change: Geophysical Fluid Dynamics Laboratory (GFDL) (Mitchell *et al.*, 1990), Max Planck Institute for Meteorology (MPI) (Cubasch *et al.*, 1992) in Hamburg, Federal Republic of Germany, and United Kingdom Meteorological Office (UK89) (Mitchell *et al.*, 1990). Depending on the scenarios, temperatures are estimated to increase 0.4-1.9°C and rainfall will change by -12.9 to +24.6% in China across different GCMs, sites, and seasons.

Three crop models were adjusted where necessary to ensure that they could be applied to China (e.g., how soil and genetic characteristics are incorporated into the models). The three models are a rice model (IRRI, 1993), a wheat model (Ritchie, 1985), and a maize model (Jones and Kinir, 1986).

A simulation was conducted using monthly GCM scenarios and daily data for present (1955 to 1985) and changed climate from the Chinese Weather Generator developed by Agrometeorology Institute, Chinese Academy of Agricultural Sciences (CAAS). Two hundred sample years were simulated for each crop to allow probability distributions to be derived for the production potential of each crop. Because of the GCMs' geographic resolution, an economic impacts assessment is feasible at a provincial level, but the physical impacts assessment may be possible at a finer geographic resolution.

2.2 IDENTIFICATION OF SENSITIVE REGIONS

Global warming is estimated to cause a general northward movement of agroclimatic regions, with certain exceptions in the south of China where moisture deficits may increase even more than in the north. Because climate change could cause both warming and drying, the agroecosystems of the extreme southern portions of current regions would have to be modified. Using a moisture deficit of 50 mm (precipitation minus evapotranspiration) as a critical value, several sensitive regions in China were identified. Additionally, areas affected by too much moisture were identified.

2.3 IDENTIFICATION OF REGIONS WITH HIGH VULNERABILITY

Five nonclimatic indicators of agricultural vulnerabilities were used to identify areas of high vulnerability. These indicators are based on readily accessible Chinese data at the provincial level for 1990 (irrigated area, cultivated area, agricultural land, available land, crop yield, cropping index, area suffering from disasters, and farmer income), and are subject to the strengths and limitations thereof (*Chinese Agricultural Yearbook 1990*). The indicators chosen were the following:

- Ratio of irrigated area to cultivated area (IC, also referred to as the irrigation coefficient). A ratio below 0.50 may be particularly vulnerable to extreme drought.

- Ratio of land used in agriculture and animal husbandry to total available land (A/L). The national average of 0.5925 is a critical value for A/L.

- Yield and cropping index (I.Y.). An I.Y. less than average in both areas should be considered to have less recoverability, hence a higher vulnerability to climate change.

- Disaster index (D/S). The D/S is calculated as the area suffering from meteorological disaster/cultivated land. The national average (0.295) of the past 10 years is used as an indicator.

- Farmer income (FI). Farmer income (average farmer income per year) reflects the ability of farmers to adapt to climate change the national average of 571 yuan in 1990 was used as the critical value.

3. Limitations

Despite the limitations of current models, understanding, and data, the need for information about broader scale responses to climate change has led to many attempts to estimate agricultural production effects. As might be expected, the variation of agricultural and climatic conditions across the world leads to very different climate impacts in different countries and regions of countries. Because regional climate predictions and agricultural impacts are highly uncertain, estimating vulnerability to potential climate change provides an approach for considering regions at relatively greater risk should climate change adversely. The economic effects of future climate change for particular localities, regions, and the world depend on how future agricultural production meets the demands of a growing population and on how climatic effects on production are transmitted among regions through international trade. Historically, agriculture has proved to be highly adaptive to changing technology, resource conditions, and increasing demand, providing evidence of the potential for agriculture to adapt to changing climate.

Variations in agriculture systems, climates, resources, and economic characteristics across and within countries may be more important in determining the effect of climate change than differences in climate scenarios themselves. Agricultural policies are an important consideration in most regions, and these policies have had many and changing goals. Climate change is generally not among top policy priorities for agricultural policy makers, but climate change could affect the cost and likelihood of achieving other policy priorities such as food adequacy and reduction in chronic hunger, improving export competitiveness, assuring regional and national economic and social development, increasing farm income and the viability of rural communities, assuring water availability and quality, reducing or reversing land degradation and soil loss through erosion, and attaining other conservation and environmental objectives.

Given the uncertainties in both magnitude and direction of impact, a key issue is vulnerability to possible climate change. Vulnerability is used here to mean the potential for negative consequences that are difficult to ameliorate through adaptive measures given the range of possible climate changes that might reasonably occur. Defining an area as vulnerable is, thus, not a prediction of negative consequences of climate change; it is an indication that across the range of possible climate changes, there are some climate outcomes that would lead to relatively more serious consequences for a region than for other regions. Improvements in the spatial and temporal resolution of GCM predictions will allow better estimates of the likelihood of detrimental climate change.

Given this information, it would seem a trivial exercise to consider the dual side of the problem that agricultural scientists and farmers have faced since cultivation began. Rather than asking "How do we limit the negative impacts of climate and weather on production?" we should be asking instead, "If climate changes, what effects would the change have on production?"

4. Results/Discussion

4.1 CROP SIMULATION RESULTS

Based on the method described above, the ranges of simulated changes in crop production under 2×CO_2 and across GCM climate change scenarios, sites, and two types of water use (rainfed and irrigated) are given in Table I. Table I shows three crops with two or three patterns each (e.g., spring and summer maize; early, late, and single rice). A preliminary conclusion is that there will be approximately a 10% decrease in production of the three main crops under climate change.

4.2 SENSITIVE AREAS

The following six areas appear to be the most sensitive to climate change (Figure 1), based on a critical value of 50 mm moisture deficit, except the last two, which would experience increased rainfall and its effects.

1. *The area along the Great Wall* lies southeast of a transition belt between agriculture and animal husbandry. The transition belt's climatic division is determined mainly by the degree to which dryland crops suffer from the rainfall deficit and from soil erosion and desertization caused by wind. The transition belt has large rainfall variability, greatly affecting the stability of arid-land agriculture, which produces bumper harvests in a limited number of water-rich years but poor harvests and greatly reduced herbage in water-deficient ones. The northwestern edge of the transition belt will become arid grassland with an expected moisture deficitof 55 mm. It will be difficult to maintain annual production of maize and millet because it is estimated that evapotranspiration will increase and rainfall may decrease.
2. *On the Hai River Plain and the Huang River Plain* the moisture deficit could increase by 70 mm, increasing dry conditions in spring (March to May) as well as hot and dry winds, which will damage wheat production and limit the extension of two crops a year. Crops that are traditionally planted three times in two years like wheat, cotton, maize, and fruit trees will suffer from drought.
3. *The area north of Huaihe River and east of Shandong* lies along the south edge of the southern temperate zone. Climatic warming could cause the northward movement of

TABLE I
Estimated changes in crop production under 2×CO_2 scenarios

Crop	Rainfed Wheat	Irrigated Wheat	Single Rice	Early Rice	Late Rice	Rainfed Spring Maize	Irrigated Spring Maize	Rainfed Summer Maize	Irrigated Summer Maize
Low§	-21%	-15.2%	-21.4%	-7.9%	-16.0%	-19.4%	-8.6%	-11.6%	-11.6%
High§	+42%	+54.7%	-3.8%	+0.1%	-4.3%	5.3%	3.6%	+0.7%	+0.7%

§ The range between low and high represents possible development scenarios for Chinese agriculture.

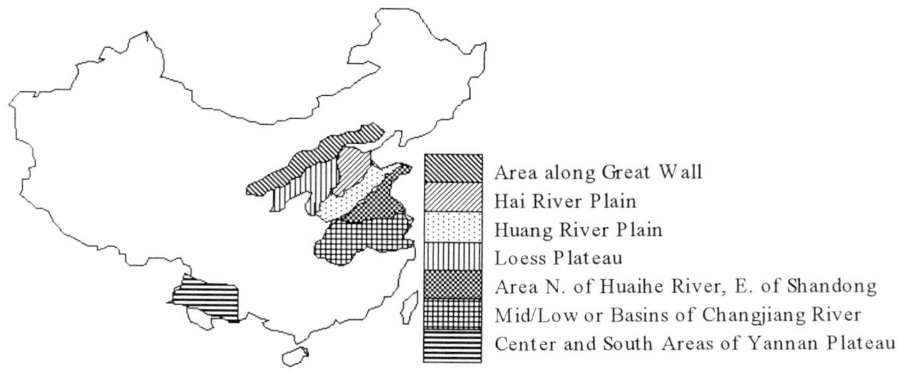

Fig. 1. Areas in China potentially sensitive to climate change.

subtropical crop areas. But frequent flooding in the south and spring dryness in the north because of growing water deficit mean that subtropical crops will not grow well in this area. The expected moisture deficit in this area is 65 mm.

4. *The central and southern areas of the Yunnan Plateau* contain regions apt to be affected by significant dryness in winter and spring under climate change. Climatic warming is estimated to increase the moisture deficit by 85 mm, and the rainfed farming system on poor red and yellow soil would be damaged. Rice and maize production, with higher water demand, would be seriously affected. The areas of trees and livestock raising might increase.
5. *The middle and lower basins of Changjiang River* are an area that would receive increasing rainfall and the largest summer variability; here alternate dog-season droughts (short droughts in the hottest season, July) and flooding often happen. Because the basins lie on the south edge of subtropical crop growth, crops planted two and three times a year like double-harvest rice, wheat, cotton, rape, bamboo, and mulberry would have higher productivity, as would fisheries and silkworm harvests. If climate change increases rainfall by 5%, and by more than 10% in summer, the agroecosystem with a high productivity will be damaged. This is because, depending on the historical data, increasing rainfall is often associated with large variability of rainfall in summer, which could cause flooding and damage the agroecosystem.
6. *The Loess Plateau* is a rainfed farming area with wheat as the main crop, which is planted as three crops in two years or two crops a year. Drought is the most damaging climate event for agriculture in this region. Because the rainfall in this region occurs mainly in summer (June to August), and is often heavy, and the spongy and fine loess is without vegetation, erosion is another damaging factor in this region. Some model results show the increase in rainfall might be 9-14% annually, and 16-32% in summer. Despite the uncertainties of GCM results, it is estimated that the agroecosystem of the area will not be able to tolerate the potential changes in rainfall.

The government and the farmer should pay attention to the sensitive areas mentioned above. Response strategies to increase adaptability, including introducing new species, using advanced technology, and increasing inputs, should be considered to reduce sensitivity to the impacts of climate change. The "anthropogenic sensitive areas" identified in this section may be sensitive to both increases and decreases in climate factors, and therefore both increases and decreases should be considered in management of the areas.

4.3 RELATIVE VULNERABILITY

Figure 2 provides a summary of "warning signals" for the vulnerabilities of the agricultural systems of China's 30 provinces to climate change. On Figure 2, "1" represents areas with one factor of higher vulnerability, "2" represents areas with two factors of higher vulnerability and continuing up to five factors of higher vulnerability. (Factors of higher vulnerability are identified in Section 2.3.) Shanxi and Inner Mongolia show vulnerabilities in all five dimensions. Nearby Gansu, Hebei, Shaanxi, Qinghai, and Ningxia are vulnerable in four indicators. These seven provinces produced 12% of China's total agricultural output value in 1990 (*Chinese Agricultural Yearbook 1990,* p. 165). In general, those provinces that are the most vulnerable are also in the areas of north and northwest China that, according to some models, are the most likely to be affected by global climate change.

5. Potential Agricultural Product Supply and Demand

5.1 DEMAND FOR AGRICULTURAL PRODUCTS

It is estimated that the population, grain production, and GNP from agriculture in China from 1990 to 2050 will increase as shown in Table II.

"The National Programme for the Reform and Development of Food Structure in China in the 1990s" set the following targets for food production in China in 2000: 500×10^6 t of grain, 39×10^6 t of meat, 14.3×10^6 t of eggs, 9.1×10^6 t of milks, 18×10^6 t of aquatic products, 32.5×10^6 t of fruits, 156×10^6 t of vegetables, 10.4×10^6 t of edible oil and 10.4×10^6 t of sugar. These targets represent estimates of China's food production needs in 2000. The estimated change in climate due to greenhouse gases will increase the difficulty in meeting these goals.

It is estimated that during the period from 2000 to 2050 the economy of China will attain the development of a medium-level developed country (Liangshu and Zhicheng, 1991a). In the first 20 years, the per capita consumption of agricultural products would have a small increase: grain by 12.5%, meat by 4%, edible oil by 20%, milks by 100%, and fruit by 100%. After 2020 the structure of food production is expected to stabilize. Apart from the increment of demand for agricultural products due to increases in population, total demand will reflect preferences for increased quality. The structure of food production is also expected

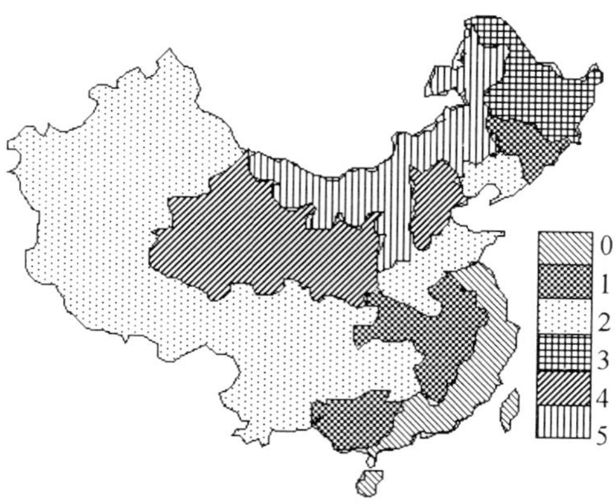

Fig. 2. Districts in China vulnerable to climate change based on nonclimatic factors.

TABLE II
Factors affecting agricultural product supply and demand

Year	Population (millions)	Grain Production (10^6 t)	GNP of Agriculture (US$ billion)	
			High	Low
1990	1,143	425	105.08	105.08
2000	1,300	500	158.84	151.81
2020	1,450	650	413.0	296.8
2050	1,500	675	993.6	463.05

Source: ADB, 1993.

to respond to the rationalization of the internal structure of agricultural production. Consequently, a total production of grain of 650 × 10^6 t in 2020 and 675 × 10^6 t in 2050 could basically meet the demand for the expected development of economy and society.

With no change in present climate and environmental conditions, but with full provision of material inputs, management, and other human factors, the maximum production of grain in China would be in the range of 840-1,040 × 10^6 t. During the past 40 years, natural disasters have brought about a 5% loss in production. It is possible that the variations and extreme events of climate would reduce the maximum grain production by 10% by 2050.

5.2 BENEFIT-COST ANALYSIS

The results of the current analysis and Tao Zhan (1993) show that with rises in temperature and changes in precipitation (mostly decreases, but some increases), the maximum production could probably drop by at most about 10%, because of the shortening of the growing period and the shortage of moisture. In 1990, the average unit yield of grain production of China was 3.975 ton ha^{-1}. To meet demand, it is estimated, considering farmland changes, that the average yield of grain production of China in the coming 50 years will change as expressed in Table III.

If the influence of climate variation on maximum production and the influence of production costs on the productive level are linear, maximum production will fall to 793-993 × 10^6 t in 2020 and 746-946 × 10^6 t in 2050. Thus in the coming 50 years, investment requirements for grain production of China would require incremental increases over 1990 investments as shown in Table IV.

The additional increments of investment for dealing with climate variation would be 4, 9, and 17% of the investment in agriculture in 1990 for each year, respectively. In the five years from 1986 to 1990, the state made a special investment of 5 billion yuan (US$1.06 billion) in the production of grain, and 20 billion yuan (US$4.26 billion) in ecological protection, technology popularization, and superior species breeding, together with some investments to support expenditures on rural production, providing disaster relief, and supporting impoverished areas, and the like. In 1990 the state invested over US$20 billion in agriculture. To adapt agricultural production to the effects of climate change in the coming 50 years, governments in China at all levels should increase the investment in agriculture every year, with the increments and their possible benefits as shown in Table V.

TABLE III
Estimated changes in the average yield of grain production to meet demand

Year	Yield (t ha^{-1})	Change
2000	4.55	1.14 times as much as that in 1990
2020	5.70	1.40 times as much as that in 1990
2050	6.14	1.54 times as much as that in 1990

Source: Erda, 1994.

TABLE IV
Incremental increase in investment for grain production required by climate change

Year	Without Climate Change	With Climate Change
2000	1.13	1.17 times as much
2020	1.42	1.51 times as much
2050	1.54	1.71 times as much

TABLE V
Potential benefits from adaptation of grain production to climate change

		Year		
		2000	2020	2050
Increased Investment		0.8	1.8	3.48
Benefit	High	1.44	16.00	79.98
or Loss§ (1990 US$ billion)	Low	1.37	11.47	32.27

§ If no newly increased investment.

6. Adaptation

Adaptation measures such as changes in crops and crop varieties, improved water-management and irrigation systems, and changes in planting schedules and tillage practices will be important in limiting any negative effects of climate change and in taking advantage of any beneficial effects. The extent of adaptation depends on the affordability of such measures, particularly in developing countries; access to knowledge and technology; the rate of climate change; and biophysical constraints such as water availability, soil characteristics, and crop genetics. The incremental costs of adaptation strategies could create a serious burden for China and other developing countries. Even though there are significant uncertainties about the capacity of different regions to adapt successfully to projected climate change, the possibility of adaptation should be considered in advance.

6.1 INCREASING INTENSIVE MANAGEMENT

It is important to maintain sown acreage and set up a system of zoning that will protect grain fields to maintain a response capacity for climate change. At present the cultivated land in China is declining by 66.6-166.6 $\times$ 10^3 ha per year (Liangshu and Zhicheng, 1991b). Also, at present, 13 million ha of arable land is held in reserve that can be opened up for agricultural use in the future. The present balance between the land acreage under cultivation and that being occupied by other uses should be maintained, and the cropping index should be increased steadily. In this way the sown acreage of all crops could reach 148.6 $\times$ 10^6 ha in 2000 and 152.7 $\times$ 10^6 ha in

2020 and thereafter, thereby allowing the sown acreage of grain to stabilize at 110×10^6 ha in order to attain the production target mentioned above. Except for Hebei Province, the provinces mentioned above where agricultural production is vulnerable to climate change have populations approaching or exceeding the carrying capacity of the existing cultivated land resources. In these areas, it is especially necessary to strictly control further occupation of farmland, and reserves should be opened up.

Strengthening irrigation capacity is one of the most beneficial means for maintaining agricultural production in the face of unfavorable climate change. The yield of grain per unit area is above the national average in one-third of the provinces and autonomous regions in China, where the irrigated farmland accounts for over 53% of the cultivated land. This ratio reaches 63% in the 6 provinces with the highest production and the 46 counties with the highest yield. The area that could be provided with irrigation facilities could probably increase by $5\text{-}6.7 \times 10^6$ ha, raising the irrigation coefficient by 5.5-6.8%. Where water resources and funding allow, increasing irrigation would be the most beneficial measure in the vulnerable areas (Liangshu and Zhicheng, 1991a).

China has 20 areas with medium or low yields that should be foci for land improvement. At least five of these are in locations vulnerable to climate change, such as the plains of the Yellow River, the Huai River, and the Hai River; the northwest area of Shandong Province; the Yellow River plain in the north Henan Province; and flood-prone clay soils in the middle and lower reaches of the Yangtze River. These medium or low-yield farmland should be provided improved irrigation and drainage capacity so that they are more adaptable to change.

Simulation exercises (see Ojima *et al.*, 1993; Duxbury *et al.*, 1993) have shown that a warming climate would have a major impact on the circulation of carbon and nitrogen in the soil. Even if precipitation were to increase, the content of organic matter in the soil would still fall by 10-30%. Therefore, to maintain the productivity of cultivated land, it is necessary to encourage the use of a more optimal fertilizer mix and to adopt the technique of subsoil application according to actual changes in soil conditions.

Other strategic intensive management measures that could reduce the vulnerability of China's agriculture sector and strengthen its adaptability are to encourage the use of superior species (e.g., those with greater drought resistance); to develop winter agriculture in southern China; to encourage the use of techniques for storing and applying water more economically; to disseminate techniques for the integrated control of pest and diseases; to improve the utilization of plastic sheeting and machinery; and to develop independent feed crop farming.

6.2 ADJUSTING THE STRUCTURE OF FARMING

The Programme for Food Development of the Chinese government stipulates that China will transform its traditional dual structure of farming (i.e., with grain crops and cash crops maintained on the majority of farm land) into a tripartite one that will entail the coordinated development of grain crops, feed crops, and cash crops, with the production of feed crops becoming relatively independent. Consequently, the proportion of farmland devoted to the three kinds of crops would change as shown in Table VI.

In time, the middle and lower reaches of the Yangtze River, north China, and southwest and northeast China would gradually supply more feed grain.

TABLE VI
Estimated proportion of crop acreage in a tripartite structure

	Year		
	1990	2000	2020
Total Sown Acreage (10^6 ha)	145	155	160
Grain Crops	76%	59%	46%
Cash Crops	20%	20%	25%
Feed Crops	4%	21%	29%

Source: Erda, 1994.

The middle and lower reaches of the Yangtze River, with 34.6% of the national production, are the major commodity grain areas in China. This area is vulnerable to probable increases in rain. Nevertheless, the increase in the production of feed maize, feed rice, and feed barley could make agricultural production less dependent on climate and probably reduce its vulnerability. But this adjustment could be sped up only if grain production increases in other areas to reduce of demand for the commodity grain of this area.

North China is vulnerable to possible droughts caused by a rise in temperature and decline in rainfall. Moreover, the social and economic resources used in agriculture are rather fragile in many places. Nevertheless, new species of wheat, maize, cotton, and pulse crops with short growing periods and high yields which are suitable for planting in this area, have been bred and put into production.

Except for the Sichuan Basin, most of southwest China has a low level of agricultural productivity. In particular, the Yunnan-Guizhou Plateau is vulnerable to drought. Adjustment in this area should rely on further improvement of the infrastructure for water conservation and transportation, in order to develop commodity production.

Northeast China has conditions that are estimated to respond positively to projected climate change with higher productivity, but we should be careful of proceeding incautiously.

6.3 USING TECHNOLOGY

While improving standardized cultural techniques, farmers should consider the possibility of climate variation so as to set up a more flexible cultivation system with stronger adaptability. Using plastic sheeting is a feasible technique to reduce the vulnerability of some arid areas and should be further encouraged.

Certain farming technologies can strengthen the adaptability of agricultural production which include techniques to save water in application and retention, and techniques for coordinated use of irrigation and fertilizer application. The technology of feed development can transform the use of land and the cropping structure in areas vulnerable to climate variation. Feed crops with low vulnerability to conditions of climate should be developed.

7. Conclusions

1. Possible climatic impacts span a wide range, depending on the climate scenario, geographic scope, and study in East Asia. For China, results show generally negative yield effects but range from less than -21% to more than +54%.

2. The general possibility of increased summer dryness in the continental midlatitudes suggests that six areas, including the area around the Great Wall, are most likely to be negatively affected by climate change in China.
3. Indices of vulnerability based on physical productivity and socioeconomic capability to adapt show that among China's 30 provinces, Shanxi, Inner Mongolia, Gansu, Hebei, Qinghai, and Ningxia are particularly vulnerable and less able to adapt to climate change.
4. The areas along the Great Wall and Huang-Huai-Hai Plains of North China are both socioeconomically and agronomically vulnerable to climate change and also are areas where climate projections suggest possible adverse changes in climate.
5. Climate change will occur against a steadily increasing demand for food in China over the next 55 years. The increased annual cost of government investment only (excluding farmers' additional costs) in agriculture due to climate change through 2050 was estimated at US$3.48 billion (17% of the cost of government investment in agriculture in 1990).

Acknowledgments

The author greatly acknowledges the contribution and support of my close colleagues Ms. Wang Jinghua, Mr. Wu Qigang, and Dr. Wu Lian Hai from Beijing Agriculture University, and Dr. Zhang Yu and Dr. Wang Xinli from the Chinese Academy of Meteorological Sciences.

References

ADB [Asian Development Bank]: 1993, National Responsible Strategy for Global Climate Change: P.R. China, Asian Development Bank Technical Assistance Project TA-1069, Manila, The Philippines.

Chinese Agricultural Yearbook 1990: 1991, Agriculture Press, Beijing.

Cubasch U., Hasselmann K., Höck H., Maier-Reimer E., Mikolajewicz U., Santer B.D., and Sausen R: 1992, *Climate Dynamics* **8**, 55.

Duxbury J.M., Harper L.A., and Mosier A.R.: 1993, Contributions of agroecosystems to global climate change, in *Agricultural Ecosystem Effects on Trace Gases and Global Climate Change*, ASA Special Publication Number 55, Madison, Wisconsin, pp. 1-18.

Erda L.: 1994, *Rural Eco-Environment* **10**, 1.

IRRI: 1993, *A Basic Model for Irrigated Low Land Rice Production*, International Rice Research Institute, Los Baños, Philippines.

Jones C.A. and Kinir J.R. (eds): 1986, *A Simulation of Maize Growth and Development*, Texas A&M University Press, College Station, Texas.

Liangshu L. and Zhicheng L.: 1991a, *Productive Structure and Developmental Prospects of Planting Industry*, Agricultural Publishing House, Beijing.

Liangshu L. and Zhicheng L.: 1991b, *Studies on the Medium and Long-Term Strategy of Food Development in China*, Agricultural Publishing House, Beijing.

Mitchell J.F.B., Manabe S., Tokioka T., and Meleshko V.: 1990, Equilibrium change, in: *Climate Change: The IPCC Scientific Assessment*, Houghton J.T., Jenkins G.J., and Ephramus J.J. (eds), Cambridge University Press, New York.

Ojima D.S., Parton W.J., Schimel D.S., Scurlock J.M.O., and Kittel T.G.F.: 1993, *Water, Air, Soil Pollution* **70**, 643.

Ritchie, J.J.: 1985, A user-oriented model of the soil water balance in wheat, in: *Wheat Growth and Modeling*, Day W. (ed), Plenum Press, New York, New York.

Smit B. (ed): 1993, Adaptation to Climatic Variability and Change, Occasional Paper No. 19, University of Guelph, Canada.

Tao Zhan: 1993, Influences of global climate change on agriculture of China, in: *Climate Biosphere Interactions*, John Wiley & Sons, Inc., New York.

THE IMPACTS OF POTENTIAL CLIMATE CHANGE AND CLIMATE VARIABILITY ON SIMULATED MAIZE PRODUCTION IN CHINA

WANG JINGHUA AND LIN ERDA

Agrometeorology Institute, Chinese Academy of Agricultural Sciences
Beijing, 100081, China

Abstract. This study assessed the impacts of potential climate change on maize yields in China, using the CERES-Maize model under rainfed and irrigated conditions, based on 35 maize modeling sites in eastern China that characterize the main maize regions. The Chinese Weather Generator was developed to generate a long time series of daily climate data as baseline climate for 51 sites in China. Climate change scenarios were created from three equilibrium general circulation models: the Geophysical Fluid Dynamics Laboratory model, the high-resolution United Kingdom Meteorological Office model, and the Max Planck Institute model. At most sites, simulated yields of both rainfed and irrigated maize decreased under climate change scenarios, primarily because of increases in temperature, which shorten maize growth duration, particularly the grain-filling period. Decreases of simulated yields varied across the general circulation model scenarios. Simulated yields increased at only a few northern sites, probably because maize growth is currently temperature-limited at these relatively high latitudes. To analyze the possible impacts of climate variability on maize yield, we specified incremental changes to variabilities of temperature and precipitation and applied these changes to the general circulation model scenarios to create sensitivity scenarios. Arbitrary climate variability sensitivity tests were conducted at three sites in the North China Plain to test maize model responses to a range of changes (0%, +10%, and +20%) in the monthly standard deviations of temperature and monthly variation coefficients of precipitation. The results from the three sites showed that incremental climate variability caused simulated yield decreases, and the decreases in rainfed yield were greater than those of irrigated yield.

Key words: China, agriculture, maize, GFDL, MPI, UKMO

1. Introduction

An enhanced greenhouse effect could lead to higher global surface temperatures and changed hydrological cycles (IPCC, 1990). China is a largely agricultural country, which increases the importance of changes in climate to the Chinese economy. By 2050, it is estimated that China would be warmer and drier because of global warming, the net balance of precipitation and evapotranspiration would be negative, and less water would be available than today; i.e., even with modest increases in precipitation, it is likely that summer soil moisture levels would decrease (Hulme *et al.*, 1992). The heavy dependence of Chinese agriculture on water supply and climate results in great sensitivity of crop yield to climate change (Lin and Wang, 1995).

Climate change due to global warming may occur in a number of ways. The changes in variability and frequency of extreme events may have greater effects on crop production than changes in mean climate; moreover, the mean and the variation of the frequency distributions of climate variables may change simultaneously (IPCC, 1990). A farmer can adapt to gradual changes in mean climate variables, but would be very vulnerable economically to increased variability (Rosenzweig and Iglesias, 1994). Therefore, sensitivity analysis of climate variability is very important. This is especially important in China because it is a developing country that is highly dependent on agriculture, and therefore economically vulnerable to variability.

Maize is one of the three most important grain crops and the most important forage crop in China. In 1992, maize production was 98.2×10^6 t, making up 22% of the total production of grain crop, and the maize crop area was 21.0×10^6 ha, or 19% of the total grain crop land. Maize yield in 1992 was 4,665 kg ha^{-1}, and the average of grain yields was 4,082 kg ha^{-1} (China Agricultural Ministry, 1993).

Water, Air, and Soil Pollution **92:** 75–85.
© 1996 *Kluwer Academic Publishers. Printed in the Netherlands.*

The objective of this study is to investigate the potential effects of climate change on maize yield in China and characterize the direction, magnitude, and degree of uncertainty of potential impacts, based on 35 maize modeling sites in eastern China that characterize the main maize regions. Additionally, this study tested maize model responses to a range of arbitrary change (0, +10, and +20%) in monthly standard deviations of temperature and monthly variation coefficients of precipitation under general circulation model (GCM) scenarios, based on three sites in the North China Plain.

2. Methodology

2.1 DESIGN AND SCOPE OF THE STUDY

The approach taken here is to compare output from the CERES-Maize model under climate change scenarios (derived from either GCM output alone or GCM output with arbitrary incremental variability) to model output from simulations of a base climate. Results for yield are analyzed as percent change from baseline simulation to provide indications of possible direction and magnitude of responses to changed climate conditions. We developed the Chinese Weather Generator to generate a daily baseline climate data sample of 200 years instead of the more conventional use of 30 years of observed climate data. The reasons are listed below:

- The impacts of climate change on crop yield may occur in a number of ways, including both the mean change and the change in the spread of yield frequency distribution. To study the latter, a period of 30 years of observed climate data is not long enough to establish a yield frequency distribution.

- A common way to create climate change scenarios is to apply mean changes (e.g., mean monthly changes) in climate variables from GCMs uniformly to observed daily climate records (Jin, 1994). Since the Chinese Weather Generator can randomly generate daily values of climate variables, this study used the Chinese Weather Generator combined with GCM output to create climate change scenarios so that mean changes from GCM were applied nonuniformly to the baseline climate. This method was applied to provide more realistic climate change scenarios.

- The Chinese Weather Generator can be used at sites in China that lack a long time series of observed daily values of climate variables, particularly solar radiation.

In this study, an important characteristic is the frequency distribution of simulated yield over 200 years. We compared results with the baseline simulation not only in the mean yield but also in different cumulative frequency yields and in the shape of frequency distribution to obtain more information on changes in yield. Establishing the frequency distribution of crop yield is also very useful in the analysis of risk and in decision making.

To analyze the possible impacts of incremental climate variability on maize yield, we specified incremental changes to variabilities of temperature and precipitation and applied these changes to the GCM scenarios to create sensitivity scenarios. This study tested CERES-Maize model responses to a range of changes (0, +10, and +20%) in monthly standard deviations of temperature (i.e., the standard deviation of daily temperature about the monthly mean) and monthly coefficients of variation of precipitation.

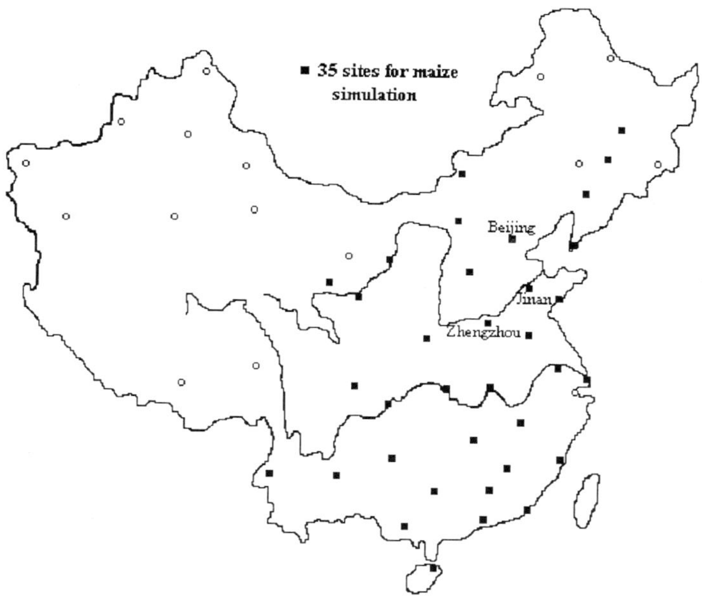

Fig. 1. Location of 51 climate data sites (○) and 35 sites for maize simulation (■) in China.

Because the maize model uses daily climate data, the Chinese Weather Generator was developed to generate daily climate data, based on 51 climate data sites in China (Figure 1). Across them, 35 sites that characterize the main maize regions were selected to simulate maize yield under three GCM scenarios. Also, three sites (Beijing, Jinan, and Zhengzhou) in the North China Plain were selected for sensitivity analysis of yield to changed climate variability.

2.2 WEATHER GENERATOR

The Chinese Weather Generator is a program for randomly generating daily maximum and minimum temperature, precipitation, and solar radiation for a long time series. The methodology taken to develop the Chinese Weather Generator is the same as that used in WGEN (Richardson and Wright, 1984). To define the model parameters in the Chinese Weather Generator, observed daily maximum and minimum temperature and precipitation for 30 years (1956-1985) and observed daily solar radiation for 18-28 years (about 1958-1985) were collected from the Monthly Report of Chinese Meteorological Record on the Surface (China Meteorological Bureau) and the Daily Solar Radiation Record of China (Beijing Meteorological Bureau). Daily solar radiation data for two sites (Xuzhou and Qingdao) were not available, and so they were calculated according to Gao and Lu (1982).

Chinese Weather Generator Validation

To validate the Chinese Weather Generator, we selected several statistics to compare the generated weather data with observed data. For all 51 climate data sites, the following statistics were compared for each month and for the year:

- mean precipitation amount
- mean number of wet days (precipitation $\geq$0.1 mm)

- mean number of days with precipitation >25 mm, or mean number days with precipitation >50 mm, depending on the study site
- mean daily maximum temperature
- mean daily minimum temperature
- mean number of days with maximum temperature >35°C
- mean number of days with minimum temperature <0°C
- mean daily solar radiation.

Most of the generated weather statistics were very comparable to the actual weather statistics at every site, but the statistics that reflect extremes of temperature and precipitation, especially mean number of days with minimum temperatures <0°C in winter months, did not compare as well with the observed data.

2.3 CLIMATE DATA

Baseline Climate
The Chinese Weather Generator was used to generate a climate data set of 200 years as baseline climate for each site, including daily maximum and minimum temperature, precipitation, and solar radiation.

GCM Equilibrium Output
Hulme *et al.* (1992) provided the results of the three equilibrium GCMs used for this study: the Geophysical Fluid Dynamics Laboratory (GFDL) model (Mitchell *et al.*, 1990), the high-resolution United Kingdom Meteorological Office (UKMO) model (Wilson and Mitchell, 1987), and the Max Planck Institute (MPI) model (Cubasch *et al.,* 1992). These results are not the original results from the three GCMs; for example, they were interpolated to the same spatial resolution and the changes in temperature and precipitation standardized by the respective global-mean warming predicted by each model (Hulme *et al.,* 1992). They include mean annual and seasonal changes in temperature and precipitation for the East Asian region (70°E to 140°E and 15°N to 60°N) with 5° × 5° resolution. All mean seasonal changes in temperature from the three GCMs show warming (0.5-1.8°C) in China; however, the changes in precipitation vary widely across the three GCMs, which show considerable seasonal and geographical variation and are very uncertain. In general, the projected changes in precipitation and other variables are less reliable than those for temperature, and mean seasonal change in temperature is less reliable than mean annual change (Hulme *et al.*, 1992).

Climate Change Scenarios
The outputs from the GCMs are not sufficiently accurate for direct use in crop models on a daily basis (Rosenzweig and Iglesias, 1994). Therefore, mean seasonal changes in temperature and precipitation from the three GCMs provided by Hulme *et al.* (1992) were applied to monthly parameters of temperature and precipitation in the Chinese Weather Generator to create climate change scenarios for each site.

Sensitivity Scenarios
Sensitivity scenarios were used to analyze the maize model's sensitivity to incremental climatic variability due to global warming. Arbitrary variability (0, +10, and +20%), including monthly standard deviations of temperature and monthly coefficients of variation of precipitation, together with output from the three GCMs, were applied to monthly parameters of temperature

and precipitation in the Chinese Weather Generator to create sensitivity scenarios. Nine fixed combinations of variability and GCM output were tested for each site.

2.4 MAIZE MODEL CALIBRATION AND VALIDATION

The CERES-Maize model (Ritchie *et al.*, 1989) was used to simulate maize yield and other parameters of maize growth. Five representative varieties were selected for the study; all are grown in the main maize regions in China. Five genetic coefficients for each variety were determined by contrasting the simulated results of phenology and yield to local field experimental data. The management variables (planting date and plant density, etc.) for each calibration site were taken from the field experiments, and local climate data for the calibration years were used. Since we had data from more than one experimental plot at each calibration site, we validated our calibrated variety with data from the second experimental plot and made small modifications to the coefficients based on the results.

3. Limitations

3.1 CLIMATE CHANGE SCENARIOS

Although GCMs are useful in studies of climate change, their outputs are unreliable at the regional level, particularly with regard to precipitation (Hulme *et al.*, 1992). Therefore, the climate change scenarios created from GCM output must not be considered as predictions, but only as examples of possible future climates for the regions under study (Rosenzweig and Iglesias, 1994).

In arbitrary climate variability sensitivity tests, the monthly standard deviations of temperature and monthly coefficients of variation of precipitation are used as a measure of climate variability. They reflect day-to-day variability. Both the standard deviation of daily values about the monthly mean (monthly SD) and the standard deviation of the interannual variation of monthly means should be considered, as appropriate (IPCC, 1990). However, to date we have not found a good way to create sensitivity scenarios with changed interannual variability.

3.2 MAIZE MODEL

In the crop model simulations, technology is held constant; nutrients are not limited; weeds, diseases, and pests are controlled; and climatic tolerances of cultivars do not change. All these assumptions tend to overestimate simulated yields (Rosenzweig and Iglesias, 1994). Therefore, output from the maize model should not be regarded as predictions under changed climate conditions. However, percent changes from baseline simulations could be used as a measure of the potential effects of climate change on maize production.

4. Results and Discussion

In this study the three GCM climate change scenarios were simulated at 35 sites and the nine sensitivity scenarios were simulated at 3 sites. Each simulation was completed under rainfed and irrigated conditions. Irrigation was simulated to provide the crop with a nonlimiting water situation (assumes no water stress and adequate nitrogen). The direct effects of CO_2 were not included in simulations.

The results of this study are listed below. They should not be regarded as predictions, but as plausible assessments of the potential effects of climate change on maize production in China.

4.1 CLIMATE CHANGE ALONE

Both the mean and the shape of the frequency distribution from 200 years of simulated maize yield under three GCM scenarios changed at all the maize modeling sites. Figures 2 and 3 illustrate these changes using the example of Jinan. The baseline distribution is skewed farther to the right than the others (Figure 2), and the interval between the mode and the mean in the baseline distribution is greater than the others, which indicates that under the GCM scenarios not only is the mean yield estimated to decrease, but also the frequency of abundant years is estimated to decrease. At different cumulative frequency levels, all simulated yields for the GCM scenarios decreased compared to baseline yields, but the yield decrease percentages varied by scenario and cumulative frequency (Figure 3). There appears to be a tendency for yields to be between 5 and 9% below baseline yields. The 90% cumulative frequency yields decreased more than the median frequency yields.

Figure 4 shows the percent changes in 90% cumulative frequency yield with the three GCM scenarios compared to the baseline scenario in China, based on 35 maize modeling sites. Both simulated rainfed and irrigated maize yields decreased at all but a few northern sites, yet simulated rainfed yields were lower and had more year-to-year variability than irrigated yields. Although irrigation did not fully compensate for the negative effects of climate change, it significantly improved simulated maize yield. Percent yield changes from baseline simulations varied by site and scenario: under rainfed conditions, yield change varied from +1 to -18% under the GFDL scenario, 0 to -19% under the UKMO scenario, and +5 to -18% under the MPI scenario; under irrigated conditions yield change varied from +3 to -7% under the GFDL scenario, +1 to -9% under the UKMO scenario, and +4 to -6% under the MPI scenario. Under the three GCM scenarios, the ranges in yield for the simulated rainfed yields were much greater than those in the simulated irrigated yields. This indicates that the simulated rainfed yield was more sensitive to changes in the climate condition.

The maize yield decreases were driven primarily by increased temperatures, which shortened the maize growth duration period, particularly the grain-filling period (Rosenzweig and Iglesias,

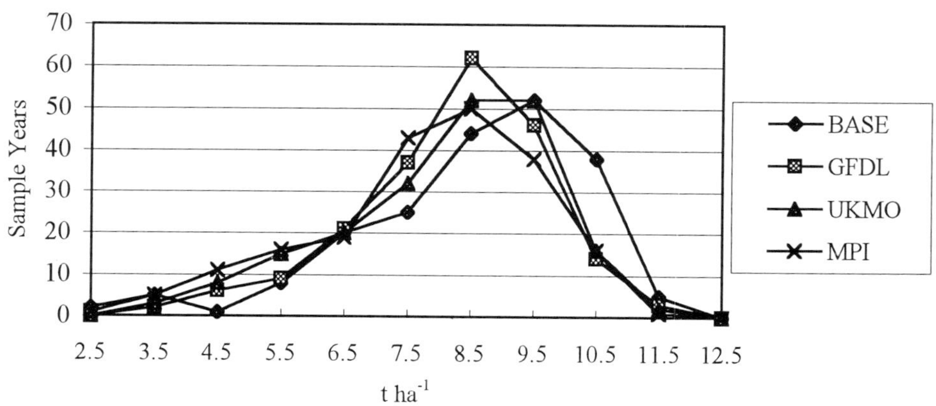

Fig. 2. Frequency distribution of rainfed maize yield with GCM climate change scenarios, Jinan.

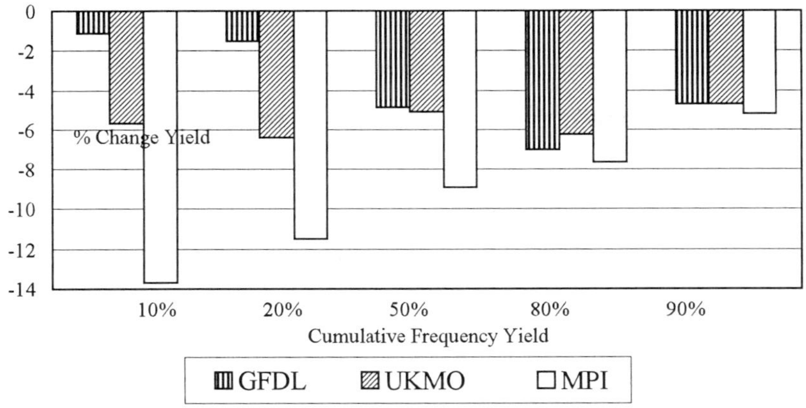

Fig. 3. Changes in yield of rainfed maize compared to base (%) at selected points along the cumulative frequency distribution with GCM climate change scenarios, Jinan.

1994). The simulation results for maize growth durations for all sites showed shortened periods under all three GCM scenarios. At the northern sites, the mean maize season length (in days) from the simulation samples of 200 years showed larger declines than at the southern sites (Table I). The change in the mean of the frequency distribution of maize season length under GCM scenarios compared to base was more obvious than the change in the shape (Figure 5). This indicates a more pronounced change in mean season length than variation in season length. The yield increases at only a few northern sites may be because maize growth is temperature-limited at these relatively high latitudes.

The simulated changes in maize yield with the GCM scenarios at different sites suggest that regional and national production may be affected by climate change in China. Based on the maize harvest area of each province (China Agricultural Ministry, 1993) and the results of the CERES-Maize model, we estimated a -3 to -6% change in annual national maize production under the GCM climate change scenarios compared to base in China (Figure 6).

4.2 CLIMATE CHANGE WITH INCREASED CLIMATE VARIABILITY

Figure 7 shows the different frequency distributions of 200 years of simulated rainfed maize yield with changed climate variability under the UKMO scenario at Jinan. The change in the shape of the frequency distribution with increased variabilities trends toward the normal distribution; in contrast, the shapes in the baseline scenario or with GCM scenarios without increased variabilities were skewed to the right.

The results of the sensitivity analysis for maize yield in 90% cumulative frequency from the simulation samples of 200 years at three sites are presented in Table II. Incremental climate variabilities with GCM climate change scenarios caused greater percent yield decreases than GCM scenarios alone at all three sites; also, simulated rainfed maize yields had greater decreases and were more sensitive to an increment in climate variability than irrigated yields. In general, the increment in variability from 0 to +10% with GCM scenarios had more negative effects on both rainfed and irrigated maize yields than the increment from +10 to +20%.

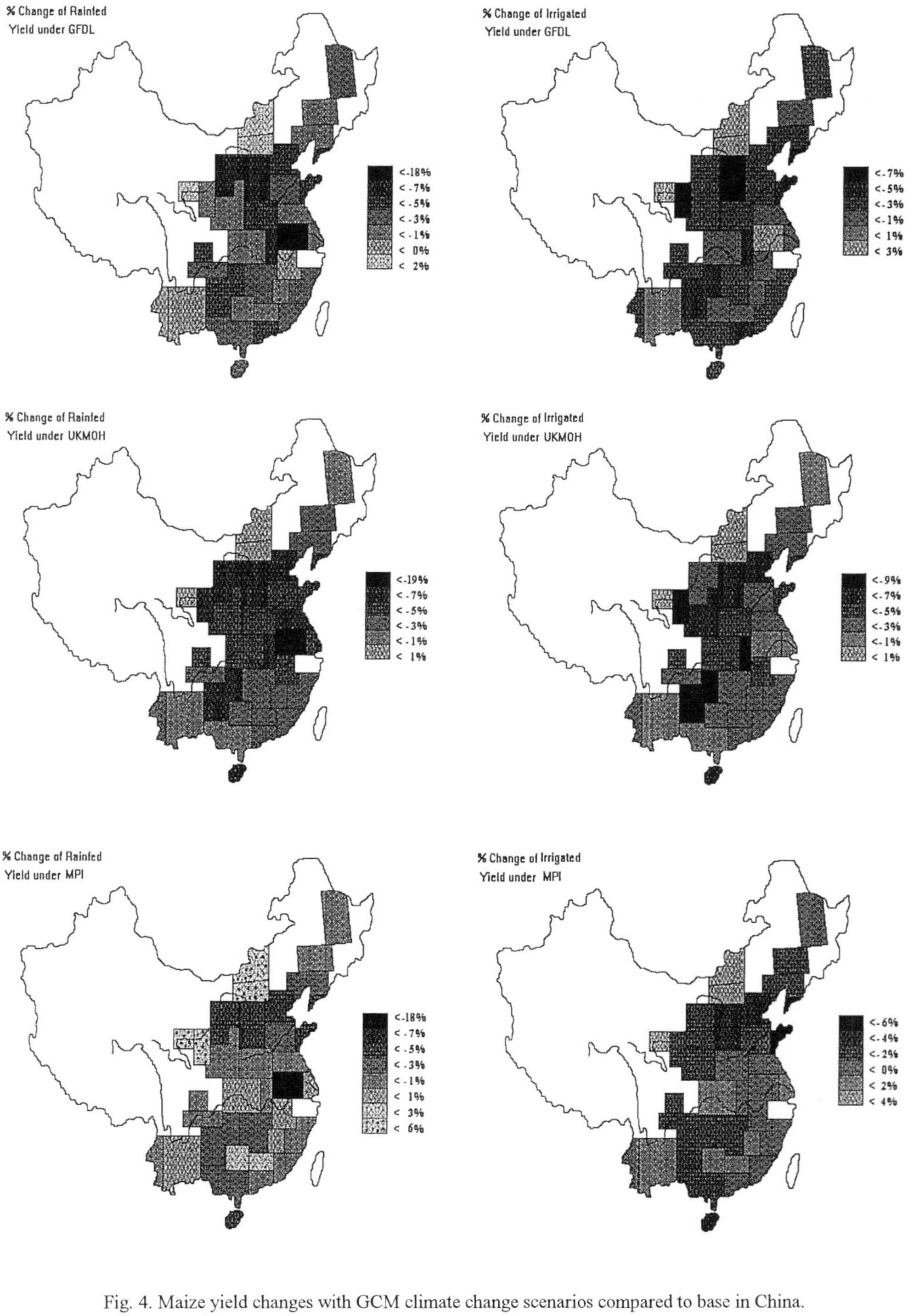

Fig. 4. Maize yield changes with GCM climate change scenarios compared to base in China.

TABLE I

Changes in average maize season length with GCM climate change scenarios compared to base at eight sites

Site	Latitude	Base (days)	GFDL (Δ in days)	UKMO (Δ in days)	MPI (Δ in days)
Haerbin	45° 41'	116	-12	-14	-8
Changchun	43° 54'	111	-8	-10	-8
Beijing	39° 56'	96	-5	-7	-5
Jinan	36° 41'	89	-3	-3	-3
Nanjing	32° 00'	93	-2	-3	-2
Changsha	28° 12'	92	-3	-3	-2
Fuzhou	26° 05'	99	-2	-3	-1
Nanning	22° 49'	106	-2	-3	-3

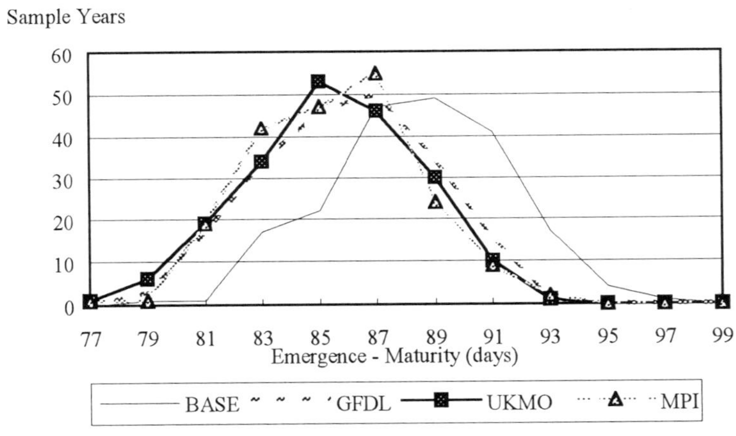

Fig. 5. Frequency diagram of maize season length with GCM climate change scenarios, Jinan.

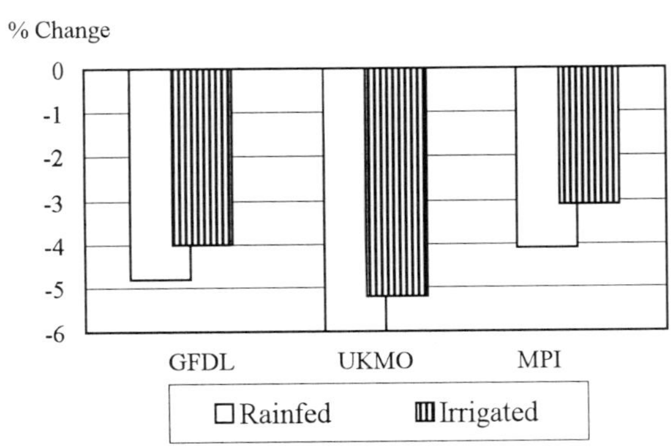

Fig. 6. Change in annual national maize production with GCM climate change scenarios compared to base in China.

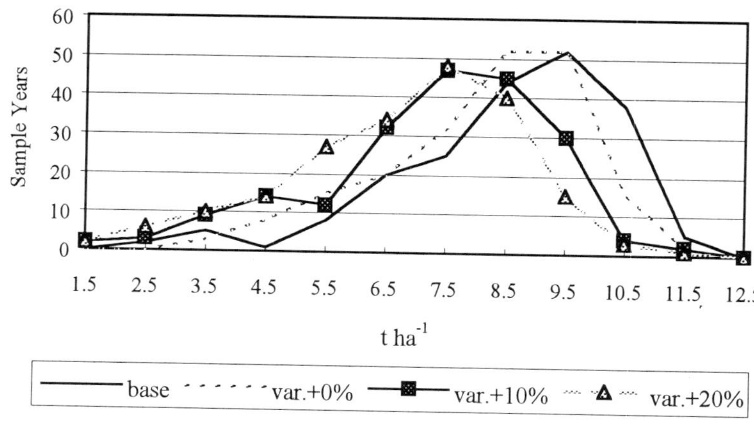

Fig. 7. Frequency distribution of rainfed maize yield under UKMO scenario with three incremental climate variabilities, Jinan.

Simulated irrigated maize yields with +20% in variability increased slightly or decreased slightly compared to those with +10% in variability, while simulated rainfed yields generally continued to decrease as variability increased.

The results for simulated maize yields showed that irrigation may be an effective adaptive strategy to improve maize yields under climate change scenarios. However, the potential for irrigation is greatly limited in the northern regions where the water stress is critical (high levels of water resource extraction already occur). The greatest barrier to establishing new irrigation systems and improving old systems is the enormous cost. Therefore, changing the methods of cultivation so as to reduce soil evaporation may be a more feasible adaptive strategy in China. In addition, earlier seeding and changing maize varieties are also effective adaptation options.

TABLE II

Sensitivity of CERES-Maize model yield to incremental variability in temperature and precipitation combined with GCM scenarios. Changes in 90% cumulative frequency yield compared to base

	Rainfed			Irrigated		
Site/Scenario	GFDL	UKMO	MPI	GFDL	UKMO	MPI
Beijing						
Var. +0%	-6.2%	-8.6%	-6.3%	-4.5%	-7.9%	-4.9%
Var. +10%	-7.0%	-9.9%	-11.9%	-6.4%	-9.3%	-7.9%
Var. +20%	-9.1%	-10.1%	-10.1%	-5.3%	-8.7%	-8.1%
Jinan						
Var. +0%	-4.1%	-5.4%	-4.7%	-4.6%	-4.0%	-3.3%
Var. +10%	-7.9%	-9.3%	-13.8%	-6.7%	-6.3%	-7.3%
Var. +20%	-10.9%	-13.1%	-11.9%	-4.7%	-8.6%	-5.6%
Zhengzhou						
Var. +0%	-6.1%	-6.1%	-1.2%	-3.0%	-6.7%	-0.1%
Var. +10%	-6.3%	-9.7%	-3.3%	-4.7%	-6.1%	-2.4%
Var. +20%	-7.1%	-10.9%	-3.9%	-3.1%	-6.8%	-3.0%

Var. = Monthly standard deviation of temperature and monthly coefficient of variation of precipitation.

5. Conclusion

Increased temperatures under the GCM climate change scenarios used in this study shortened maize growth duration, especially in northern China, which generally decreased simulated maize yields except in areas where maize was originally temperature-limited. Under the GCM climate change scenarios, simulated rainfed maize yields are lower and more sensitive to an increment in climate variability than irrigated yields. Based on these findings, Chinese maize production could be negatively affected by climate change. Decreases in simulated maize yield under GCM scenarios occurred not only in the mean yield but also in the frequency of abundant years. Because the shapes of the frequency distribution of simulated maize yield are non-normal, it is better to use the median of the distribution rather than the mean in analyzing the yield change.

Acknowledgments

We gratefully acknowledge the contribution and support of our close colleagues Wu Qigang and Chen Shanxian, and especially acknowledge Zhao Zongci for providing the results of the GCM scenarios.

References

China Agricultural Ministry (eds): 1993, *Agricultural Statistics of China 1992*, Agriculture Press, Beijing.

Cubasch U., Hasselmann K., Höck H., Maier-Reimer E., Mikolajewicz U., Santer B.D., and Sausen R.: 1992, *Climate Dynamics* **8**, 55.

Gao G. and Lu Y.: 1982, *Radiation and Heat Balance on the Ground Surface in China*, Scientific and Technological Publishing House, Beijing.

Hulme M., Zhao Z., Leemans R., Wang F., Markham A, Wigley T., Ding Y., and Jiang T.: 1992, Climate Change due to the Greenhouse Effect and its Implications for China, World Wide Fund for Nature, Gland, Switzerland.

IPCC WG1: 1990, *Climate Change: The IPCC Scientific Assessment*, Cambridge University Press, Cambridge.

Jin Z., Ge D., Chen H., and Fang J.: 1994, Effects of climate change on rice production and strategies for adaptation in Southern China, in: *1994, Implications of Climate Change for International Agriculture Crop Modeling Study*, Rosenzweig C. And Iglesias A. (eds), EPA 230-B-94-003, U.S. Environmental Protection Agency, Office of Policy, Planning and Evaluation, Climate Change Division, Adaptation Branch, Washington, DC.

Lin E. and Wang J.: 1995, *Advance in Earth Sciences* **10**, 597.

Mitchell J.F.B., Manabe S., Tokioka T., and Meleshko V.: 1990, Equilibrium change, in: *Climate Change: The IPCC Scientific Assessment*, Houghton J.T., Jenkins G.J., and Ephramus J.J. (eds), Cambridge University Press, New York.

Richardson C.W. and Wright D.A.: 1984, WGEN: A Model for Generating Daily Weather Variables, ARS-8, U.S. Department of Agriculture, Agricultural Research Service, Washington, DC.

Ritchie J. *et al.*: 1989, A User's Guide to CERES-Maize — V2.10, International Fertilizer Development Center, Muscle Shoals, Alabama.

Rosenzweig C. and Iglesias A. (eds): 1994, Implications of Climate Change for International Agriculture: Crop Modeling Study, EPA 230-B-94-003, U.S. Environmental Protection Agency, Office of Policy, Planning, and Evaluation, Climate Change Division, Adaptation Branch, Washington, DC.

Wilson C.A and Mitchell J.F.B.: 1987, *Journal of Geophysical Research* **92**, 13315.

VULNERABILITY OF SRI LANKA TEA PRODUCTION TO GLOBAL CLIMATE CHANGE

M.A. WIJERATNE
Tea Research Institute
Research Advisory and Extension Centre
Ratnapura, Sri Lanka

Abstract. The tea industry is Sri Lanka's main net foreign exchange earner and source of income for the majority of laborers. Tea yield is greatly influenced by weather, and especially by droughts, which cause irreparable losses because irrigation is seldom used on tea plantations. At the other extreme, heavy rains erode top soil and wash away fertilizers and other chemicals. In the recently published Sri Lanka country report on climate change, it was reported that the island will experience extreme rainfall intensities and warmer temperatures as a result of climate change. The possibility of a 10% increase in the length of dry and wet seasons per year in the main plantation area was also indicated. Thus both drought damages and soil losses in tea production areas will increase in the years to come. An analysis of the results of field experiments with weather data shows that increases in temperature, soil moisture deficit, and saturation vapor pressure deficit in the low elevations will adversely affect growth and yield of tea. Reports have also shown that about 30 cm of soil has already been eroded from upland tea plantations. Under these circumstances, the tea industry in Sri Lanka is clearly vulnerable to predicted climate changes, and subsequently greater economic, social, and environmental problems. This paper discusses the various aspects of the adverse effects of climate change on Sri Lanka's tea industry.

Key words: Sri Lanka, agriculture, tea

1. Introduction

Although industrial exports such as textiles and garments bring in a higher percentage of foreign exchange, agriculture is the highest net foreign exchange earner in Sri Lanka. Of the agricultural exports, tea alone contributes about 15-25% of the total exchange earnings and, hence, plays a key role in the Sri Lankan economy. About 30% of the employees of the public sector (government and semigovernment) are manual workers on estates, and the majority of these are employed on tea plantations. Moreover, about 239,000 tea small holdings also generate a large proportion of employment opportunities in the country. Accordingly, more than 700,000 workers and their families are dependent on the tea industry. These figures show the importance of the tea industry to the social and economic stability of Sri Lanka.

In spite of the expansion of the small holdings sector, in Sri Lanka, the total area of tea plantations has decreased since the 1930s, i.e., from about 0.44 to 0.23 × 10^6 ha. The decline in the estate sector tea lands began early in the 1960s, because of low productivity brought about by soil and bush debilitation. Many of the midcountry (600-1,200 m elevation) tea plantations have now become marginal, warranting crop diversification. Adversities of weather and poor management practices have been blamed for this situation; changes in the microclimate in tea plantations after shade removal during the early 1960s have also affected the productivity of the tea bushes (Fuch, 1989). Some adverse effects of the nationalization of the plantations in the 1970s, such as the neglect of agricultural practices, contributed to the decline in the estate sector production (Fuch, 1989).

Being a rainfed plantation crop in Sri Lanka, tea depends greatly on weather for optimal growth. Therefore, changes in weather conditions would undoubtedly affect tea production. The relationship between tea yield and weather has been discussed by many researchers (e.g., Devanathan, 1975; Kandiah and Thevadasan, 1980; Carr and Stephens, 1992). The

Water, Air, and Soil Pollution **92:** 87–94.
© 1996 *Kluwer Academic Publishers. Printed in the Netherlands.*

findings of this study differ from those of the Sri Lanka country report (ADB, 1994). This could be due to the different models used to predict tea yield. It is generally accepted that an increase in temperature increases tea yield, and this relationship has been used in many models. But recent findings show that at higher temperature regimes (greater than 25-26 °C), the increase in temperatures reduces tea yield.

Droughts inflict irreparable losses to the tea industry and hence to the economy of Sri Lanka. The consequences of the droughts in 1983 and 1992 are good examples (Central Bank, 1983; 1992). The decline in production due to drought in early 1983 was about 4% over the previous year, and the drought in 1992 reduced tea production by about 26% compared to that of 1991. It also increased the costs of production by 19%, depriving the country of about 3 billion rupees (US$70 million) of foreign exchange. Total production in 1991 and 1992 was 240.7 and 178.9×10^6 kg, respectively, and the latter was the lowest production recorded since the end of the 1950s (Figure 1). These figures show the magnitude of the loss that could be incurred by adversities of weather.

Heavy rainfall also causes considerable damage to tea plantations through soil erosion, poor growth due to lack of sun, and increases in disease incidence. Poorly covered old seedling tea fields, pruned tea fields, and young tea fields during the first two years are more vulnerable to soil erosion due to inadequate ground cover. It has been estimated that more than 30 cm of top soil has already been lost from Sri Lanka's tea plantations, especially in the uplands (Krishnarajah, 1985). Landslides also adversely affect plantations and endanger the lives of workers on the hilly slopes.

2. Climate Change and Methods

According to climate change scenarios, the increases in global atmospheric CO_2 concentrations and temperatures by 2100 could be in the range of 600-700 ppm and 1.0-3.5 °C, respectively,

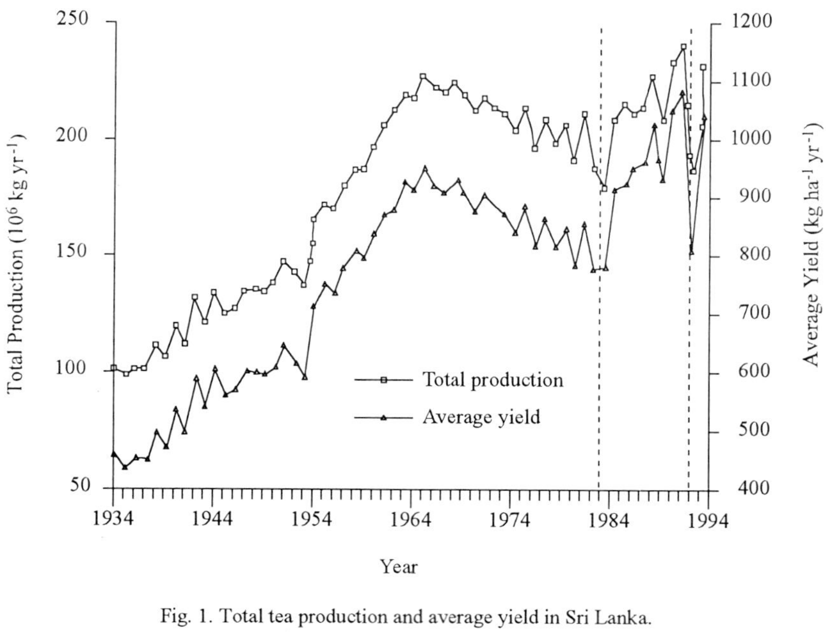

Fig. 1. Total tea production and average yield in Sri Lanka.

compared to 1990, depending on different scenarios of variations in greenhouse gas emissions (CO_2, CH_4, N_2O) and oceanic changes (Houghton *et al.*, 1996).

The consequences of climate change will differ from one country to another. In the recently published Sri Lanka country report (ADB, 1994), it was shown that Sri Lanka will experience frequent droughts, warmer spells, and extreme rainfall events as a result of the climate change. Scenarios of temperature changes for Sri Lanka show an increase in temperature of 0.4-3.0°C by 2070 (ADB, 1994). Moreover, climate change scenarios presented for Sri Lanka have shown that the frequency and severity of such extreme weather conditions may increase, and thus affect the agriculture sector. It is also indicated that there will be a 10% increase in the length of dry and wet seasons per year in the main tea plantation area. Although an increase in rainfall is predicted, any significantly favorable impact on tea plantation agriculture is unlikely because of increased evaporation losses brought about by high temperatures and the possibility of the distribution of rainfall being erratic or uneven (ADB, 1994).

In many other countries, vulnerability assessments on different economically and socially important areas are being conducted to inform policy makers about suitable adaptation measures that could be implemented. Hence, it has become a current need to address the possible impacts of global climate change on the tea industry in Sri Lanka.

In this study, the effects of environmental factors on growth and yield of tea were studied using data on annual variation of climatic factors and yield parameters of tea. The relationship between the climatic factors and tea yield was analyzed using linear regression analysis.

3. Results and Discussion

As described previously, since there is no irrigation, tea yield is greatly influenced by weather. Tea grows well under air temperatures in the range of 18-25°C (Carr, 1972; Watson, 1986). A well-distributed rainfall of about 1,300-1,400 mm per year is generally considered adequate for the growth of tea in Sri Lanka. It is also reported that an annual rainfall of about 2,500-3,000 mm is optimum for tea cultivation (Fuch, 1989; Watson, 1986). There is a wide variation in temperature and rainfall in the different tea growing regions in Sri Lanka.

Although the relationship between weather and tea yield has shown that increases in rainfall and temperature increase tea yield (Devanathan, 1975; Squire, 1990), recent observations have shown that at higher temperature regimes (>25-26°C), the yield components of tea (shoot population density, shoot weight, and shoot extension rate) tend to decrease with increasing temperatures:

$$SW = 0.647\ (\pm 0.059) - 0.017\ (\pm 0.002)\ T$$
$$R^2 = 40\%,\ p < 0.001\ ,$$

and

$$SER = 225\ (\pm 38) - 6.62\ (\pm 1.37)\ T$$
$$R^2 = 29\%,\ p < 0.001\ ,$$

where SW, SER, and T are the shoot dry weight (g/shoot), shoot extension rate (mm/week), and temperature (°C), respectively. Low R^2 values were obtained because this experiment was conducted under field conditions where none of the environmental factors were controlled.

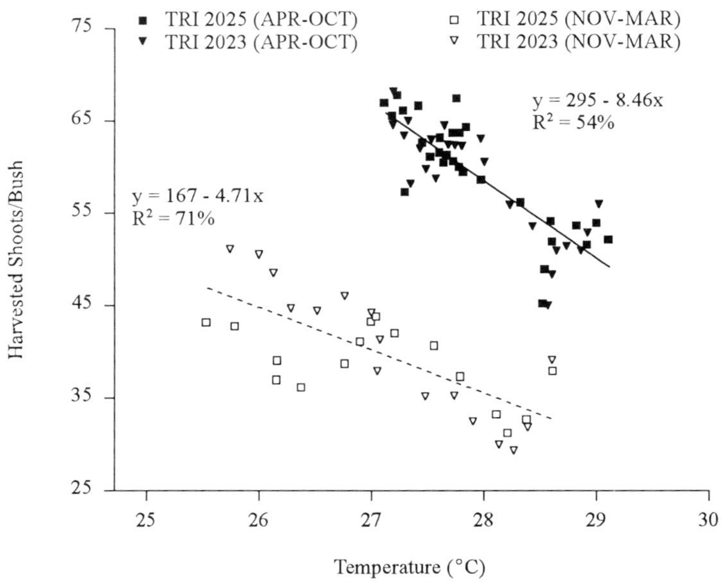

Fig. 2. Effect of temperature on population density of tea shoots.

Figure 2 shows the relationship between tea shoot population density and temperature for TRI 2025 and TRI 2023, two common tea clones (cultivars) in Sri Lanka.

Experiments under controlled environments have shown that shoot extension rate increases with increasing temperatures up to 22°C, and further increases in temperature up to 34°C result in a decline in the extension rate (Figure 3; the dashed lines below 15°C in the figure are extrapolations of the linear relationship to obtain the base temperature above which tea grows).

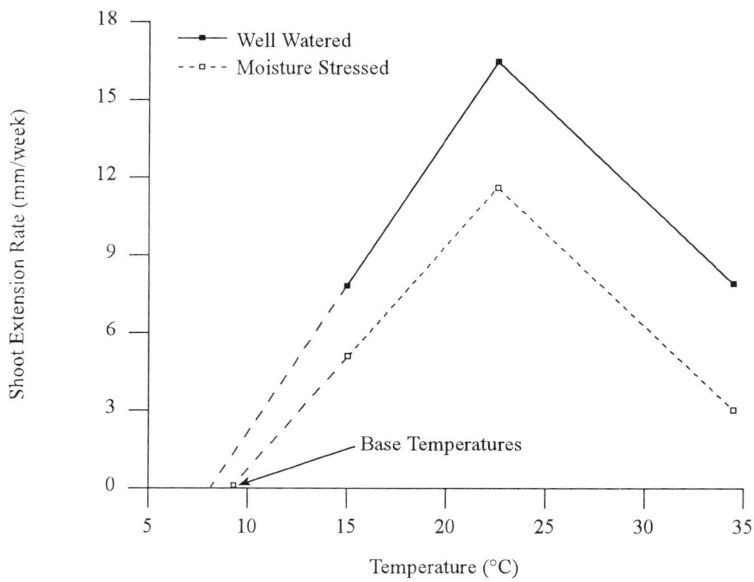

Fig. 3. Effect of temperature on shoot extension rate under controlled conditions.

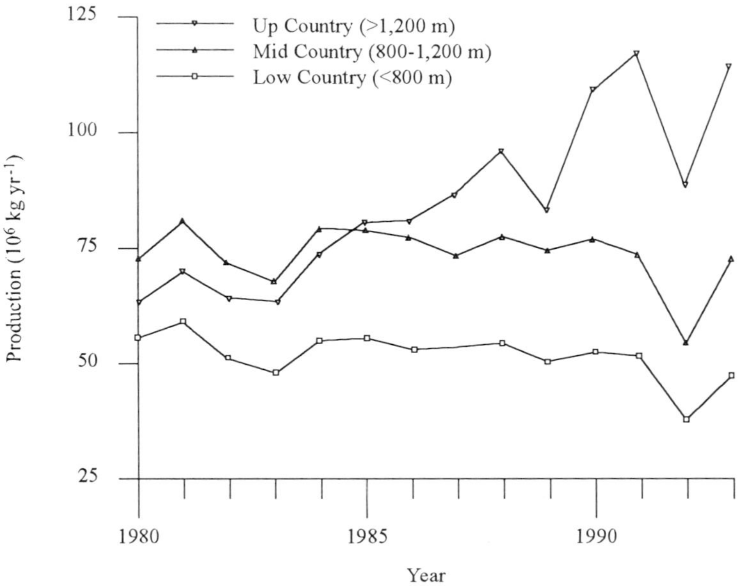

Fig. 4. Annual tea production in different tea growing regions in Sri Lanka.

The adverse effects of climate change are expected to be greater in the low country tea growing regions (<600 m above mean sea level) where the mean air temperatures are usually higher than 25°C and drought damages are greater. However, this is the region where the majority of tea production enters the market (Figure 4). Further, most of the small holdings are concentrated in the low country districts.

The effects of drought on tea plantations are well known (Wijeratne and Ekanayake, 1990; Navaratne, 1992; Yatawatte, 1992). Increases in temperature, soil moisture deficit, and vapor pressure deficit create a plant water deficit, which leads to growth retardation. Experimental results in the low country have shown that clonal tea yields could be adversely affected at temperatures >26°C, soil moisture deficits >30-50 mm, and saturation vapor pressure deficits >1.2 kPa (Wijeratne, 1994). Given these results, it could be assumed that the predicted climate change for Sri Lanka, i.e., higher temperatures and drier weather (ADB, 1994), will be unfavorable for tea production.

Extreme rainfall events within a short period of time also cause considerable damage through erosion. Loss of fertility, reduction in water holding capacity because of an increase in graveliness, exposure of hard pans, exposure of root systems, and reduction in microbial activities due to loss of organic matter are some of the negative consequences of soil erosion, adversely affecting the growth and yield of tea. With the loss of top soil, cultivable lands may become barren or unproductive and can be used neither for replanting with tea nor for any other agricultural purpose. Moreover, rehabilitation of such lands takes a long time, and may even be uneconomical. In addition, the growth of tea bushes could be reduced because of heavy rainfall due to lack of sun and an increase in the incidence of diseases such as blister blight.

When these facts are considered, it is clear that the predicted climate change in Sri Lanka, i.e., recurrent warm seasons, droughts, and heavy rains with erratic distribution, will undoubtedly affect the tea industry. Tea needs an even distribution of rainfall to ensure continuous production. Although increased atmospheric CO_2 levels can enhance photosynthesis, this beneficial effect cannot be expected to offset the negative impacts expected because of limitations such as high temperatures and poor soil conditions.

Currently, more than 50% of Sri Lanka's tea fields are seedling fields with poor ground cover. Although these seedlings are generally more adaptable to environmental stress, their ability to withstand dry weather is being degraded by poor soil and poor management. Most of these fields are planted along the slope (but not on contours) and hence are also vulnerable to soil erosion. Recent observations have shown that even seedling tea fields are subject to drought damage due to poor shade and poor soil (Yatawatte, 1992). Hence, seedling tea fields will also suffer considerable damage from climate change.

Shade is an essential requirement in tea plantations: shade trees reduce temperature, conserve soil, and increase relative humidity. After the abrupt removal of shade trees in the early 1960s (shade was removed with the hope that an increase in light intensity would enhance the assimilation of tea), shade management in Sri Lanka's tea plantations was very poor. Under such conditions, vulnerability of the tea industry to global warming and soil erosion will be much greater.

As a result of the poor management of shade trees and of green manure crops, the addition of organic matter to Sri Lanka's tea lands is currently inadequate. Higher temperatures in the years to come will quickly degrade organic matter, leaving unfertile soil with poor physical and chemical properties, especially in the low elevations. Further, increased soil erosion will also reduce the soil's organic carbon levels. The consequences of the loss of organic matter, such as reduced efficacy of applied nutrients, reduced microbial activity, and increased erodibility, will weaken the tea bushes and reduce yield. An increase in the susceptibility of the debilitated bushes to various pests and diseases may further decrease production.

As mentioned above, the small holding tea sector in Sri Lanka is expanding; it now makes up about 40% of the land used for tea. Hence, the present conditions of these small holdings should also be considered. In small holdings, shade management and soil conservation measures are given less priority because of their smaller size and the lack of resources for adopting costly practices such as building drains and terraces. Furthermore, infilling programs (planting tea in patches vacant because of dead plants) are also lacking in the small holding sector and hence ground cover is poorer. A number of reports have highlighted that these lands are more vulnerable to heavy soil erosion (e.g., Stocking, 1992). Therefore, tea production in this sector could be badly affected by the predicted change in climate.

Environmental changes such as global warming could increase outbreaks of dry weather pests or even new pests. Furthermore, the susceptibility of tea plants to diseases such as canker could be increased because of stress. Planting high yielding, vegetatively propagated tea clones is now the common practice. Because of their high production, their moisture and nutrient requirements are also high, and they are usually more susceptible to adverse weather. Growers are reluctant to plant some hardy tea clones because of their moderate or low yielding potential. Hence, a greater percentage of Sri Lanka's tea plantations and small holdings may suffer from climate change in the years to come.

Rehabilitation of the land before planting is unlikely given the high cost and the extension of the unproductive period of tea cultivation. Hence, the soil conditions of the tea lands may degrade markedly and the establishment of young plants in the field may be poor. In addition, the quality of present agricultural practices such as plucking, pruning, weed management, and fertilizer application is quite marginal.

According to the Sri Lanka country report (ADB, 1994), the tea industry in Sri Lanka would benefit from climate change. However, it is clear from the above discussion that climate change could cause considerable damage to tea cultivation, and this matter warrants further investigation.

If the tea plantations are adversely affected by global warming, the country will lose a greater amount of foreign exchange earnings, leading to numerous economic problems. Not only the yield per hectare but also the area under cultivation may be reduced. As a result, social problems such as unemployment may arise. To prevent such a catastrophic situation, it will be necessary to implement suitable adaptation measures, after careful evaluation, because changes in plantation agriculture take a long time.

4. Adaptation

Awareness of climate change is inadequate in Sri Lanka. Therefore it is of utmost importance to enlighten the relevant parties such as policy makers and agriculturists on the imminent effects and ensuing problems. In this context, the advisory and extension services of various institutions and organizations should play a key role. In addition, new research programs should be designed to find and assess socioeconomically acceptable adaptation measures. In a strategy to minimize the adverse effects of global warming on the tea industry in Sri Lanka, the following suggestions should be considered.

- Use of hardy tea clones that are resistant to drought, pests, diseases, and adaptable to changing weather conditions (Wijeratne, 1996).
- Improvement and implementation of soil conservation measures (e.g., sloping agricultural land technology, or SALT) to arrest soil erosion, and addition of organic matter content to enhance the soil's physical and chemical properties.
- Proper shade management to provide a better environment and thus minimize stress conditions.
- Expansion of multicropping systems (e.g., tea and rubber intercropping) to reduce the risks of monocropping.

Studies should also be initiated to assess the magnitude of the impact of climate changes on the tea industry and to identify the most vulnerable areas, and these should then be given high priority. Selected measures to be implemented should be carefully evaluated to suit the availability of resources in the country. In the case of financial constraints in implementing the selected measures, subsidy schemes may be required, and hence, the involvement and assistance of policy makers are also necessary.

Acknowledgment

I express my sincere gratitude to Dr. J. Ratnasiri, Coordinator, U.S. Country Studies Program, Ministry of Transport, Environment and Women's Affairs, Colombo, Sri Lanka, for helpful suggestions.

References

ADB: 1994, Climate Change in Asia: Sri Lanka Country Report, Asian Development Bank, Colombo, Sri Lanka.

Carr M.K.V.: 1972, *Experimental Agriculture* **8**, 1.

Carr M.K.V. and Stephens W.: 1992, Climate weather and the yield of tea, in: *Tea: Cultivation to Consumption*, Willson K.C. and Clifford M.N. (eds), Chapmann and Hall, London, pp. 87-135.

Central Bank: 1983, *Annual Report*, Central Bank of Sri Lanka, Colombo.

Central Bank: 1992, *Annual Report*, Central Bank of Sri Lanka, Colombo.

Devanathan M.A.V.: 1975, *Tea Quarterly* **45**, 43.

Fuch H.-J.: 1989, *Tea Environments and Yield in Sri Lanka*, Margraf Scientific Publishers, Weikershelm, Federal Republic of Germany.

Houghton J.T., Meira Filho L.G., Callender B.A., Harris N., Kattenberg A., and Maskeu K.: In press, Climate Change 1996: The Science of Climate Change, Contribution of Working Group I to the Second Assessment Report of the Intergovernmental Panel on Climate Change, Cambridge University Press, New York.

IPCC: 1994, *IPCC Technical Guidelines for Assessing Climate Change Impacts and Adaptations*, University College London and Center for Global Environmental Research, London.

Kandiah S. and Thevadasan T.: 1980, *Tea Quarterly* **49**, 2533.

Krishnarajah P.: 1985, *Sri Lanka Journal of Tea Science* **54**, 91.

Navaratne D.K.: 1992, *Tea Bulletin* **12**, 34.

Squire G.R.: 1990, *The Physiology of Tropical Crop Production*, CAB International, Wallingford, Oxon, United Kingdom.

Stocking M.: 1992, Soil Erosion in the Upper Mahaveli Catchment, Technical Report No.14, Environment and Forest Division, Mahaveli Authority of Sri Lanka, Polgolla.

Watson M.: 1986, Soil and climatic requirements, in: *Hand Book on Tea*, Sivapalan P., Kulasegaram S., and Kathiravetpillai A. (eds), Tea Research Institute of Sri Lanka, pp. 3-5.

Wijeratne M.A.: 1994, Effect of Climatic Factors on the Growth of Tea (*Camellia sinensis* L.) in the Low Country Wet Zone of Sri Lanka, PhD Thesis, University of London.

Wijeratne M.A.: 1996, Some adaptations of the tea plant to dry environments, in: *Adapting to Climate Change: Assessments and Issues,* Smith J.B., Bhatti N., Menzhulin G., Benioff R., Budyko M.I., Campos M., Jallow B., and Rijsberman F. (eds), Springer-Verlag, New York, pp. 174-1187.

Wijeratne M.A. and Ekanayake P.B.: 1990, *Tea Bulletin* **10**, 15.

Yatawatte S.T.: 1992, *Tea Bulletin* **12**, 87.

THE POTENTIAL IMPACTS OF CLIMATE CHANGE ON PASTURE AND CATTLE PRODUCTION IN MONGOLIA

B. BOLORTSETSEG and G. TUVAANSUREN
Hydrometeorological Research Institute
Khudaldaany 5, Ulaanbaatar 11, Mongolia

Abstract. The objective of this study was to assess the impact of climate change on plant and livestock production in several natural regions of Mongolia. The Simulation of Production and Utilization of Rangelands model was used for the assessment of the climate impacts. Long-term (1961-1990) climatic data and biophysical and physiological parameters of pasture grassland and cattle were used in the study. The selection of simulation sites for the study was based on regions where there are many cattle. Climate change scenario data were obtained by combining historical weather data from each site with predicted output from general circulation models. Results from baseline runs were compared with four climate change scenarios and a scenario with baseline climate conditions and doubled carbon dioxide ($2\times CO_2$). The impact of climate change on pasture production is estimated to be negative in the Gobi desert area and favorable in colder regions. Livestock intake and livestock weight are estimated to generally decline in late summer when digestibility is lower. Average daily weight gain of Mongolian steers that have only pasture forage is estimated to be lower, whereas it is estimated that there will be no negative effects on the weight gain of steer that are provided supplemental feed.

Key words: Mongolia, grasslands/livestock, cattle production, pasture production, CCCM, GFDL, GISS, SPUR2, UKMO

1. Introduction

Average rates of carbon dioxide (CO_2) increase during the 1980s were 0.4% or 1.5 ppm per year (IPCC, 1995). Predictions from general circulation models (GCMs) indicate that with a doubling of atmospheric carbon dioxide, the average global temperature may rise by 1.5 to 4.5°C (IPCC, 1995). Changes in global climate, as predicted by the GCMs, could have profound implications for agriculture production worldwide, and for rangeland production and animal husbandry in Mongolia.

The Mongolian economy relies heavily on livestock production. Products such as meat, butter, wool and hair, and hides and furs account for one half the country's output and almost 90% of its exports. The pastures that support livestock are strongly affected by climate and weather conditions, so, ultimately, climate affects the economic condition of the country. In Mongolia there are many different types of pasture and, within these types, different amounts of standing biomass. Pasture production is unstable from year to year, and in some years it is quite low because of prolonged dry and warm weather during the growing season. The mean of peak standing biomass varies from 100 to 1,000 kg ha^{-1}, and decreases from north to south in Mongolia, depending on the features of the regions. In the north part of the country, there are high mountains and partially wooded steppe, in the east and southeast is steppe, and in the south is the Gobi Desert. Because cattle and pasture production are economically important to Mongolia, the potential effects of climate change on production is an important issue to study. The study of the vulnerability of cattle and pasture production to potential climate change is also important for future policy development.

Research has been conducted to examine the response of rangeland ecosystems to greenhouse warming. For example, Baker *et al.* (1993) used the Simulation of Production and Utilization of Rangelands (SPUR2) enhanced model, developed by the USDA-ARS Great Plain Systems

Water, Air, and Soil Pollution **92:** 95–105.
© 1996 *Kluwer Academic Publishers. Printed in the Netherlands.*

Research Unit and Colorado State University, to examine the potential effects of climate change on ecosystem processes and cattle production on U.S. rangelands. This model was used in the study presented here.

The objective of the study described here was to assess the impacts of potential climate change under several different climate change scenarios on rangeland production and beef cattle in Mongolia using SPUR2.

2. Methods

The SPUR2 model (Hanson *et al.*, 1992), an enhanced version of SPUR (Wight and Skiles, 1987), is a general grassland ecosystem model that simulates the cycling of carbon and nitrogen through several components, including standing green, standing dead, live roots, dead roots, seeds, litter, and soil organic matter. Competition between plant species and the impact of grazing on vegetation are also simulated. SPUR2 has been modified to simulate the direct effects of CO_2 on plant production. SPUR2 also contains herd-wide, life-cycle beef cattle production and steer simulation models.

Ten sites that represent five different natural regions were selected for the simulations of rangeland and cattle production of Mongolia (Figure 1, Table I).

The five functional groups simulated by the model include warm and cool season grasses, warm and cool season forbs, and shrubs. Julian dates for the beginning of seed production and

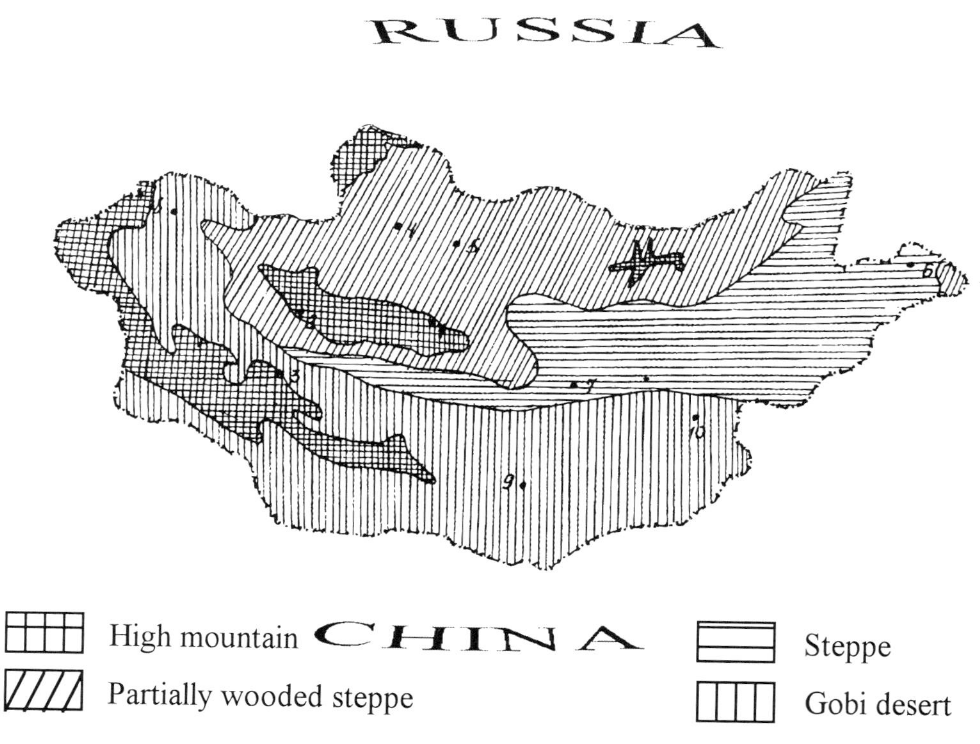

Fig. 1. Natural zones of Mongolia and site locations (site names are given in Table I).

TABLE I
Characteristics of study sites

Natural Region	Map Reference	Site Name	Latitude	Longitude	Elevation (m)
High Mountain (Khangai Mountains)	1	Tsetserleg	47°27'	101°28'	1,695
	2	Uliastai	47°51'	96°51'	1,751
High Mountain (Altai Mountains)	3	Altai	46°24'	96°15'	1,751
Partially Wooded Steppe	4	Moron	49°38'	100°10'	1,288
	5	Hutag	49°23'	102°42'	938
Steppe	6	Halhgol	47°37'	118°31'	688
	7	Mandalgov	45°46'	106°17'	1,393
Gobi Desert	8	Ulaangom	49°59'	92°04'	934
	9	Dalanzadgad	43°23'	104°25'	1,469
	10	Sainshand	44°54'	110°07'	915

the beginning and end of senescence, and maximum root to shoot ratio for each functional group, were also used. Soil texture class and the representative composition of each site were obtained from the Institute of Land Use Policy.

The steer simulation model was used in this study to determine the effects of climate change on animal intake and weight for animals approximately 6-18 months old. Three steer genotypes (Selenge, Halhgol, and Mongolian) were simulated in this study. Mongolian steers are pastured in all territories of Mongolia. Selenge and Halhgol steers are primarily raised for meat production and are bred in the northern part of Mongolia where pasture plant biomass is higher. In the model, the Selenge and Halhgol genotypes are fed a 2 kg daily supplement from January to May, but no supplemental feed is provided to the Mongolian genotype. The weight parameters of steer genotypes are shown in Table II.

The SPUR2 model is driven by daily inputs of precipitation, maximum and minimum temperatures, solar radiation, and daily wind. Six climate scenarios were used in the study: a baseline scenario, a baseline climate with $2\times CO_2$ scenario, and climate change scenarios derived from four GCMs. The baseline scenario was developed using 30 years of historical climate data (1961-1990) from each simulation site. The daily maximum and minimum temperatures, precipitation, and wind speed for that period were obtained from the Climatic Data Base made available by the Hydrometeorological Research Institute. Daily solar radiation was calculated using observed daily sunshine duration. The GCM scenarios used for the climate change runs were produced by the following models: Goddard Institute for Space Studies (GISS) (Hansen *et al.*, 1983), Geophysical Fluid Dynamics Laboratory (GFDL) (Mitchell *et al.*, 1990), United Kingdom Meteorological Office (UKMO) (Wilson and Mitchell, 1987), and Canadian Climate Centre Model (CCCM) (Boer *et al.*, 1992). Additionally, a scenario that maintained current climatic variables but with a doubled atmospheric concentration of carbon dioxide was considered ($2\times CO_2$ scenario). Other researchers have found that the UKMO $1\times CO_2$ scenario results provide the closest fit to Mongolia's current climate (e.g., Bayasgalan *et al.*, 1995).

The main limitations in using the SPUR2 model are that management practices remain constant over the simulated period, therefore adaptive responses were not considered (Benioff *et al.*, in press).

TABLE II
Weight parameters of steer genotypes

Genotype	Birth Month	Birth Weight (kg)	Weight at 8 Months (kg)	Weight at 18 Months (kg)	Mature Weight (kg)
Selenge	March-May	25.0	175	275	560
Halhgol	March-May	20.5	163	253	510
Mongol	March-May	19.5	145	216	480

3. Results

The GCMs predicted various climate changes for different locations and times in Mongolia. The estimated temperature increases range from 1.7 to 12.5 °C depending on the location and the scenario, and precipitation increases range from 40 to 167%. In the mountainous north, average precipitation could increase by 8% in the growing season, but in the south precipitation averages are estimated to increase by only 2%. Consequently, the estimated effects of climate change on plant and animal production varied by month, scenario, and location.

Under the scenario of current climate conditions and doubled CO_2, the peak standing biomass did not increase greatly at the 10 sites (Table III). But when CO_2 doubling was combined with changes in temperature and precipitation, the peak standing biomass increased significantly at most study sites under most scenarios. At the sites in mountainous regions, the increase in peak biomass may be due to increased temperatures. The results show that in the Khangai mountains and in the partially wooded steppe, the peak standing biomass would increase under climate conditions estimated by the GCMs for most scenarios at most sites (Figure 2a). At Tsetserleg (Khangai mountains) and Moron (partially wooded steppe), the GFDL and GISS scenarios predicted decreases in precipitation during the growing season, so the decrease in peak standing biomass under these climate scenarios would be expected. In the cold and elevated regions of the country, climate change could prove to be favorable for rangeland production because precipitation would be higher under global warming than it is today.

TABLE III
Peak standing biomass changes under different scenarios
(as percent of current standing biomass)

Natural Region	Site Name	Actual Peak Standing Biomass (kg ha^{-1})	Simulated Peak Standing Biomass (kg ha^{-1})	Peak Standing Biomass Changes under Different Climate Scenarios				
				2×CO_2	CCCM	GFDL	GISS	UKMO
High Mountain (Khangai Mountains)	Tsetserleg	590	569.4	1.5%	9.2%	-6.5%	-9.4%	2.3%
	Uliastai	260	394.1	3.3%	15.0%	3.9%	14.1%	13.3%
High Mountain (Altai Mountains)	Altai	110	189.3	4.7%	2.8%	3.1%	8.5%	3.4%
Partially Wooded Steppe	Moron	390	307.9	6.2%	310.2%	-27.0%	-39.7%	25.1%
	Hutag	330	468.1	4.6%	12.2%	7.2%	4.6%	8.7%
Steppe	Halhgol	1,110	388.5	4.2%	24.0%	-13.9%	5.6%	12.9%
	Mandalgov	240	235.5	4.8%	8.1%	3.1%	-21.3%	5.6%
Gobi Desert	Ulaangom	120	221.7	1.9%	21.5%	3.2%	-4.2%	14.2%
	Dalanzadgad	90	170.3	4.5%	8.3%	6.2%	3.9%	6.3%
	Sainshand	190	178.2	2.9%	5.4%	3.7%	-2.4%	4.5%

The south steppe region and the Gobi region of Mongolia are warmer and drier. For the sites in southern Mongolia, the UKMO predicted a drop in precipitation in the first part of the growing season, and the CCCM and GISS showed decreased rainfall in last part of growing season. At some sites in the southern part of the country, the UKMO and GISS scenarios predicted decreased rainfall throughout the growing season. Thus, generally peak standing biomass in these regions is not estimated to increase significantly; see, for example, Dalanzadgad (Figure 2b). Under the GCM scenarios, the time of peak standing biomass is simulated as 6-37 days earlier than under current conditions and the vegetative growing period is simulated as 13-49 days longer than under current climate conditions. Plant growth could start early because of precipitation and temperature increases for a short period in the spring. This result is similar to results from Hunt *et al.* (1991) and Baker *et al.* (1993), which indicated that changes in temperature increased the length of the growing season in the United States. Because of climate change, standing biomass accumulation on pastures may begin in March-April at most of sites, but from June onward, standing biomass would be close to or lower than the baseline scenario (Figure 2).

Simulated potential evapotranspiration generally increases under climate warming. For Mongolia, larger increases of potential evapotranspiration are simulated for April, May, and June. In the Gobi desert areas, potential evapotranspiration increases start from March.

Soil evaporation projections for Mongolia are different under the different GCM scenarios. The differences depend on precipitation projections and on plant cover. Available water for plants generally decreases for the entire growing season (April to October) at the study sites.

With increasing temperature, evapotranspiration and transpiration are estimated to be higher and water available to plants is estimated to be lower. The combination of these factors could have an unfavorable effect on plant growth. Because of early plant growth, estimated plant nitrogen contents were higher from March to May and lower later in the year under the climate change scenarios.

Increases in soil carbon were predicted under all GCM scenarios for sites in the northern regions; therefore, conditions would be suitable for an increase in carbon sequestration and an increase in potential nutrient storage. At the same time, decreases in soil carbon were predicted for the southern regions. Soil nitrogen deficits could increase desertification. However, in the simulations soil inorganic nitrogen did not significantly change.

The simulations also found that in the spring the simulated C:N ratio was higher than in the baseline scenario, but that after flowering the C:N ratio would potentially be decreased. The C:N ratio of the above-ground biomass is used to indicate the change in plant tissue quality. A decline in the C:N ratio may indicate lower plant quality.

Steer production generally declined under the climate change scenarios. In first period of the growing season (beginning in April) when crude protein and digestibility were greater, livestock intake was increased. During summer (June to August), temperature increased, so estimated livestock intake was decreased (Figures 3 and 4). Figure 5 is an example of intake for Mongolian steer. Intake for all genotypes was similar.

Under the current climate in Mongolia, the weight of the Mongolian genotype of steer, which graze year-round, declines in winter and spring and increases in summer and autumn (Figure 6). The simulation results showed that average daily gain decreased at Tsetserleg under the GISS, GFDL, and UKMO scenarios, and at Halhgol under the CCCM, GFDL, and $2{\times}CO_2$ scenarios. The simulation results also showed that average daily gain increased at Hutag under all scenarios. The study results show that meat production of the Mongolian genotype of steer could generally decline.

(a)

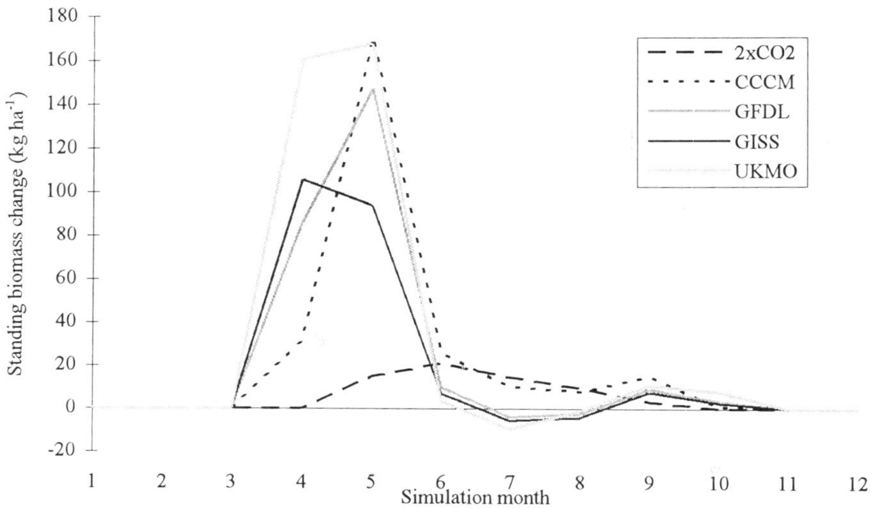

(b)

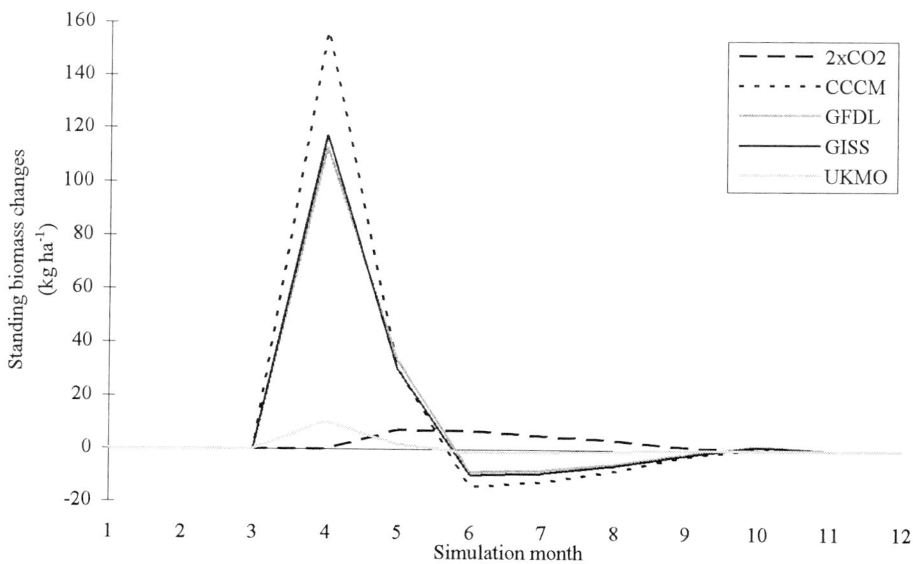

Fig. 2. Change from baseline of standing biomass (kg ha^{-1}) under different scenarios for (a) Hutag and (b) Dalanzadgad.

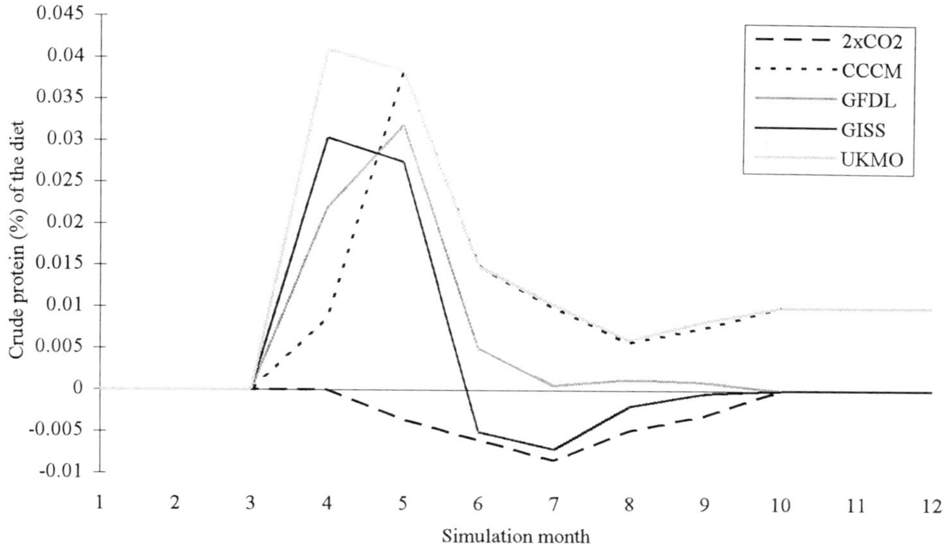

Fig. 3. Change from baseline of crude protein (%) of the diet selected for different scenarios for Hutag.

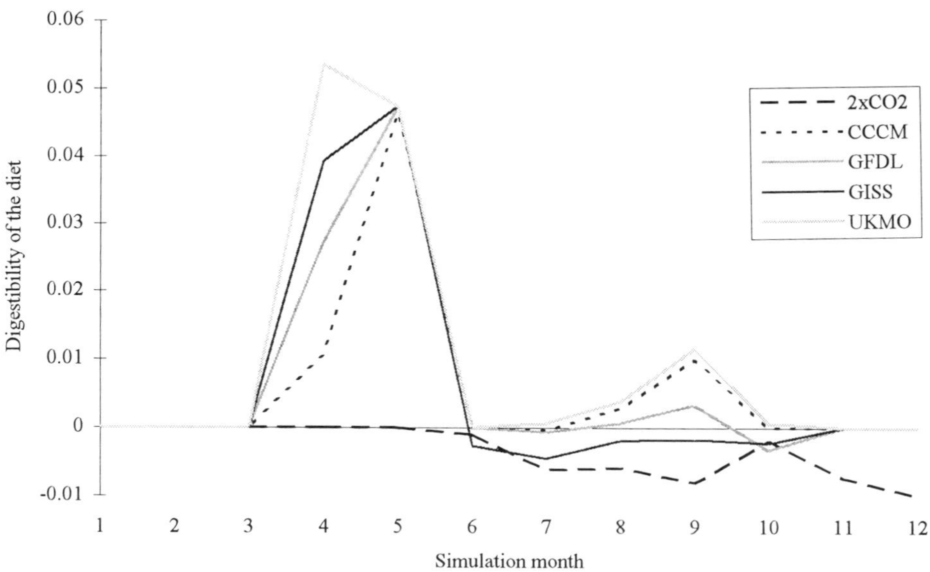

Fig. 4. Change from baseline of digestibility (%) of the diet selected for different scenarios for Hutag.

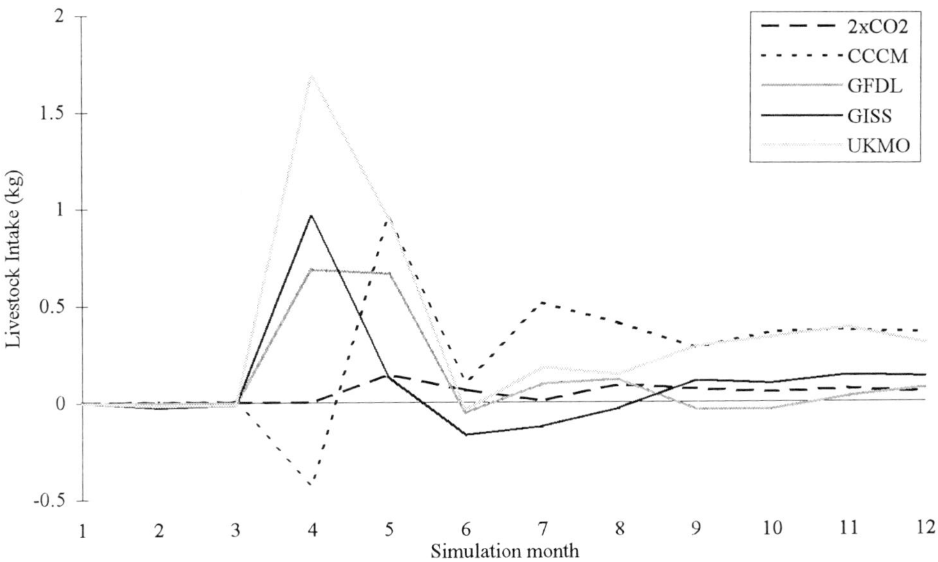

Fig. 5. Change from baseline of Mongolian steer intake (kg) for difference scenarios for Hutag.

In summer and autumn, the daily weight gain of the Selenge and Halhgol genotypes of steer, which receive supplemental feed in winter and spring, increased under the climate change scenarios (Figure 7).

4. Conclusions

Results from this study demonstrate that changes in global climate could affect pasture and livestock production in Mongolia. The impact of climate change on pasture production in the Gobi desert areas may be negative. In colder regions, climate change could have favorable effects on plant production. Because precipitation would increase in spring (March to May) under global warming, plant biomass accumulation could start early, the length of growth stages would be changed, the time of peak standing crop would occur earlier, and the growing season would be longer. Consequently, pasture plant biomass and its tissue quality are estimated to be higher in the first period of the growing season and lower in the second period. With global warming, potential evapotranspiration would rise, and the amount of water available to plants would decrease. The soil organic carbon pool is estimated to increase in the north and decrease in the south under the GCM scenarios. After flowering, with decreased above-ground standing biomass, plant nitrogen content, C:N ratio, and crude protein would be potentially decreased. With increased temperatures and lower plant quality, livestock intake and livestock weight could decline in late summer when digestibility is lower. Average daily weight gain of Mongolian steers that have only pasture forage, is estimated to be lower by 0.2 kg in the second part of the summer. All of these effects would mean a decrease in meat production of Mongolian steer. Steer weight could be lower by approximately 15-30 kg at the end of the autumn. However, the simulation indicated that there would be no negative effects on the weight gain of steer that are provided supplemental feed.

(a)

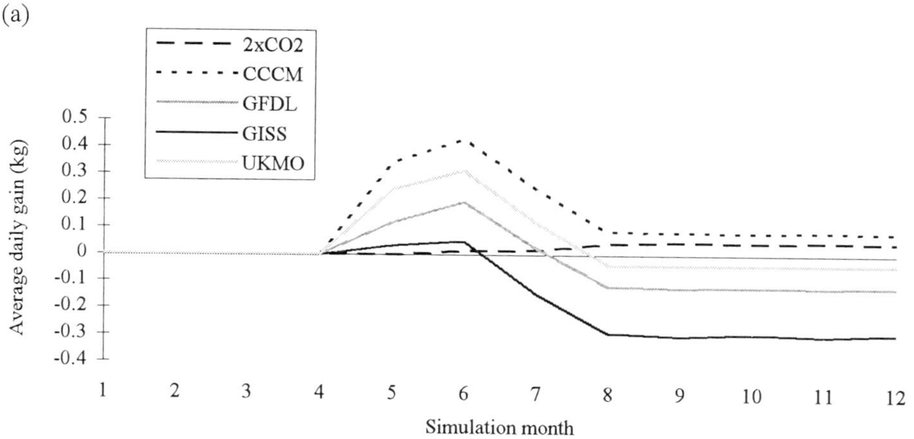

(b)

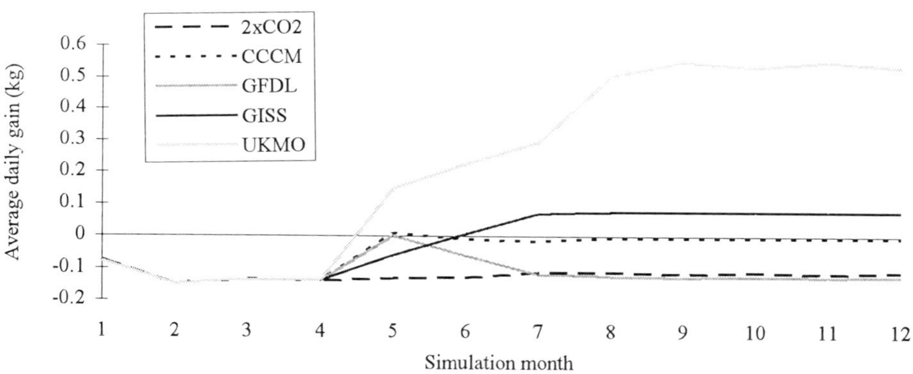

(c)

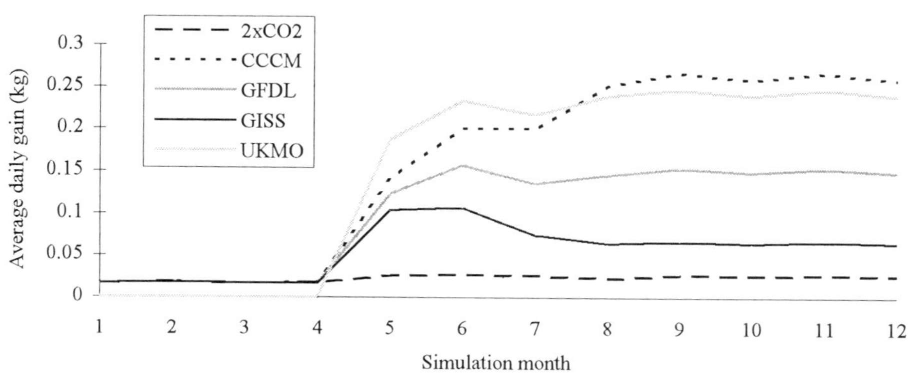

Fig. 6. Change from baseline of average daily gain of Mongolian genotype steers under different scenarios for (a) Tsetserleg, (b) Hutag, and (c) Halhgol.

(a)

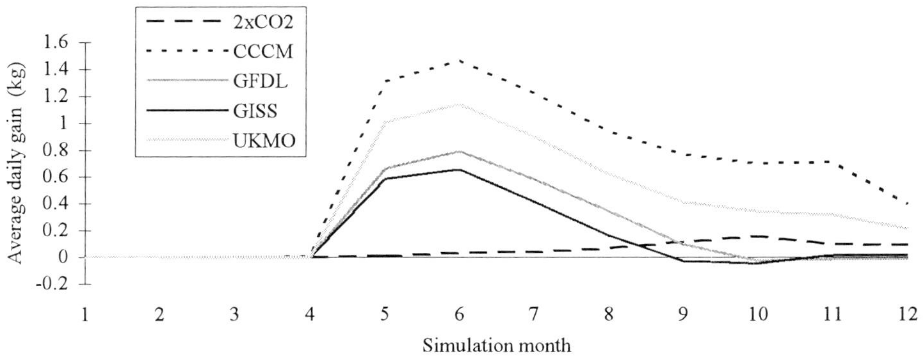

(b)

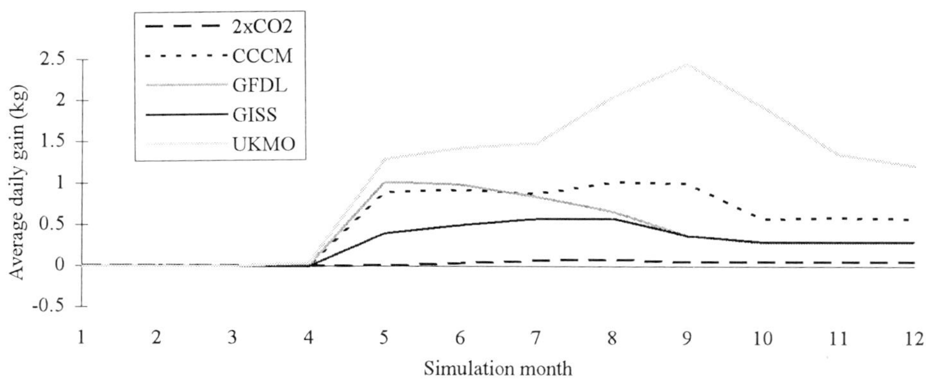

(c)

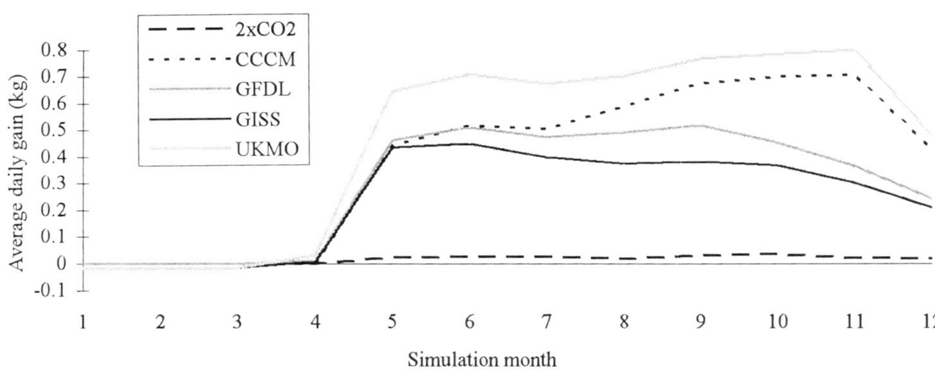

Fig. 7. Change from baseline of average daily gain of Selenge genotypes for (a) Tsetserleg and (b) Hutag, and Halhgol genotypes for (c) Halhgol.

To keep livestock production at its present level, adaptation options must be considered, such as strengthening of forage reserves that are set aside at the end of summer to provide a supply of fodder.

References

Baker B.B. *et al.*: 1993, *Climatic Change* **25**, 97.

Bayasgalan Sh., Mijiddorj R., Namkhai A., and Ulziisaikhan V.: 1995, Some changes in the climate integral characteristics and future trends for natural zones, in: *Proceedings of the Second Mongolia-China Symposium on Climate Change in Arid and Semi-Arid Regions of the Central Asia*, September 12-16, 1995, Ulaanbaatar, pp. 102-113.

Benioff R., Guill S., and Lee J. (eds): In press, *Vulnerability and Adaptation Assessments: An International Guidebook*, Kluwer Academic Publishers, Dordrecht, The Netherlands.

Boer G.J., McFarlane N.A., and Lazare M.: 1992, *Journal of Climate* **5**, 1045.

Hansen J., Russell G., Rind D., Stone P., Lacis A., Lebedeff S., Ruedy R., and Travis L.: 1983, *Monthly Weather Review* **3**, 609.

Hanson J.D. *et al.*: 1992, *SPUR2 Documentation and User's Guide,* Great Plains Systems Research Technical Report 1, U.S. Department of Agriculture, Fort Collins, Colorado.

Hunt H.W., Trlica M.J., Redente E.F., Moore J.C., Detling J.K, Kittel T.G.F., Walter D.E., Fowler M.C., Klein D.A., and Elliott E.T.: 1991, *Ecol. Mod.* **53**, 205.

IPCC: 1995, *Climate Change 1994: Radiative Forcing of Climate Change and an Evaluation of the IPCC IS92 Emission Scenarios,* Houghton J.T., Meira Filho L.G., Bruce J., Hoesung Lee, Callender B.A., Haites E., Harris N., and Meskell K. (eds), Cambridge University Press, Cambridge, England.

Mitchell J.F.B., Manabe S., Tokioka T., and Meleshko V.: 1990, *Climate Change: The IPCC Scientific Assessment,* Houghton J.T., Jenkins G.J., and Ephramus J.J. (eds), Cambridge University Press, New York.

Wight J.R. and Skiles J.W. (eds): 1987, *SPUR: Simulation of Production and Utilization of Rangelands. Documentation and User Guide,* U.S. Department of Agriculture, Agricultural Research Service, ARS 63, 372 pp.

Wilson C.A. and Mitchell J.F.B.: 1987, *Journal of Geophysical Research* **92**, 13315.

MODELING POTENTIAL CHANGES OF FOREST AREA IN THAILAND UNDER CLIMATE CHANGE

K. BOONPRAGOB and J. SANTISIRISOMBOON
Thailand Environment Institute, 210 Sukumvit 64
Bangkok 10600, Thailand

Abstract. The forest cover of Thailand has been characterized according to the Holdridge Life Zone Classification, a model that correlates climatic features with vegetation distribution. Six Holdridge life zone types of forest cover are found in Thailand: subtropical dry forest, subtropical moist forest, subtropical wet forest, tropical dry forest, tropical moist forest, and tropical wet forest. Climate change scenarios were simulated by three general circulation models: two United Kingdom Meteorological Office models (the low and high resolution versions) and the Goddard Institute for Space Studies model. These scenarios were used to simulate the effects of future climate change on Thai forests. The ratios of precipitation and the absolute values of temperature changes were incorporated into a baseline climate scenario from the International Institute of Applied Systems Analysis. Under the climate change scenarios simulated by the three general circulation models, the subtropical dry forest could potentially disappear, and areas of tropical very dry forest would appear. In general, the area of subtropical life zone would decline from about 50% to 20%-12% of total cover, whereas the tropical life zone would expand its cover from 45% to 80%. All three general circulation model scenarios suggest that the tropical dry forest has the greatest potential to extend into the subtropical moist forest. This analysis suggests that global climate change would have a profound effect on the future distribution and health of Thai forests.

Key words: Thailand, forest, GISS, Holdridge Life Zone Classification, UKMO, UK89

1. Introduction

Forests respond to prevailing environmental conditions, including climate and topography, which are the primary factors that control their distribution at global and regional scales. It is generally accepted that the broad-scale patterns of forest distribution are at equilibrium with present soil and climate conditions (Smith *et al.*, 1992b). On a local scale, many environmental factors such as soil characteristics, altitude, and slope aspect are involved in the distribution of vegetation patterns (Rundel and Boonpragob, 1995).

Increasing greenhouse gases in the atmosphere are expected to lead to changes in climate. The changes could include temperature rise, changes in seasonality, changes in precipitation patterns, and accelerated sea level rise. The warming rate is estimated to be about 0.3 °C per decade, and there is great uncertainty about how species will respond to this rapid change (Melillo, 1990). Changes in the seasonality of climate variables could have a marked impact on forests (Melillo *et al.*, 1990). Evidence from the past suggests that the potential for ecosystem changes in a warmer future is large (Warrick *et al.*, 1986).

Tropical forests are complex and highly productive. They have suffered and continue to suffer directly from human disturbance, due largely to poverty and population pressure. A CO_2-induced climate change, an indirect effect of human activity, will impose even greater pressure on tropical forests in the future. It is important to understand the possible impact of changing climate on potential forest distribution in the tropics to bring about sustainable utilization of forest resources.

The simplest and most general approach to examining the possible responses of the world's ecosystem types to climate change is to use hypothetical relationships between climate and vegetation derived in present-day conditions, and apply these to scenarios of changed climate

Water, Air, and Soil Pollution **92:** 107–117.
© 1996 *Kluwer Academic Publishers. Printed in the Netherlands.*

model widely used to examine global distribution of terrestrial vegetation by relating annual precipitation and biotemperature with ecosystem characteristics. General circulation models (GCMs) are mathematical formulations of atmospheric, ocean, and land surface processes that are based on classical physical principles used to simulate climatic patterns under changing climate (Gates *et al.,* 1990). Using Holdridge and GCMs, the extent of the overall impact of climate change can be characterized by shifts in the boundaries of major forest types.

This study characterizes potential changes of forest cover in Thailand based on simulated climate change scenarios from three GCMs utilizing climate data from local weather stations and global climate data. The trend of future forest areas in Thailand are also characterized.

2. Methods

2.1 FOREST CLASSIFICATION SYSTEM

The Holdridge Life Zone Classification system is a bioclimatic model relating the distribution of natural vegetation associations to climate indices. The life zones are depicted by a series of hexagons formed by intersecting intervals of climate variables on logarithmic axes in a triangular coordinating system. Two variables, average biotemperature and average annual precipitation, determine the classification. Average biotemperature is the average temperature over a year where the unit temperature values (i.e., daily, weekly, or monthly) are used in computing the average which is set to 0°C if they are less than or equal to 0°C. The Holdridge classification assumes, on the basis of studies of several ecosystems, that potential evapotranspiration (PET) is proportional to biotemperature. The PET ratio in the Holdridge system is therefore dependent on the two primary variables, annual precipitation and biotemperature, with an additional frost killing factor along a critical temperature line that divides hexagons between 12 and 24°C into warm temperate and subtropical zones. (Holdridge, 1967; Smith *et al.,*1992b).

2.2 FOREST AREA UNDER CURRENT CLIMATE

The global climate data of mean monthly precipitation and temperature at 0.5° latitude × 0.5° longitude resolution from the International Institute for Applied Systems Analysis (IIASA) developed by Leemans and Cramer (1990) are in good agreement with the magnitude and pattern of climate data from local weather stations (Boonpragob, unpublished). Utilizing IIASA data, present distributions of forest types under current climate based on Holdridge Life-Zone Classification were mapped at the 0.5° × 0.5° resolution using ARC/INFO software.

2.3 FOREST AREA UNDER CLIMATE CHANGE SCENARIOS

Climate change scenarios of doubling CO_2 were constructed from selected GCMs whose baseline climate variables of temperature and precipitation most resemble observations from local meteorological stations. The UK89 model from the United Kingdom Meteorological Office (UKMO) is a high resolution model (Mitchell *et al.,* 1989) that best duplicates both the magnitude and the patterns of local climate, whereas the Goddard Institute for Space

TABLE I
General circulation models used to construct climate change scenarios in Thailand

GCM	Model Resolution (lat. × long.)	Base $1\times CO_2$ (ppm)	Change in Mean Global Temp. (°C)	Change in Mean Global Precip. (%)
UK89	2.50° × 3.75°	320	+3.5	+9
UKMO	5.00° × 7.50°	320	+5.2	+15
GISS	7.83° × 10.0°	315	+4.2	+11

Studies model (GISS; Hanson *et al.*, 1988) and the older version of the UKMO model (Wilson *et al.*, 1987) have a lower resolution (Boonpragob, unpublished). The GISS and UKMO models do not depict baseline climate as well as the UK89 model. However, they are better than the other GCMs (Geophysical Fluid Dynamics Laboratory and Canadian Climate Centre Model), whose magnitude and pattern vary greatly from the observed baseline climate. Table I describes the characteristics of these three models.

Simulations of current ($1\times CO_2$) and $2\times CO_2$ climates from the UK89, GISS, and UKMO models were used to construct climate change scenarios. Changes in mean monthly precipitation and temperature were calculated for each GCM scenario for each 0.5×0.5 grid using linear interpolation. Temperatures were expressed as absolute difference ($2\times CO_2$ minus $1\times CO_2$) and precipitation as the ratio of $2\times CO_2$ to $1\times CO_2$. Changes in monthly precipitation and temperature were then applied to the IIASA climate data to provide a change scenario. The altered data bases corresponding to each of the three GCM scenarios were then used to reclassify the grid cells (0.5° resolution) using the Holdridge classification.

3. Results

Current theoretical types and boundaries of forest cover in Thailand according to the Holdridge classification based on biotemperature and precipitation data from IIASA are illustrated in Figure 1. There are six types of forests in Thailand: subtropical dry forest, subtropical moist forest, subtropical wet forest, tropical dry forest, tropical moist forest, and tropical wet forest. Under current climate, subtropical moist forest dominates, occupying 48% of the forest cover, and tropical dry forest and tropical moist forest contribute 32 and 15%, respectively. The other forest types are found in small areas, ranging from 5% to less than 1%.

The potential distributions of forests under the climate change scenarios simulated by the UK89, GISS, and UKMO models as classified according to the Holdridge model are illustrated in Figures 2, 3, and 4, respectively. The area and proportion of each forest type are shown in Table II. Subtropical forest is reduced under all three models' scenarios and is replaced by tropical dry forest. A new type of forest, tropical very dry forest, appears in the GISS and UKMO simulated climates. All three models predict a decline in areas of the subtropical moist forest and subtropical wet forest, and an increase in tropical dry forest, tropical moist forest, and tropical wet forest (Table II).

The relative extent of major forest types under current climate (Figure 1) and under the UK89-simulated changing scenario (Figure 2) reveals the potential extension of tropical dry forest into subtropical moist forest in the north and northeastern region and tropical moist forest expansion into subtropical moist forest in the south. Tropical wet forest replaces subtropical wet forest in the south and central-west region. The areas of subtropical moist forest and subtropical wet forest decline from 48 to 18% and 5 to 1% of the total area, respectively. In addition, the

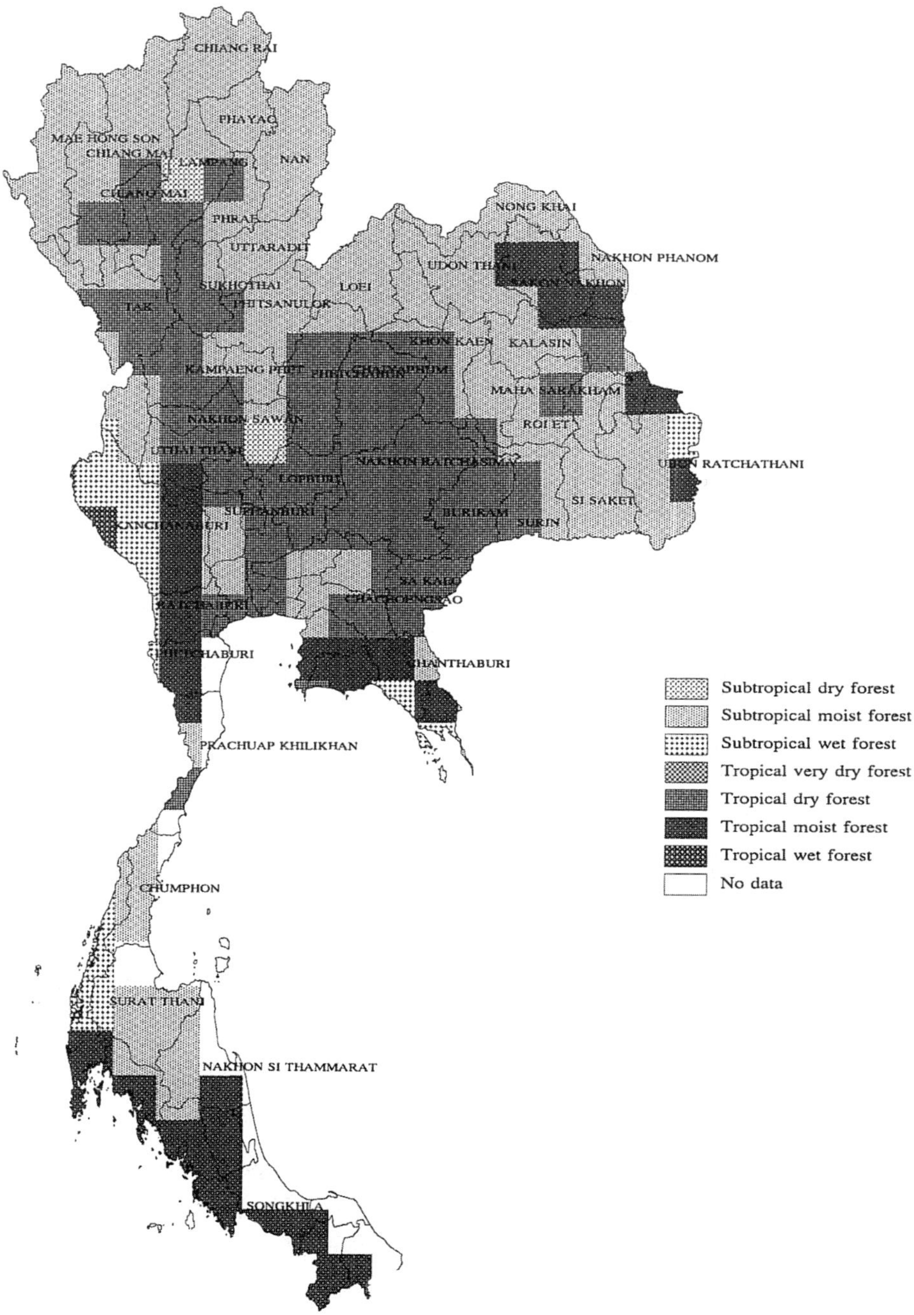

Fig. 1. Map of potential forest distribution in Thailand based on Holdridge life zones under current climate condition from International Institute for Applied System Analysis. The resolution is 0.5° latitude × 0.5° longitude.

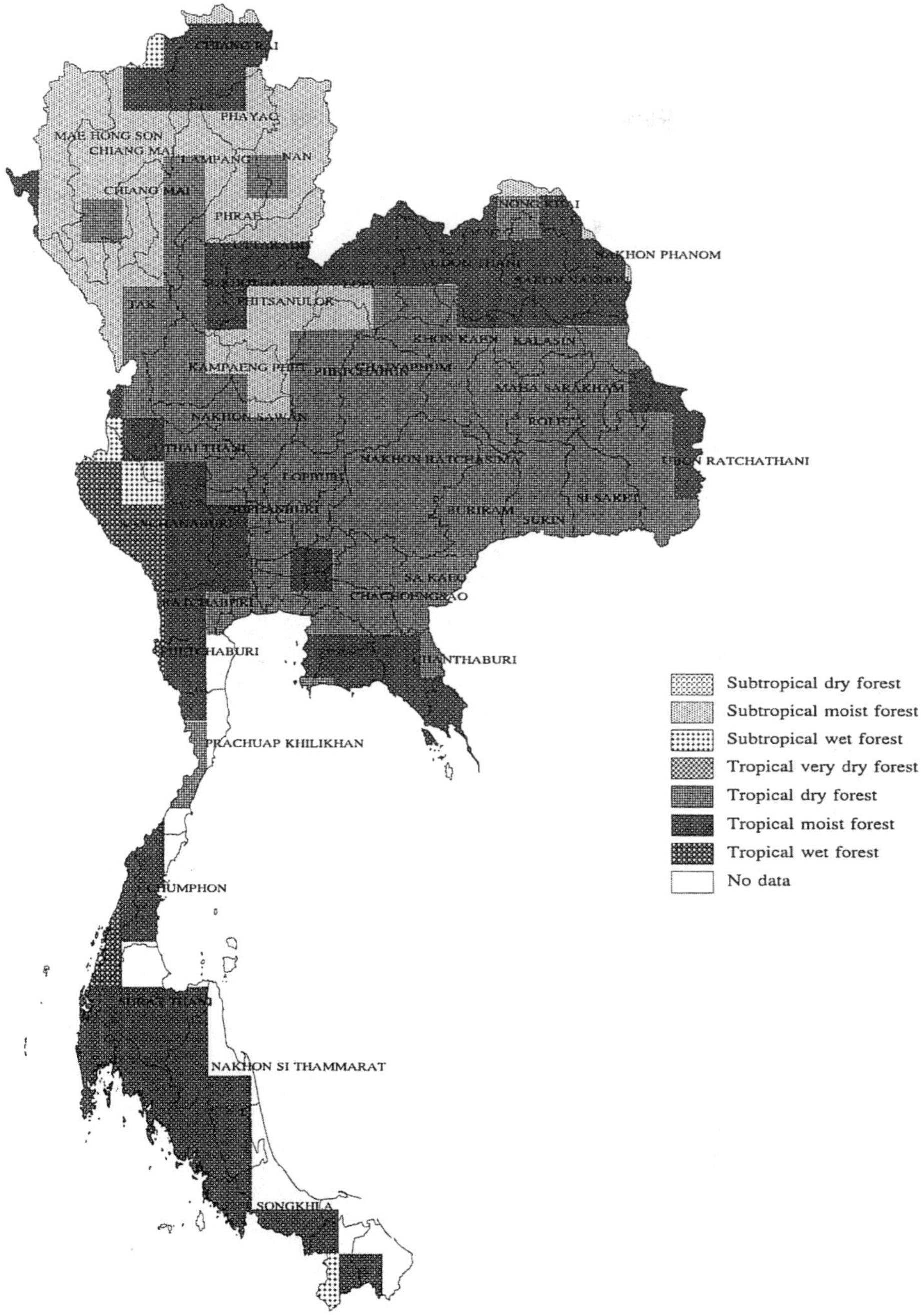

Fig. 2. Map of potential forest covers in Thailand based on Holdridge life zones under UK89-derived climate change scenario. The resolution is 0.5° latitude × 0.5° longitude.

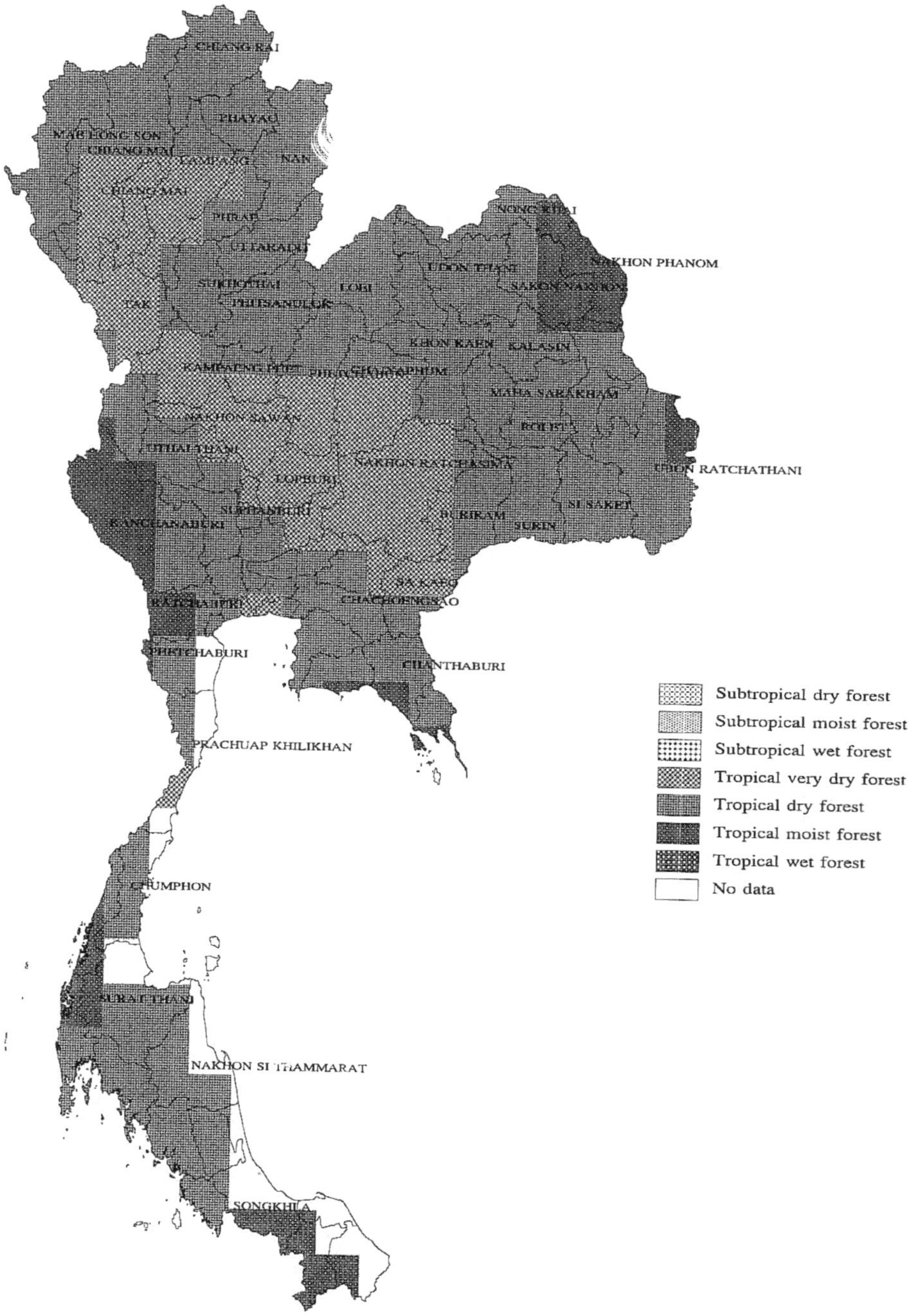

Fig. 3. Map of potential forest covers in Thailand based on Holdridge life zones under GISS-derived climate change scenario. The resolution is 0.5° latitude × 0.5° longitude.

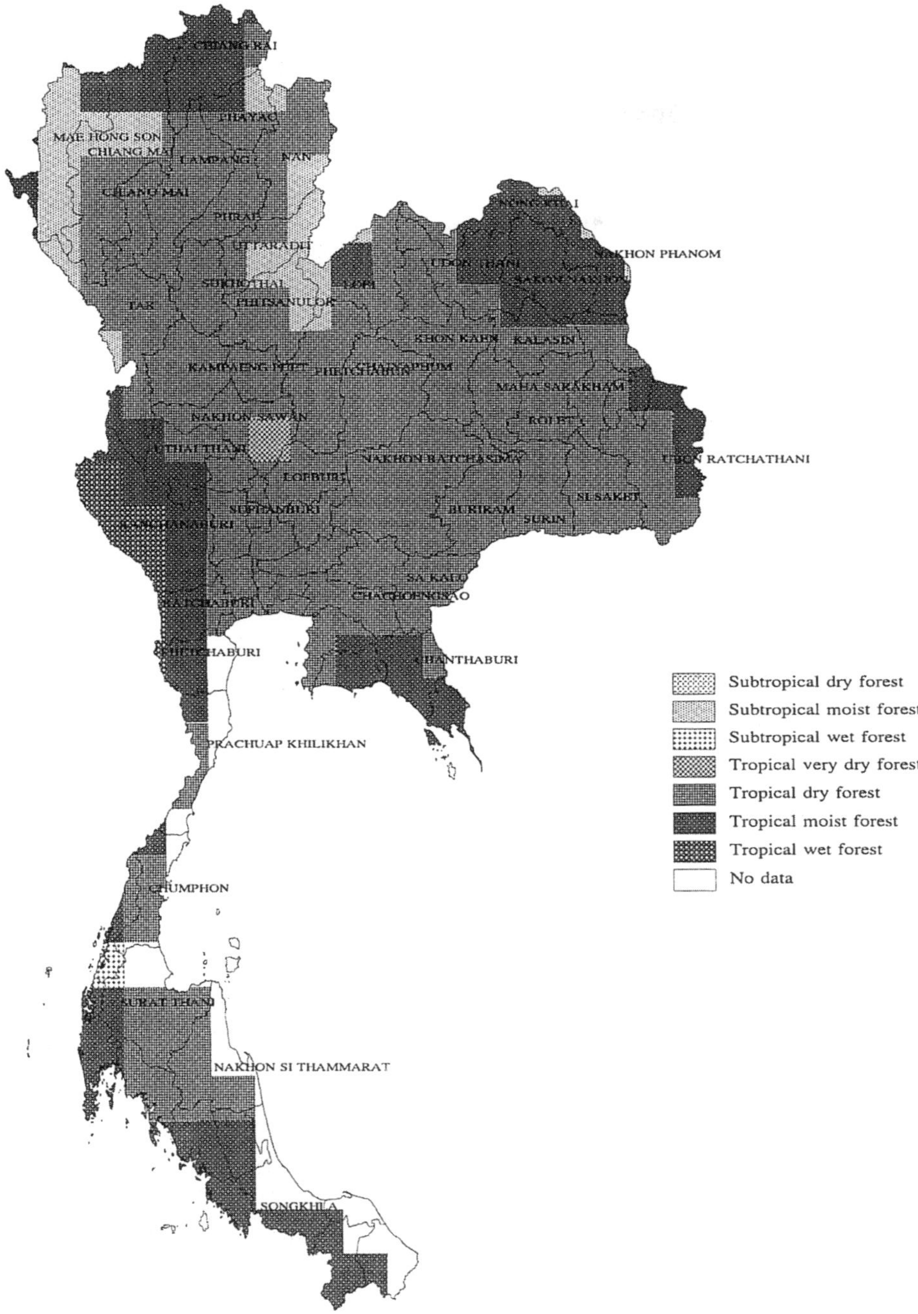

Fig. 4. Map of potential forest covers in Thailand based on Holdridge life zones under UKMO-derived climate change scenario. The resolution is 0.5° latitude × 0.5° longitude.

TABLE II
Areas of major forest types in Thailand under current and changed climate conditions according to three GCM scenarios

Forest Type	Current Area $km^2 \times 10^3$ (%)	Climate Change Area $km^2 \times 10^3$ (%) UK89	GISS	UKMO
Subtropical Dry Forest	5.9 (1.2)	—	—	—
Subtropical Moist Forest	234.5 (47.7)	87.7 (17.8)	9.5 (1.9)	59.4 (12.1)
Subtropical Wet Forest	22.2 (4.5)	6.6 (1.3)	5.3 (1.1)	1.8 (0.4)
Tropical Very Dry Forest	—	—	11.9 (2.4)	3.0 (0.6)
Tropical Dry Forest	156.5 (31.8)	218.6 (44.4)	341.3 (69.3)	290.1 (58.9)
Tropical Moist Forest	71.5 (14.5)	166.8 (33.9)	120.5 (24.5)	128.0 (26.0)
Tropical Wet Forest	1.6 (0.3)	12.6 (2.6)	3.7 (0.8)	9.9 (2.0)

areas of tropical dry forest and tropical moist forest increase from 32 to 44% and 15 to 34% of the total area (Table II). Although the extent of tropical wet forest, present in small areas of less than 1% under the current climate, has the highest percent change with an estimated increase to about 3% of total area under the climate change scenario (Table II).

The simulated changes in biotemperature and precipitation from the GISS scenarios indicate that subtropical moist forest, previously dominant in the area under $1\times CO_2$, is almost entirely replaced by tropical dry forest. Its area declines from 48 to 2%, whereas the area of tropical dry forest increases from 32 to 70% (Table II, Figures 1 and 3). There is a lesser increase of tropical moist forest, from 15 to 25%, mostly replacing subtropical wet forest. The small area previously designated subtropical dry forest turns into tropical dry forest and tropical very dry forest.

Potential forest cover simulated by the UKMO scenario demonstrates an extension of tropical dry forest and tropical moist forest into subtropical moist forest, whereas tropical wet forest has the potential to replace subtropical wet forest. The increases in tropical dry forest, tropical moist forest, and tropical wet forest are from about 32 to 59%, 15 to 26%, and 0.3 to 2% of the total area, respectively. Alternately, subtropical moist forest is subject to a tremendous decline, from 48 to 12%, whereas subtropical wet forest has the potential to decline from 5 to less than 1%. The previous subtropical dry forest is replaced by tropical dry forest and tropical very dry forest (Table II, Figures 1 and 4).

4. Discussion

A number of studies have been conducted to simulate the potential impact of global climate change on terrestrial vegetation (e.g., Emanuel, 1985; Prentice *et al.*, 1990; Smith *et al.*, 1992a; Smith *et al.*, 1992b; Neilson, 1993). These studies concentrate on temperate regions, are based on broad-scale global climate patterns, or utilize only temperature as a determining factor. Nevertheless, they predict a poleward shift in vegetation boundaries. This study attempts to investigate spatial changes in the forest cover of Thailand based on a finer scale of IIASA data. Although the mean annual temperature from IIASA is slightly lower than the observed values from the Thai Meteorological Department, the patterns are compatible (Boonpragob, unpublished). However, there are many uncertainties in the use of GCM scenarios to estimate changes in forest boundaries.

Because of limited data availability and a lack of understanding of factors controlling ecosystems, it is appropriate to discuss the theoretical changes in forest types that might result from only climate without taking into consideration other edaphic factors (i.e., soil type, elevation). However, these results should not be viewed as predictions and should be interpreted with caution. Given the difficulty in comparing local forest classification with Holdridge life zones, the comparison of life zones in this study is determined explicitly from the Holdridge classification system. However, a preliminary comparison based on temperature and precipitation of the ecosystems indicates relative compatibility between the two systems.

Preliminary caution relating to model biases must be taken into account in evaluating future climate changes. Additional caution arises from GCMs' relatively crude treatment of the ocean and their neglect of other potentially important elements of the climate system such as the upper atmosphere, atmospheric chemical processes, and surface biological processes. There will be a range of uncertainties associated with their results (IPCC, 1990). However, substantial uncertainty in the models simulating climate change scenarios does not prevent us from elucidating that global climate change leads to change in forest ecosystems. Looking at a number of plausible scenarios at least tells us the potential trend of changes. The predictions of forest covers based on changing climate simulated by the three GCMs, UK89, GISS, and UKMO, consistently suggest that global change will potentially lead to the replacement of the subtropical life zone with the tropical life zone. The difference lies in the extent of the boundaries of changes. The major distinction between the two life zones is based on a critical temperature line at 24 °C in the Holdridge scheme. All three GCMs simulated elevated temperature under climate change scenarios, resulting in the shift of these life zones.

As simulated by UK89, the expansion of tropical moist forest into subtropical moist forest indicates a shift into a higher temperature zone, but remaining at the same annual precipitation boundary of 2000-4000 mm. This condition might possibly lead to drier conditions because increasing precipitation cannot compensate for increasing evapotranspiration driven by higher temperatures. These potential changes are found in the upper northeastern region of the country and the tip of the northern region (Figures 1 and 2). The highest increase in temperature in the country, about 3.5 °C, is found in this area along with the lowest increase in precipitation, less than 20% of normal (Boonpragob, unpublished). An extension of tropical wet forest into subtropical wet forest also indicates a change to a higher temperature zone, with shifts in annual precipitation to a wetter condition from 2000-4000 mm to 4000-8000 mm. This change might possibly lead to a wetter habitat. This potential change is found in a small part of the central-west and southwest where simulated climate change indicates the lowest increase in temperature, about 2.5 °C, along with the highest increase in precipitation, of about 40% of normal.

In general, forest changes simulated by GISS indicate shifts to warmer life zones, but not the wetter precipitation zones of the Holdridge scheme. This trend might be due to a larger increase in temperature but a smaller increase in annual precipitation under the GISS climate change scenario. The changes in forest cover based on climate simulated by UKMO indicate a larger shift to a drier conditions and a lesser change to higher precipitation. This response might be due to a slightly higher magnitude of temperature change than that simulated by UK89.

The small areas of subtropical wet forest and tropical wet forest found under current climate and under changing climate are due to incomplete data in the southern peninsula of the country. This area is a narrow strip of land that extends into the ocean. It is separated by mountain ridges running north-south, resulting in different climate zones for the east and the west coasts. Actual climate monitoring is not being conducted in this area. Interpolation of the data would be inappropriate because of the great distances of the neighboring points from the continent.

However, this region is recognized as having high rainfall, typical of the wet tropics similar to Malaysia and Indonesia. It is covered with tropical wet forest or tropical rain forest (Rundel and Boonpragob, 1995).

Plant climate-related models are not sufficiently sophisticated to predict changes in local vegetation with great confidence. Seasonality, an important factor influencing the regeneration processes of tropical forest, is not captured by the Holdridge model. Any change in seasons has far-reaching ecological effects (Western, 1991). Additionally, soil characteristics are another nonclimate factor of importance in determining the local vegetation of the region. Although many properties of soils are closely related to climate and are reflected by the Holdridge classification in terms of their influence on vegetation, physical and geological factors such as parent material are also important soil-forming factors that are not treated in the purely climatic scheme of the Holdridge classification.

This study lays an important and fundamental foundation of the potential impact of climate change on forest ecosystems in Thailand based on available data and technology. Adaptation policies as well as research to reduce the uncertainty associated with models are prudent and necessary to minimize the catastrophic effects of climate change. This is especially true in natural ecosystems that have only limited possibilities to adapt rapidly to such change. It is not currently possible to prescribe the ways and means of relating the research to a policy process in a generic fashion. All this, however, leads us to the process of identifying future adaptation strategies by considering economic feasibility, technical feasibility, social and legal acceptability, spatial linkages, and environmental sustainability. Perhaps the single most important factor is flexibility in choosing options for adaptation, allowing the possibility of future choice while meeting present needs.

5. Conclusions

Many scientists believe that climate change is real and the impact is likely to be irreversible and catastrophic. Simulations of climate change scenarios in the tropics showing the vulnerability of tropical forests to these changes are rarely performed. This study is the first of its kind in Thailand to assess the potential impact of climate change on forest cover. The simulation models consistently predict the expansion of tropical dry forest into subtropical moist forest in the upper part of the country, whereas subtropical forests in the southern portion will be replaced by tropical forests. National action plans and adaptation policies can be primarily drawn from the results of this study. For example, the reforestation project under way in Thailand should take into consideration the potential for future climate change to ensure the survival of planted trees. An appropriate strategy would be to select drought adapted species for the northeast and north and wet tolerant species for the south.

Acknowledgments

The authors wish to thank Surat Sarasup and Paul Hesting for their help in producing the map of forest classification using ARC/INFO software. Additional thank are extended to the U.S. Country Studies Program for providing financial and technical support.

References

Emanuel W.R., Shugart H.H., and Stevenson M.P.: 1985, *Climatic Change* **7**, 29.

Gates W.L., Rowntree R.R., and Zeng Q.C.: 1992, Validation of climate models, in: *Climate Change: The IPCC Scientific Assessment,* Houghton J.T., Jenkins G.J., and Ephramus J.J. (eds), Cambridge University Press, Cambridge, pp. 99-130.

Hansen J., Fung I., Lacis A., Rind D., Lebedeff S., Ruedy R., Russel G., and Stone P.: 1988, *J. Geophys. Res.* **93**, 9341.

Holdridge L.R.: 1967, *Life Zone Ecology,* Tropical Science Center, San Jose, Costa Rica.

IPCC: 1990, *Climate Change 1990: The IPCC Scientific Assessment,* Houghton J.T., Jenkins G.J., Ephramus J.J. (eds), Cambridge University Press, Cambridge, England.

Leemans R. and Cramer W.: 1990, The IIASA Climate Database for Land Area on a Grid of 0.5 Resolution, WP-41, International Institute for Applied Systems Analysis, Laxenburg, Austria.

Melillo J.M., Callaghan T.V., Woodward F.I., Salati E., and Sinha S.K.: 1990, Effects on ecosystems, in: *Climate Change: The IPCC Scientific Assessment,* Houghton J.T., Jenkins G.J., and Ephramus J.J. (eds), Cambridge University Press, Cambridge, pp. 282-310.

Mitchell J.F.B., Senior C.A., and Ingram W.J.: 1989, *Nature* **341**, 132.

Neilson R.P., King G.A., and Lenihan J.: 1994, Modeling forest response to climate change: The potential for large emissions of carbon from dying forests, in: *Carbon Balance of World's Forested Ecosystems: Toward a Global Assessment,* Kanninen M. (ed), Academy of Finland, Helsinki, Finland, pp. 150-162.

Prentice K.C.: 1990, *J. Geophys. Res.* **95**, 11811.

Rundel P.W. and Boonpragob K.: 1995, Dry forest ecosystems of Thailand, in: *Seasonally Dry Tropical Forests,* Bullock S.H., Moony H.A., and Medina E. (eds), Cambridge University Press, New York, pp. 93-123.

Smith T.M., Leemans R., and Shugart H.H.: 1992a, *Climatic Change* **21**, 367.

Smith T.M., Shugart H.H., Bonan G.B., and Smith J.B.: 1992b, *Advances in Ecological Research* **22**, 93.

Warrick R.A., Shugart H.H., Antonovsky M.Ja., Tarrant J.R., and Tucker C.J.: 1986, The effect of increased CO_2 and climate change on terrestrial ecosystems, in: *The Greenhouse Effect, Climate Change and Ecosystems*, Bolin B., Doss B., Jager J., and Warrick R. (eds), SCOPE 29, John Wiley and Sons, New York, pp. 368-392.

Western D.: 1991, Climate change and biodiversity, in: *A Change in the Weather: African Perspectives on Climate Change*, Omnide S.H. and Jurna C. (eds), African Center for Technology Studies, Nairobi, Kenya, pp. 87-96.

Wilson C.A. and Mitchell J.F.B.: 1987, *J. Geophys. Res.* **92,** (D11) 13315.

VULNERABILITY AND ADAPTATION OF THE LARCH FOREST IN EASTERN SIBERIA TO CLIMATE CHANGE

K.I. KOBAK,[1] I.YE. TURCHINOVICH,[1] N.YU. KONDRASHEVA[1]
E.-D. SCHULZE,[2] W. SCHULZE,[2] H. KOCH,[2] and N.N. VYGODSKAYA[3]

[1] *State Hydrological Institute*
St. Petersburg, Russia

[2] *Bayreuth University*
Bayreuth, Germany

[3] *A.N. Severtzov Institute of Animal Evolutionary Morphology and Ecology*
Moscow, Russia

Abstract. The most widely distributed coniferous forests in the world are the larch forests. In the Russian Federation they occupy 27.6×10^6 ha. In Siberia, the larch species *Larix russica* generally grows west of the Yenissei River, and *Larix gmelinii* grows to the east. The morphological and physiological features of *L. gmelinii* make it possible for this species to grow in the far north of eastern Siberia, where climate conditions are more severe: The range of air temperature fluctuations in this region is more than 100 °C, from 38 °C down to 64 °C below zero. One of the major adaptions to unfavorable soil conditions is provided by a specific feature of root formation in *L. gmelinii*, in which the apex central root dies off at the permafrost border and a root system develops in upper soil layers. The major larch vulnerability factors are natural and anthropogenic fires and damage caused by insects, which become more frequent with hot and dry weather. The consequences of projected global warming could be both positive and negative for larch forests. Permafrost melting may result in improved soil nutrition in the areas the larch forests occupy, yet the frequency of forest fires and damage by pathogens are likely to increase. Global warming is expected to cause forest dieback and increased areas of steppe in the southern regions of eastern Siberia.

Key words: Siberia, forest

1. Introduction

Global climate warming due to the increase in greenhouse gas concentration in the atmosphere is likely to cause changes in terrestrial biota (Budyko *et al.*, 1990; Houghton *et al.*, 1990; IPCC, 1995). Of critical importance is the study of potential changes in forest ecosystems and peatland due to climatic fluctuations. These ecosystems contain much more carbon in phytomass and soil pools than other ecosystems. The changes in their carbon pools and fluxes between ecosystems and atmosphere noticeably affect the carbon cycle and the future chemical composition of the atmosphere (Dixon *et al.*, 1994).

The larch forests are the most widely distributed coniferous forests in the world. In the Russian Federation, they occupy 27.6×10^6 ha, and the growing stock equals 25.2×10^9 m^3. According to the recent forest inventory data (Forest Fund of the USSR, 1990; Forest Fund of Russia, 1995), larch forests contain 32% of the total growing stock of Russia, or 35% of the total carbon storage of forests. The shares of other species are as follows: 20% for *Pinus* sp., 15% for *Picea*, 10% for *Pinus sibirica*, 3% for *Abies*, 13% for *Betula*, 5% for *Populus*, and 2% for *Quercus* and other trees and shrubs.

There are three main larch species in Siberia: *Larix russica* (Endl.) Sab.ex Trautv. (formerly *L. sibirica* Ledeb.), *L. gmelinii* (Rupr.) Rupr. (formerly *L. dahurica* Turcz.ex Trautv.), and *L. sukaczewii* Dyl. *Larix russica* occupies western Siberia, *L. gmelinii* grows eastward of the

Water, Air, and Soil Pollution **92:** 119–127.
© 1996 *Kluwer Academic Publishers. Printed in the Netherlands.*

TABLE I

Mean long-term meteorological data (1937 to present) from Yakutsk weather station (62°08'N, 129°75'E)

	Temperature (°C)					Precipitation (mm)		
	Mean			Absolute				
Month	Daily	Max	Min	Max	Min	Mean	Max	Min
1	-42.9	-39.4	-46.4	-8	-64	7	—	—
2	-35.8	-30.6	-40.5	-7	-62	6	—	—
3	-22.4	-14.6	-29.2	6	-53	4	—	—
4	-7.4	-1.1	-14.0	21	-41	8	—	—
5	5.7	11.5	-0.1	33	-21	14	—	—
6	15.3	22.0	8.9	34	-5	27	—	—
7	18.8	25.5	12.1	38	0	37	—	—
8	14.9	21.8	9.0	35	-7	36	—	—
9	6.3	12.4	1.2	27	-10	21	—	—
10	-7.8	-3.7	-11.3	18	-40	15	—	—
11	-27.9	-23.7	-31.6	2	-54	12	—	—
12	-39.5	-35.8	-43.1	-2	-58	8	—	—
Annual	-10.2	-4.6	-15.4	38	-64	195	350	134

Source: *Basic Data on the Climate of the USSR*, 1976.

Yenissei River, and *L. sukaczewii* grows in northwestern Siberia. *Larix gmelinii* occupies more than 85% of larch forest area, *L. russica* about 15%, and *L. sukaczewii* less than 0.5%.

Systematic studies of larch forests started more than 150 years ago. The taxonomy of *Larix genus* has been changed by different botanists many times. *L. gmelinii (dahurica)* was taxonomically separated from *L. decidua* and *L. russica (sibirica)* more than 100 years ago. Later, the larches of northern Yakutia, the Kuril Islands, the coast of the Sea of Okhotsk, and the coast of the Amur river were identified as separate species. Dylis (1961) called the larches of Yakutia, Chukotka, Baikal, parts of Amur, and the northern China regions *L. dahurica*. In his opinion, the differences in morphological structure of these species are not large compared to the differences between the larches in the western and eastern parts of the area occupied with *L. dahurica*. He identified two subspecies (*ssp. dahurica* and *cajanderi*). The areas occupied with these species never overlap, and they border at 120°E.

In spite of numerous investigations of *L. gmelinii* (Dylis, 1961; Utkin, 1965; Pozdnyakov, 1975, 1983), our knowledge about its physiological processes and adaptation mechanisms to the severe environmental conditions in Siberia (Table I) is limited. However, we need to know more about these processes to forecast the reaction of larch forest to future climate warming. This paper focuses on *L. gmelinii* because it is the predominant species in eastern Siberia and significant changes in the phytomass of this species could noticeably affect the global carbon cycle.

2. Methods

2.1 MORPHOLOGICAL AND PHYSIOLOGICAL FEATURES OF *L. gmelinii*

To investigate *L. gmelinii,* an expedition to Central Yakutia, headed by Prof. D.-E. Schulze from Bayreuth University, was organized in the summer of 1993 (Schulze *et al.,* 1995).

The research was done in pristine forest on the Aldan plateau 155 km south of the city of Yakutsk (60°51'N, 128°16'E). We investigated the composition, structure, biomass, and nitrogen dynamics of trees and stands of *L. gmelinii*. Carbon and water fluxes were also measured.

The Aldan plateau is an extensive limestone plate south of a prominent bow in the region's major water course, the Lena River. The land is gently undulating, with broad, shallow valleys dissecting most of the region. The study sites were located along a south and east-facing hillside. The top of this hill was only 50 m above the valley bottom. At all sites, the soils were deep alluvial silt loam on limestone. The larch forest covered the slopes and hilltops in almost pure stands.

To identify various stages of development of a *L. gmelinii* forest after fire and its relation to other tree species, mainly *Pinus silvestris* and *Picea obovata*, measurements were made along six representative transects, each 50 m long and 2.5 to 8 m wide, and in a separate 0.5 ha plot.

Tree measurements included species, stem diameter at a height of 1.3 m (dbh), total height, height of the lowest live branch, and estimation of crown cover in the stand. At 1 m intervals along each transect, the percent cover of species of the forest floor was recorded along the 50 m line transect. Soils for each transect and the 0.5 ha plot were described. A total of 66 trees were selectively harvested from the six transects and the plot. The trees represented the range of age class and canopy positions (subcanopy, canopy, and emergent) found in the forest. We identified six stand types:

- young stand (50 years old), an early stage of development with high density of trees (about 50,000 per ha)
- pole stand (130 years old), a later stage of development with a less, but rather high, density of trees (4,800 trees per ha)
- *Vaccinium* stand, an open mature forest (125 years old) with *Vaccinium* and *Arctostaphyles* understory
- *Alnaster* stand, a dense mature forest (125 years old) with an understory of *Alnaster*
- *Ledum* stand, an open mature forest (130 years old) with *Ledum* and *Alnaster* understory
- Lichen stand, an open, very old-growth stand (380 years old) with lichen understory.

Fresh (and dry) weights of the main stem, coarse branches (>2 cm diameter), twigs (<2 cm diameter), and current year shoots were determined. The total lengths were measured separately for the main and lateral branches as well as for current year shoots. Needle and twig biomass was determined for each tree. Tree age was counted on stem segments originating from close to the soil surface of 66 felled trees.

A photographic print exposure technique was used to estimate areas of fresh needle samples. Leaf area index was calculated from the size distribution of trees on each plot. The roots distribution was studied on the plot with *Vaccinium* (*Laricetum uliginoso-vaccinosum*) (Schulze *et al.*, 1995).

2.2 EFFECTS OF CLIMATE CHANGE ON NATURAL ZONES AND ECOSYSTEMS

Based on the concepts and methods of Budyko (1971) and Zubenok (1976), we used some climatic parameters to determine the relationship between the distribution of natural zones and climate factors. These parameters are potential evapotranspiration (E_o), aridity index (E_o/r, where r is the annual precipitation) and continentality index (a function of the annual temperature amplitude and latitude) for mountainous regions. Our investigations have shown that natural zone distribution in the former Soviet Union and North America corresponds to particular ranges of selected climatic parameters. The maps of zonal distributions developed using modern values of these parameters do not differ from those developed by geobotanists (Kobak and Kondrasheva, 1992).

To forecast possible changes in the location and areas of natural zones with temperature increases of 1.4 and 2.2°C, we used maps of predicted evapotranspiration, developed by the complex method of Budyko, and precipitation anomaly maps for warming of 1-2°C, developed by Borzenkova (Budyko and Izrael, 1987). Using this method, we reconstructed the temperature of the warm periods of the Holocene. Aridity index values were calculated and maps of natural zone distribution in the former Soviet Union territory were developed with global warming of 1.4 and 2.2°C. The result is a hypothetical zonality pattern which is optimum with changed climate conditions. We did not take into account the succession process rate, possible increase in productivity and biomass due to CO_2 concentration and temperature increase, and some other factors.

3. Results

3.1 CHARACTERISTICS OF *L. gmelinii*

Our species cover estimates indicated that we examined the main types of *Larix gmelinii* forests. The young and mature stands are essentially the same plant community at different stages of tree crown closure. More radiation reaches the forest floor in the relatively open stands (the lichen stand, for example), leading to increases in lichen cover (*Cladina rangiferina*, *Cetraria laevigata*), and an increase in cover or invasion by other light-demanding herbaceous species (e.g., *Poa pratensis*).These light-demanding species were also found along the edges of the young stand. In the more closed stands, herbaceous species disappeared, and lichens accounted for a lower percentage cover.

The forest structure investigation of the six stands suggests that regeneration takes about 40-50 years. The age classes in our study suggest that these stands regenerated after at least three major fires that burned various areas over the past 400 years. The analysis of the growth parameters we obtained showed that the diameter of main stems tapered by less than 10 mm per 1 m. The limited variability in stem form from age 50 to 380 years suggests that the straight, columnar stems of *L. gmelinii* develop from strong competition for light at an early age, which was evident in the young stand. The above-ground biomass of *Larix* forest increases up to about 125-130 years. Our data from the 380-year-old stand showed a decrease in stand density, but above-ground biomass remained unchanged (Table II). For all stands, tree volume and biomass increased curvilinearly with stem diameter. In the largest tree, the stem biomass comprised 96% of tree biomass, branches contributed 3%, and needles 1%.

The average water content for all stands was 0.4 kg H_2O kg^{-1} dry wood, and wood density (dry weight per volume) was equal to 471 kg m^{-3}.

Tree height increased initially linearly with increasing stem diameter up to 20 cm, but then height appeared to reach an upper limit of about 24 m. The over-bark stem diameter included an increasing quantity of bark because there was a positive relation between stem diameter and bark thickness. Bark comprised 12-14% of the stem volume.

TABLE II
Some characteristics of the *Larix gmelinii* stands (adapted from Schulze *et al.*, 1995)

Characteristics	Stand Types					
	Young	Pole	*Vaccinium*	*Alnaster*	*Ledum*	Lichen
Tree Age (years)	49	130	125	125	130	380
Tree Height (m)						
Max.	7.5	15.0	20.5	23.0	16.5	20.0
Average	2.1±0.1	7.8±0.6	12.3±0.9	12.9±0.9	12.1±0.7	11.0±1.1
Tree Diameter (mm)						
Max.	79	140	236	344	235	551
Average	23±1	67±5	112±10	133±12	128±8	163±21
Tree Crown Cover (%)	100	87	89	88	55	55
Above-Ground Biomass (kg m^{-2})						
Total	4.36	7.17	9.26	12.02	7.23	8.91
Wood	3.93	6.94	9.06	11.79	7.07	8.77
Needle	0.43	0.23	0.20	0.23	0.16	0.14
Leaf Area Index (m^2 m^{-2})	5.70	1.68	1.47	1.63	1.14	1.03
C/N Ratio (mol mol^{-1})						
Litter	42±2.0	63±7.0	55±4.4	47±4.5	50±6.1	56±9.8
Topsoil	21±0.5	35±1.6	38±4.6	45±9.4	41±0.9	36±1.4

With a relatively constant sapwood thickness, sapwood area increased linearly with stem diameter. The average sapwood thickness of 9 mm was about 20% of the average stem-wood radius of 48 mm, excluding about 6 mm of bark. A significant linear relationship between total needle biomass and sapwood area was suggested. This relationship showed increasing variation with tree size. On the average, about 5 g (or about 7,000 needles) are supported by each 1 mm circumference of the stem.

Larch roots are characterized as shallow or anchoring (Utkin, 1965; Pozdnyakov, 1975). Most root mass in highly productive types of forest is distributed at a depth of 40% of thawing layer (the layer of melting permafrost), and in less productive forests at a depth of 30% of this layer. *Larix gmelinii* roots very seldom reach the permafrost layer. According to Pozdnyakov (1975, 1983), mass root distribution zone coincides with the upper accumulation, alluvial soil layer (75-85% of total mass), whereas only 15-25% of total root mass reaches the lower soil layers.

Our studies of *Laricetum uliginoso-vaccinosum* roots show that about 90% of big and medium root mass (80% of the total root mass) and about 60% of fine (contractile) root mass (6.7% of total root mass) are distributed in the upper soil layer (0-40 cm). Only individual, mainly fine roots reached a depth of 70 cm. Fine roots showed comparatively uniform distribution along the depth of soil profile. Their mass decreased considerably only at a depth of 60-70 cm (5% of fine root mass). We found that tree root systems extend far beyond individual projected crown areas. Three (out of five) big lateral roots distributed to the north and south, and two to the east. Their lengths ranged from 3 to 5 m. There was a predominance of densely overlapping tree roots in the thin top soil. The relationship between above-ground and below-ground biomass of *L. gmelinii* is presented in Table III. The results show that above-ground biomass contains more than 75% of total tree biomass. Big and medium roots contain about 17% of total biomass.

The shallow limit of roots partly reflects the unfavorable thermal regime of the deeper soil during the growing season in eastern Siberia. In addition, with a limited depth of soil for root

TABLE III

Above-ground and below-ground biomass of *L. gmelinii* from *Laricetum uliginoso-vaccinosum* (height = 18.1 m, dbh 23.2 cm, crown beginning (level of the lowest branches) is 8.4 m, crown length is 9.7 m, about 130 years old)

Parts of Trees	Dry Weight (kg)	Percentage of Total Biomass (%)
Above Ground		
Stem	210.0	68.9
Needle and Branch	20.1	6.6
Subtotal	230.1	75.5
Under Ground		
Base of Rootage	16.0	5.3
Roots		
1st Order (main lateral roots)	45.0	14.8
2nd Order (diameter ~3.8 cm)	4.0	1.3
3rd Order (diameter ~2.9 cm)	2.0	0.7
4th Order (diameter <2.5 cm)	1.0	0.3
Fine Roots (diameter <2 mm)	6.5	2.1
Subtotal	74.5	24.5
Total	304.6	100.0

activity, the availability of nitrogen and other mineral nutrients per tree and stand declines with increasing tree age.

3.2 CHANGES IN NATURAL ZONES AND ECOSYSTEMS INDUCED BY CLIMATE CHANGE

According to the results of estimated future climate developed using the concepts of Budyko (1971) and Zubenok (1976), noticeable changes in the locations and areas of zones and ecosystems are likely to occur with global warming of 1-2 °C. First, the main natural zones tend to shift northward. The tundra and boreal zones are expected to experience significant changes. The southern boundary of tundra would shift toward the Arctic Ocean. Tundra and forest-tundra areas would contract to be equal to only 3% of the former Soviet Union territory (they now equal 19.5%). Their current area would be replaced with coniferous forest of northern taiga. Only small areas of coastal continental tundra and forest-tundra would be preserved between 80 °E and 180 °E. The area of coniferous forest is expected to decrease by 20-25%. The northern boundary would shift northward and the southern boundary would reach 67 °N with global warming of 2.2 °C. The steppe area would double in the former Soviet Union, and would increase four or fivefold in the territory between 100 °E and 150 °E. Small forest-steppe and steppe areas currently found among taiga in Yakutia are expected to increase significantly and shift northward and eastward.

These results are generally in good agreement with those of modeling the reaction of Siberian vegetation to future warming (Monserud *et al.*, 1993). A regional model of Siberian vegetation and different scenarios of climate change with the doubling of CO_2 concentration in the atmosphere, obtained by four GCMs, predict considerable changes in tundra, taiga, and steppe vegetation. Tundra and northern taiga are predicted to shift northward by 600-1,000 km. Middle and southern taiga subzones would shift northward by 1,000-1,500 km. Boreal steppes and temperate forest-steppes are likely to shift northward by 2,500 km. Only steppes and tundra would also shift eastward. Forest-tundra is expected to be preserved on 5% (central part) to 50% (western part) of its present area. Forest ecosystems of the eastern part of northern and middle taiga are likely to disappear completely, and the forest-steppe and steppe area would increase by six times.

The results, obtained by various scenarios of global warming, show certain discrepancies. For example, distances of shift of certain zones differ by 1.5 times. However, all the estimates indicate the same tendency: natural zones and subzones in Siberia are expected to shift northward (with some northeastward movement) due to global climate change.

Most results obtained by the available models of vegetation response to climate change show that the most significant changes are expected to occur in the high and middle latitudes (IPCC, 1995). In spite of certain disagreements in the results, many specialists predict northward shift of natural zones and ecosystems. The results of the forecast depend mainly on the selection of global warming scenarios.

4. Discussion

The large area of *Larix gmelinii* distribution is evidence of the excellent adaptation of this species to a very severe continental climate with very widely distributed permafrost. The northern boundary of *L. gmelinii* is the northernmost forest boundary in the world. There are larch woodlands not far from the Polar Circle at 72°38'N (Ary-Mas, Taimyr peninsula); their average height is only 4-5 m. In the river valleys of southern Siberia, *L. gmelinii* grows to the dry steppe zone. Its western and eastern boundaries coincide with an isotherm of -30°C in January and isohyet line of precipitation of 200-350 mm yr^{-1}.

Certain anatomical features of the stem and roots (very thick bark layer, well-developed systems that produce turpentine in the bark and wood, thick-walled cells of a late wood) make it possible for this species to exist under the conditions of low snow cover and air temperature fluctuations of more than 100°C (temperature maximum is 38°C and minimum is 64°C below zero).

Our investigations showed that above-ground biomass in larch forests increased in trees up to 125-130 years old. And although the data obtained in the 380-year old stand showed a decrease in stand density, its above-ground biomass stayed constant. The distribution of pure larch forest on very large areas over a long time is evidence of the great resistance of *L. gmelinii* to severe environmental conditions and to fires. If there are no fires and climatic conditions are less severe, light-needle forest (larch and pine) can be replaced with dark-needle species (spruce and silver fir). A significant proportion of *Picea obovata* in the lichen stand and the mature stands indicates possible succession toward *Picea*.

The C/N ratio in topsoil can be considered as an indicator of carbon and nitrogen utilization by the ecosystem. Our data show that the C/N ratio was 21 in the young stands (Ah-horizon; the top horizon of soil under the leaf litter horizon), but it was about twice that value for the older stands. These differences in topsoil C/N ratio may be attributable to differences in tree age (Schulze *et al.*, 1995). But they also can result from differences in microclimatic conditions. This phenomenon could also be considered as evidence of the resistance of *L. gmelinii* to insufficient nutrients, and especially nitrogen.

The major vulnerability factors of Siberian forests are natural and anthropogenic fires, and insect and disease damage. According to official statistics, only 60% of forest inventory (approximately 600 × 10^6 ha) are under some form of forest-fire protection. Forest fires in the rest of the territory are not registered, and nothing is done to control or extinguish them. An estimated 10,000-15,000 fires destroy Siberian forests every year (Shvidenko and Nilsson, 1994). Based on the 1989-1990 forest fire statistics, it can be concluded that each fire destroys about 100 ha of forest area and 2,000-5,000 m^3 of timber. In Siberia, 26.5×10^6 ha of unforested land were created from forest fires, and a further 5×10^6 ha were caused by harvesting. About 20% of the harvested volume for all Russia is wasted; and an estimated 40% of the potential removals are left in the forests.

The damage caused by insects and diseases in Siberia is another serious problem. Some 1×10^6 ha of forested areas are reported to be seriously affected by insects and diseases annually. In 1991, about 1.2×10^6 ha in the Far East were damaged by the insect *Lymantria dispar* alone. About 70% of larch forests in this region have a large amount of rotten wood (root and stem rot). Anthropogenic factors, mainly industrial pollutants, have a negative effect on natural regeneration and forest stability. Human settlements have negatively affected 6.5 million ha of forest (Shvidenko and Nilsson, 1994).

The effects of projected global warming could be both positive and negative for larch forests. Permafrost melting may result in improved soil nutrition. Our data suggested that *L. gmelinii* ecosystems will be sensitive to disturbances in the nitrogen cycle. We would expect increases in tree productivity and stand biomass with increases in temperature and in external input of factors that favor additional nitrogen fixation. The predicted warming in the northern latitudes would increase active soil depth and the ecosystem nutrient pool due to permafrost melting.

In contrast, it is possible that significant changes in many ecosystems and natural zonality maybe caused by future climatic fluctuations. Paleobotanic reconstructions showed that the Pliocene conditions of what is now called Yakutia were favorable for dark-needle taiga. In the late Pliocene and during most of the Pleistocene, mixed forests (*Picea*, *Tsuga*, *Larix russica*, *Betula*, *Corylus*) occupied the areas of recent *L. gmelinii* forests. *L. gmelinii* appeared in northeastern Siberia in the late Pleistocene when glaciation and mountain-forming processes developed (Dylis, 1961). Gradually, *L. gmelinii* areas increased toward the west, south, and southwest. The *L. russica* areas decreased and shifted westward. This process occurred quite rapidly in the late Pleistocene and in the early Holocene because of increased areas of continental climate and of permafrost.

The forecast of the future changes in ecosystems and natural zone locations depends generally on regional climatic forecasts for central Siberia. A decrease in continental climate conditions could cause a significant expansion of dark-needle taiga to the territory of Yakutia. Similar conditions existed in the late Atlantic time of the Holocene (Monserud *et al.*, 1993). According to another forecast, increasing potential evapotranspiration could cause an expansion of steppes areas and a decrease in forest areas (Kobak and Kondrasheva, 1992).

5. Conclusion

Recent *Larix gmelinii* ecosystems are well adapted to the severe climatic conditions found in Siberia. The major vulnerability factors of these forests are anthropogenic impacts (e.g., industrial pollutants, intensive harvesting), fires (both anthropogenic and natural), and insect and disease damages. Global warming is expected to cause significant changes in the locations of ecosystems and natural zones. Even if projected climate changes are favorable for forest stability, we have to bear in mind that the forest resources of Siberia are deteriorating, slowly but significantly. To protect forests we must take prompt actions to improve fire control and prevention, to enhance disease and pest control, and to decrease harvested areas. Economic resource allocation should include measures providing reforestation and afforestation, and increase in forest productivity.

Acknowledgments

The authors thank the U.S. Country Studies Management Team for the opportunity to present the results of the research at the Regional Workshop on Climate Change Vulnerability and Adaptation in Asia and the Pacific (Manila, Philippines, 15-19 January 1996) and for valuable comments during preparation of the manuscript.

References

Basic Data on the Climate of the USSR: 1976, VNIIGMI-MCD, World Data Center, Obninsk, Russia.

Budyko M.I.: 1971, *Climate and Life*, Gidrometeoizdat, Leningrad, Russia.

Budyko M.I. and Izrael Yu.A. (eds): 1987, *Anthropogenic Climatic Change,* Gidrometeoizdat, Leningrad, Russia.

Budyko M. *et al.* (eds): 1990, *Prospects for Future Climate: A Special U.S./U.S.S.R. Report on Climate and Climate Change*, Lewis Publishers, Chelsea, Michigan.

Dixon R.K., Brown S., Houghton R.A., Solomon A.M., Trexler M.C., and Wisniewski J.: 1994, *Science* **263**, 185.

Dylis N.V.: 1961, *Larch in Eastern Siberia and Far East*, Acad. of Sci., Moscow, Russia.

Forest Fund of the USSR: 1990, Sci. Institute of Forest Resources, Moscow, Russia.

Forest Fund of Russia: 1995, Sci. Institute of Forest Resources, Moscow, Russia.

Houghton *et al.* (eds): 1990, *Climate Change. The IPCC Scientific Assessment,* Cambridge University Press, Cambridge, England.

IPCC: 1995, *IPCC Second Scientific Assessment of Climate Change*, Cambridge University Press, Cambridge, England.

Kobak K.I. and Kondrasheva N.Yu.: 1992, *Ecology* **3**, 9.

Monserud R.A., Denissenko O.A., Tchebakova N.M.: 1993, *Climate Res.* **3** (3), 143.

Pozdnyakov L.K.: 1975, The productivity of forest in Siberia, in: *Resources of the Biosphere,* issue **1**, Nauka, Leningrad, Russia, pp. 43-55.

Pozdnyakov L.K.: 1983, *Forest Sci.* **6**, 28.

Schulze E.-D., Schulze W., and Hillonger D.Y.: 1995, *Canadian Jour. of Forest Res.* **25**, 943.

Shvidenko A. and Nilsson S.: 1994, *Ambio* **23**, 396.

Utkin A.I.: 1965, *The Forests of Central Yakutia*, Acad. of Sci., Moscow, Russia.

Zubenok L.I.: 1976, *Evaporation on the Continents*, Gidrometeoizdat, Leningrad, Russia.

POTENTIAL IMPACT OF GLOBAL CLIMATE CHANGE ON FOREST DISTRIBUTION IN SRI LANKA

S. SOMARATNE[1] and A.H. DHANAPALA[2]
[1] *Division of Botany, The Open University of Sri Lanka Nawala, Nugegoda, Sri Lanka*
[2] *Participatory Forestry Project Forest Department, Battaramulla, Sri Lanka*

Abstract. The potential impact of climate change on forest distribution in Sri Lanka was evaluated. The Holdridge Life Zone Classification was used along with current climate and climate change scenarios derived from two general circulation models, the Geophysical Fluid Dynamics Laboratory model and the Canadian Climate Centre Model, at a 0.5° × 0.5° resolution. Current and future distributions of life zones were mapped with a Geographic Information System. These maps were then used to calculate the extent of the impact areas for the climate change scenarios. The current distribution pattern of forest vegetation includes tropical very dry forest (6%), tropical dry forest (56%), and tropical wet forest (38%). Results obtained using the Geophysical Fluid Dynamics Laboratory model show an increase in tropical dry forest (8%) and decrease in tropical wet forest (2%). The Canadian Climate Centre Model scenario predicted an increase in tropical very dry forest (5%) and tropical dry forest (7%), and a decrease in tropical wet forest (11%). Both models predicted a northward shift of tropical wet forest into areas currently occupied by tropical dry forest. The application of general circulation models such as the Geophysical Fluid Dynamics Laboratory model and the Canadian Climate Centre Model, as well as the Holdridge Life Zone Classification, to estimate the effect of climate change on Sri Lankan forests in this paper indicates that these methods are suitable as a tool for such investigations in Sri Lanka.

Key words: Sri Lanka, forest, CCCM, GFDL, Holdridge Life Zone Classification

1. Introduction

The impact of global climate change on terrestrial vegetation distribution patterns has been a major issue over the last decade, and has been analyzed by numerous scientists (e.g., Emmanuel *et al.*, 1985; Woodward, 1987; Smith *et al.*, 1990; Smith *et al.*, 1992). It is a well-known fact that the distribution pattern of terrestrial vegetation is a result of bioclimatic interactions. Thus changes in the global climate due to increased atmospheric greenhouse gases could alter the present patterns.

A number of studies have been carried out to investigate the potential impacts of climate change on terrestrial vegetation distribution on regional scales (e.g., Leemans, 1989; Neilson *et al.*, 1989; Neilson *et al.*, 1991; Halpin and Smith, 1992). The potential impacts of such climate changes on the distribution patterns of Sri Lankan vegetation, and especially its forests, have not yet been examined in detail. There are, however, many detailed studies on classification of Sri Lankan climate and terrestrial vegetation that resulted from the equilibrium of vegetation and climate interaction (Holmes, 1956; Koelmayer, 1957, 1958; Mueller-Dombois, 1968). This paper presents a preliminary investigation to evaluate the response of Sri Lankan forests to climate change. The objectives of this report were to evaluate the potential impacts of climate change on forests of Sri Lanka, evaluate the suitability of general circulation models in predicting Sri Lankan climate, verify the applicability of the Holdridge Life Zone Classification to Sri Lanka, and document potential climate change impacts on Sri Lankan forests.

Water, Air, and Soil Pollution **92:** 129–135.
© 1996 *Kluwer Academic Publishers. Printed in the Netherlands.*

2. Methods

The Holdridge Life Zone Classification (Holdridge, 1967) is a climate classification scheme that relates the distribution of major ecosystem complexes to the climatic variables of biotemperature, mean annual precipitation, and the ratio of potential evapotranspiration (PET) to precipitation. The life zones are depicted by a series of hexagons in a triangular coordinate system.

To construct baseline climate scenarios for Sri Lanka, mean monthly climate data of a 30 year period, 1951-1980, from 21 climate stations were used. These data were interpolated to $0.5° \times 0.5°$ to estimate the gridpoint data on an area defined by longitude 79.5°E- 81.5°E and latitude 6.5°N-9.5°N. A comparison was made between observed $1{\times}CO_2$ climate data and estimated $1{\times}CO_2$ data from three general circulation models (GCMs), the Geophysical Fluid Dynamics Laboratory (GFDL) model (Mitchell *et al.*, 1990), the Canadian Climate Centre Model (CCCM) (Boer *et al.*, 1992), and the United Kingdom Meteorological Office (UKMO) model (Wilson and Mitchell, 1987), to select the most suitable model or models that best simulated the Sri Lankan climate. Based on the comparison, the CCCM and the GFDL model were selected.

Estimated current distributions of life zones were mapped using observed climate data of mean monthly precipitation and temperature at a $0.5° \times 0.5°$ (latitude and longitude) resolution. Simulations of current ($1{\times}CO_2$) and $2{\times}CO_2$ climates from two GCMs (Table I) were used to construct climate change scenarios. Changes in mean monthly precipitation and temperature were calculated for each GCM scenario for each computational grid element by taking the difference between simulated current and $2{\times}CO_2$ climates. Temperatures were expressed as absolute difference ($2{\times}CO_2$ - $1{\times}CO_2$) and precipitation as the ratio of $2{\times}CO_2$ to $1{\times}CO_2$. These data from each GCM were interpolated to $0.5° \times 0.5°$ using the same technique as applied in the development of the data base for current climate (Leemans and Cramer, 1990). The technique used was a triangulation of all data points. Changes in monthly precipitation and temperature were then applied to the observed climate data to provide a change scenario. The altered data bases corresponding to each of the two GCM scenarios were then used to reclassify the grid cells ($0.5° \times 0.5°$) using the Holdridge classification and to create a map of the estimated distribution of life zones under $2{\times}CO_2$ conditions.

TABLE I

Adjustment statistics for monthly precipitation and temperature used in the construction of CCCM and GFDL scenarios for Sri Lanka

	Rainfall (%)		Temperature (°C)	
Month	CCCM	GFDL	CCCM	GFDL
January	-35	-14	1.6	1.7
February	-28	-21	1.5	2.1
March	-46	+04	1.4	2.0
April	-09	-08	1.7	2.1
May	-01	+10	1.6	2.0
June	-35	-24	1.8	2.1
July	-08	+05	2.1	2.3
August	-09	-10	2.2	2.5
September	+17	-24	2.2	2.7
October	-10	-23	2.1	2.6
November	-03	-01	1.9	2.3
December	-51	-15	1.7	2.3

3. Results

As a consequence of the need for food, fuel, fiber, and forest products, the indigenous forest cover of the island decreased from approximately 70% at the beginning of twentieth century to less than 25% today. Based on remote sensing data supplemented by field observations, the latest figures for closed-canopy indigenous forest cover indicate that it now occupies only 23.87% of the land area. It is generally expected that to maintain normal local climate conditions and provide ecological security, it would be desirable to have a forest cover of at least 25% of the land area. Under the current climate, of the 24% of existing closed canopy forest, approximately 18% is in the tropical dry forest zone and 6% is distributed in the tropical mesic forest zone (Figure 1).

In the CCCM scenario, 12% of the closed canopy forest is in the tropical mesic forest, 10% in the tropical dry forest, and 2% in the tropical very dry forests (Figure 2). Under the GFDL scenario, the closed canopy forest is distributed in the ratio of 3:1 in tropical mesic forests and tropical dry forest, respectively (Figure 3).

Current forest distribution patterns (Figure 1) include four different life zones, i.e., tropical very dry forest, tropical dry forest, tropical moist forest, and tropical wet forest. The changed distribution pattern of forest under 2×CO_2 climate scenarios predicted by CCCM and GFDL are represented in Figures 2 and 3, respectively. Both GCM scenarios show different degrees of scatter distribution of forest vegetation under the changed climate.

Under the CCCM scenario (Figure 2), the dry forest area increases and there is a shift of mesic forest toward the north and northeast part of the island. Further, a strip of a very dry forest that extends from the northwestern province is projected under the CCCM scenario. A tropical dry forest occupies the rest of the island, and extends diagonally from northwest to southeast.

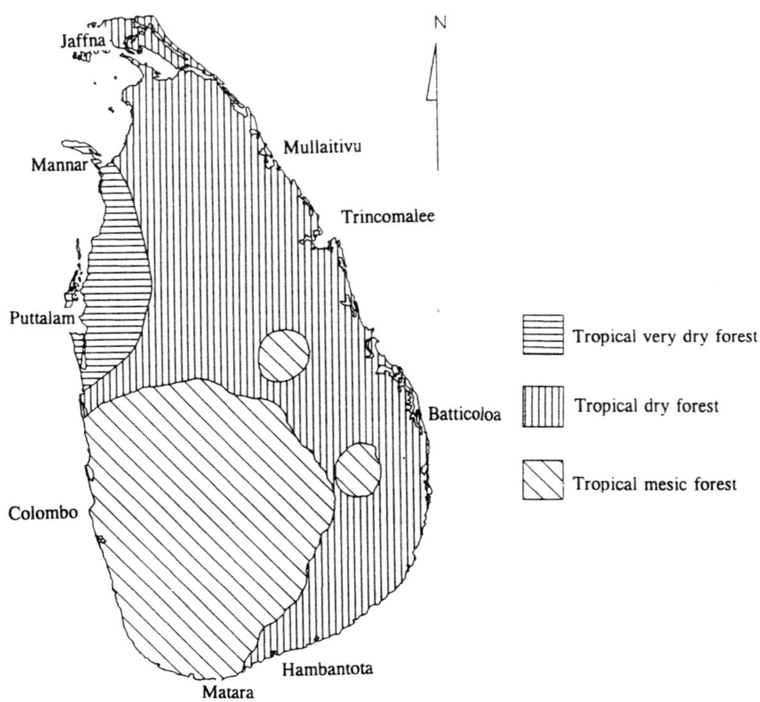

Fig. 1. Holdridge life zones of Sri Lankan forests under current climate conditions.

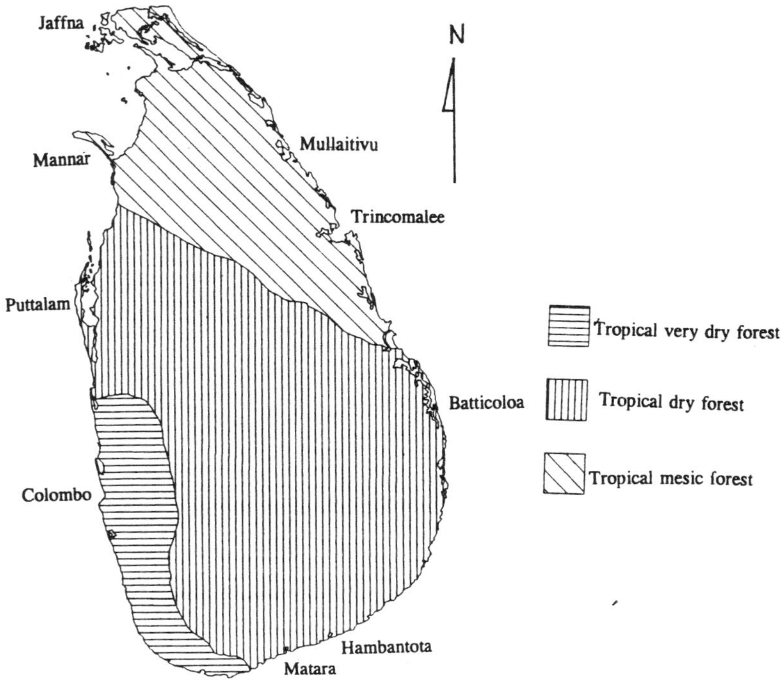

Fig. 2. Holdridge life zones of Sri Lankan forests under CCCM derived climate change scenarios.

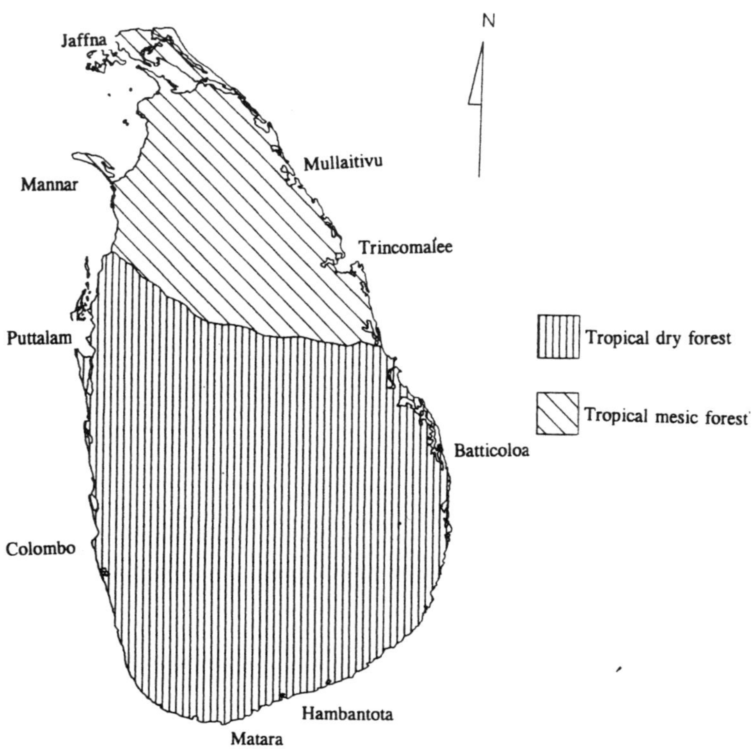

Fig. 3. Holdridge life zones of Sri Lankan forests under GFDL derived climate change scenarios.

TABLE II
Approximate areas of potential life zones under current and changed climate conditions in Sri Lanka

		GFDL		CCCM	
Vegetation Type	Baseline (ha)	Area (ha)	% Change from Baseline	Area (ha)	% Change from Baseline
Tropical very dry forest	387,099	0	-100%	715,149	+85%
Tropical dry forest	3,674,160	4,199,040	+14%	4,100,625	+12%
Tropical mesic forest[§]	2,460,375	2,355,399	-4%	1,738,665	-29%

[§] Tropical mesic forest = Tropical moist forest + tropical wet forest + sub tropical moist forest.

The forest distribution pattern under the GFDL scenario is different from that of the CCCM scenario. Under the GFDL scenario, tropical very dry forest is not indicated, and most of the area in the south is covered by tropical dry forest (Figure 3). Intermediate forest types are not reflected on either of the maps developed from the Holdridge classification and the GCM scenarios.

The approximate areas of current and potential life zones under climate change conditions are shown in Table II. The areas of potential life zones under climate change conditions were estimated using the Holdridge classification and $2\times CO_2$ climate scenarios generated from the GFDL and CCCM. Percent change from baseline is also shown for each life zone type.

4. Discussion

The following trends were identified as potential results of climate change in Sri Lanka. A northward shift of mesic forest over existing dry forest, indicated by both the GFDL and CCCM scenarios, would result from increased temperature and rainfall in the northern region of the island. Most of the forest area is currently distributed in the northern and southeastern parts of the island, where population density is low. Climate change would convert the current forest in these areas to a mesic forest. Tropical mesic forests in the south part of the island would be replaced by tropical dry forests.

The most vulnerable forest areas would be the Sinharaja Forest Reserve, a national heritage of Sri Lanka, and Peak Wilderness Forest Reserve. Most of the Sri Lankan endemic species find their refuge in these forest patches, and change in the forest type due to climate change would probably eliminate most of these species from Sri Lanka.

These findings indicate that, although there are minor differences, the CCCM and GFDL could be used in assessment of climate change impacts on different sectors other than forest in future studies. The Holdridge Life Zone Classification is useful, but for future studies in-depth investigations on carbon cycling, biomes, biodiversity, and other factors should be undertaken.

The current life zones identified in this study based on climate data (Figure 1) to a certain extent agreed with Thornthwaite (1931) and the vegetation map of Mueller-Dombois (1968). In addition, the findings associated with current climate conditions do not deviate significantly from findings by the Asian Development Bank (1994), which pointed out that climate change scenarios developed for Sri Lanka indicate significant change in rainfall and temperature in the island. As a consequence, the Asian Development Bank (1994), showed an increase in rainfall and temperature in both the dry and wet zone of Sri Lanka during the northeast monsoon. However, in this study both CCCM and GFDL scenarios showed a transformation of existing dry vegetation in just the northern part of the island.

Although the GCM (CCCM and GFDL) data simulate the current climate with a certain extent of accuracy, these models are limited due to their inherent characteristics. Further, the low resolution of the GCM data and the assumption that terrain is smooth during GCM funs pose difficulties in generating scenarios for a small country like Sri Lanka. Topographically Sri Lanka is heterogenous and that heterogeneity could not be incorporated in the GCM scenarios. Thus these limitations of the GCM data should be taken into account in the interpretation of climate change scenarios and the results of this study.

When the results from the Holdridge Classification are considered it must be noted that this methodology is suitable for exploration of broad-scale vegetation distribution pattern and the influence of climate changes on the suitability of a region to support different vegetation types. However, the resolution of the data used in the generation of Holdridge life zones are very low, and such resolution gives an approximate estimation of life zones. Further studies on the forest sector, especially studies combined with a Gap model, would add to the understanding of the extent of potential climate change impacts on distribution of Sri Lankan forests.

Acknowledgments

The authors would like to express their sincere thanks to Dr. Robert Dixon and Joe Wisniewski for their critical review of the manuscript and valuable suggestions. Thanks are also due to Ms. Liz Madden, U.S. Country Studies Program Library, Washington, DC, for providing relevant reference materials. The assistance given by Mr. S.R. Krishnarajah, Division of Zoology, Mr. B.K.L. Wickremasinghe, Division of Botany, Mr. A.M.P.B. Abeysinghe, Elementary Computer Laboratory, all of the Open University, and Mr. P.L.M. Liyanage and Ms. M. Withanage, Department of Geography, University of Colombo, is appreciated.

References

Asian Development Bank: 1994, *Climate Change in Asia: Executive Summary*, Asian Development Bank, Manila, Philippines.

Boer G.J., McFarlane N., and Lazare M.: 1992, *Journal of Climate* **5**, 1045.

Emmanuel W.R., Shugart H.H., and Steven M.P.: 1985, *Climate Change* **7**, 29.

Halpin P.H. and Smith T.M.: 1992, Potential Impacts of Climate Change on Forest Protection in the Humid Tropics: A Case Study of Costa Rica, Climate Change Division, OPPE, U.S. Environmental Protection Agency, Washington, DC.

Holdridge L.R.: 1959, *Science* **130**, 572.

Holdridge L.R.: 1967, *Life Zone Ecology*, Tropical Science Center, Costa Rica.

Holmes C.H.: 1956, *Ceylon Forester* **II**, 209.

Koelmeyer K.O.: 1957, *Ceylon Forester* **III**, 144.

Koelmeyer K.O.: 1958, *Ceylon Forester* **III**, 265.

Leemans R.: 1989, Possible changes in natural vegetation patterns due to global warming, in: *Der Treihauseffekt Das Problem — Mogliche Folgen — Erforderliche Masshamen*, Hackle A. (ed), Akadamie für Umvelt und Energie, Laxenburg, Austria.

Leemans R. and Cramer W.: 1990: The IIASA Climate Database for Land Area for 0.5° Resolution, WP-41, International Institute for Applied System Analysis, Laxenburg, Austria.

Mitchell J.F.B, Manabe S., Tokioka T., and Meleshko V.: 1990, Equilibrium change, in: *Climate Change: The IPCC Scientific Assessment*, Houghton J.T., Jenkins G.J., and Ephramus J.J. (eds), Cambridge University Press, New York.

Mueller-Dombois D.: 1968, *Ceylon Forester* **VIII**, 39.

Neilson R.P. and King G.A.: 1992, Continental scale biome response to climate change, in: *Ecological Indicators,* McKenzie D., Hayat E., McDonald H.J., and McDonald J.C. (eds), Proceedings of International Symposium, Elservier Science Publication, Ltd., Ft. Lauderdale, Florida, October 16-19.

Neilson R.P., King G.A., DeVelice R.L., Lenihan J., Marks D., Campbel W., and Glick G.: 1989, Sensitivity of Ecological Landscapes and Regions to Global Climatic Change, U.S. EPA Environment Research Laboratory, EPA/600/3-89 1073, NTIS-PB90-120-072/AS, Corvallis, Oregon.

Smith T.M., Leemans L., and Shugart H.H.: 1992, *Climate Change* **21**, 367.

Smith T.M., Shugart H.H., and Halpin P.N.: 1990, *Global Forests,* U.S. EPA/OPA, Washington, DC.

Thornthwaite C.W.: 1931, *Geo. Rev.* **21**, 633.

Wilson C.A. and Mitchell J.F.B.: 1987, *Journal of Geophysical Research* **92**, 13315.

Woodward F.I.: 1987, *Climate and Plant Distribution,* Cambridge University Press, New York.

POTENTIAL IMPACTS OF SEA LEVEL RISE ON THE COASTAL RESOURCES OF MANILA BAY: A PRELIMINARY VULNERABILITY ASSESSMENT

R.T. PEREZ,[1] R.B. FEIR[2]
E. CARANDANG,[2] and E.B. GONZALEZ[1]

[1] *Philippine Atmospheric, Geophysical and Astronomical Services Administration (PAGASA), Department of Science and Technology (DOST), Quezon City 1100, Philippines*
[2] *Coast and Geodetic Survey Department (CGSD), National Mapping and Resources Information Authority (NAMRIA), Department of Environment and Natural Resources (DENR), Binondo, Manila, Philippines*

Abstract. This study considers the Manila Bay coastal area of the Philippines for evaluation of possible consequences of accelerated sea level rise in the context of climate change, and suggests adaptive responses to such threats. The semi-enclosed Manila Bay is bounded by the provinces of Bataan, Pampanga, Bulacan, and Cavite, and some of the towns and cities of Metro Manila along the eastern side. This region is important to the commercial, industrial, agricultural, and aquacultural activities of the Philippines, with Manila as the seat of the national government and the rest among the political constituents of the National Capital Region. An increasing trend in the mean sea level has been observed since 1965 and continues today. The bay area is already subject to several hazards including floods and storm surges during tropical cyclones. The shoreline has changed greatly in the last 5 to 10 years due to reclamation for housing, ports, coastal roads, buildings, and other urbanized developments, adding to the threat of inundation. Selection of appropriate responses is looked at in terms of expected vulnerability, costs, land use, and other sociopolitical and legal considerations. Partial results show that parts of Cavite and Metro Manila Bay areas are especially vulnerable to accelerated sea level rise.

Key words: Philippines, Manila Bay, coastal resources, coastal zone management, sea level rise, AVVA

1. Introduction

Accelerated sea level rise is one of the most certain outcomes of global warming (National Research Council, 1987). With its long coastline of 18,000 km, the Philippines is taking a critical look at the potential effects and possible responses. The physical effects of sea level rise include inundation (submergence) of low-lying wetland and dryland areas, erosion, salt-water intrusion into estuaries and aquifers, higher water tables, and increased risk of flooding and storm damages. In turn, these physical changes may lead to substantial socioeconomic losses such as loss of coastal structures, both natural and artificial; dislocation of the population; and changes of livelihood. The same physical changes may bring about certain ecological consequences such as redistribution of wetlands, destruction of coral reefs, reduction in biological diversity, wildlife extinction, and changes in biophysical and biochemical properties of the coastal zones. This may necessitate a range of possible responses such as construction of seawalls and dikes, upgrading of coastal infrastructure to accommodate higher water levels, relocation of coastal populations, mangrove regeneration, and others.

A detailed impact assessment is currently being undertaken by the Philippine Country Study to Address Climate Change through one of its study elements: Vulnerability Assessments and Evaluation of Adaptations on Coastal Resources Due to Accelerated Sea Level Rise. The objectives of this study element are the following:

- to study the possible effects of accelerated sea level rise on the coastal ecosystem
- to examine the possible effects of accelerated sea level rise on socioeconomic structures and activities in the coastal areas

Water, Air, and Soil Pollution **92:** 137–147.
© 1996 *Kluwer Academic Publishers. Printed in the Netherlands.*

- to formulate response strategies by identifying potential costs and benefits
- to identify needed actions to develop a long-term coastal zone management plan
- to develop the capability to apply the methodology to other areas in the Philippines
- to brief policy and decision makers on the vulnerability of the Philippines to accelerated sea level rise and the urgent need to mitigate or adapt to climate change.

This paper addresses the first three objectives by presenting the partial results of the Philippine vulnerability and adaptation assessment.

2. Methods

The methodology being followed is essentially that of the seven step common methodology of the Coastal Zone Management Subgroup of the Intergovernmental Panel on Climate Change (IPCC) Working Group II (IPCC, 1992). The common methodology consists of seven steps. These include (1) the delineation of the study area and accelerated sea level rise boundary conditions; (2) inventory of physical and socioeconomic data; (3) identification of relevant development factors; (4) assessment of physical changes and natural system responses; (5) formulation or response strategies; (6) assessment of vulnerability and interpretation of results; and (7) identification of needs and action.

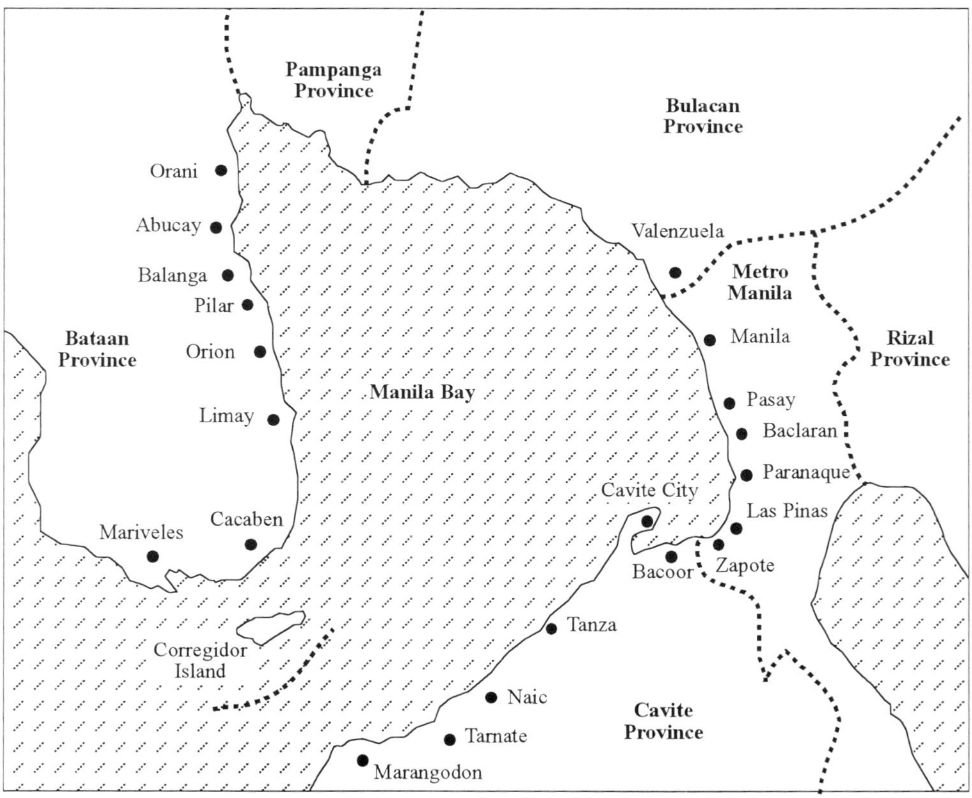

Fig. 1. Area of study: Bataan, Bulacan, Metro Manila, Cavite (Manila Bay area).

2.1 DELINEATION OF THE STUDY AREA AND SPECIFICATION OF SCENARIOS

2.1.1 Study Area

For this study, the Manila Bay area in Western Philippines, which is between 120°30'E to 121°E and 14°15'N to 14°50'N (Figure 1), is considered to evaluate the possible consequences of accelerated sea level rise, and to draw adaptive responses to such threats. The semi-enclosed bay is bounded by the provinces of Bataan, Pampanga, Bulacan, and Cavite, and some of the towns and cities of Metro Manila along the eastern side.

The Manila Bay coastal area contributes significantly to the Philippines' economy through trade and commerce. The North and South Harbors in the city of Manila serve both the international and domestic shipping industries. The bay also supports the agriculture, aquaculture, and tourism needs of the country.

Even without the anticipated effects of climate change, the bay coastal area is already subjected to several hazards such as flooding and storm surges during tropical cyclones. Reclaimed sections were often built higher than the former land areas, thereby retarding the natural seaward flow of water. Hence, floodwaters resulting from rainstorms take longer to subside, especially during high tides. In addition, coastal floods are not uncommon along low-lying areas during storm surges. These floods impede water discharges from upstream areas of rivers and thus resulting in further accumulation of water.

2.1.2 Accelerated Sea Level Rise Scenarios

For this study, the three accelerated sea level rise scenarios used were 0.3, 1.0, and 2.0 m by 2100. The first two values are the low and high estimates of the IPCC (IPCC, 1990), and the 2.0 m rise represents a worst case scenario.

2.2 INVENTORY OF THE STUDY AREA CHARACTERISTICS

2.2.1 Physical Environment

From tide gauge data taken at a station at Manila South Harbor (14°35'N, 120°58'E), an increasing trend in the mean sea level has been observed since 1965 and continues to manifest itself today (see Figure 2). [The tide gauge stations are operated and maintained by the Coast and Geodetic Survey Department (CGSD) of the National Mapping and Resource Information Authority (NAMRIA), Department of Environment and Natural Resources (DENR).] Previous investigations (Carandang, 1989) made at the tide gauge stations showed no evidence of land subsidence.

The bay experiences a tidal regime ranging from a low level of -0.475 m to a high level of 0.528 m (mean sea level was used as the datum). Water circulation is largely influenced by rivers entering the bay, tidal fluctuations, and prevailing winds. The coastline of Manila Bay is 190 km long; the shores are generally gently sloping, with the highest elevation less than 5.0 m above mean sea level. These shores are covered by mud, soft clay, and alluvial deposits. The groundwater level averages from 50 to 150 m below sea level.

The bay area experiences two pronounced seasons: dry (from December to May) and wet (from June to November). The period of maximum rain is from June to September. The area is generally exposed to the southwest monsoon, as indicated by climatological wind values from the Climatological and Agrometeorological Branch of PAGASA presented in Table I. According to a record of 45 years (1948-1992), 5.64% of the total tropical cyclones that crossed

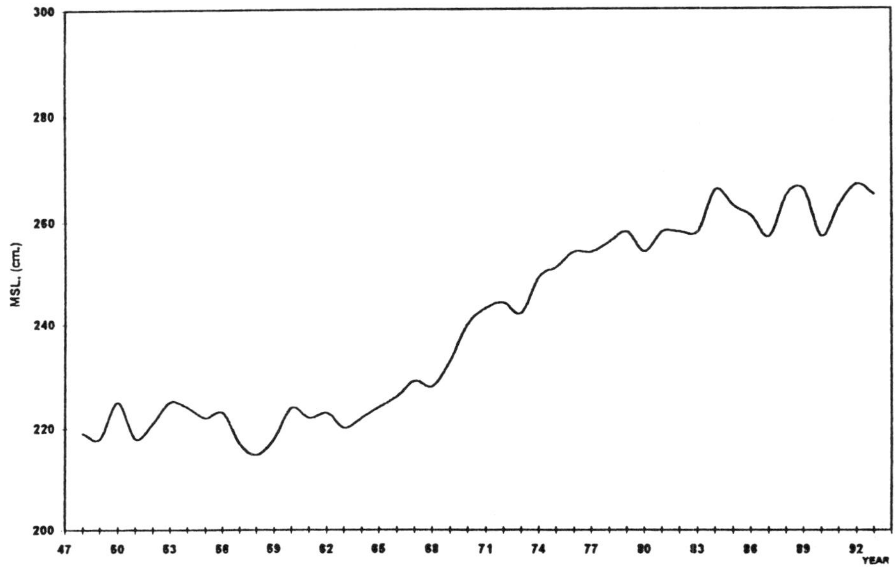

Fig. 2. Monthly mean sea level for Manila Bay.

the country passed over this area (Metro Manila, Bulacan, Cavite), a frequency of about five cyclones in three years. Storm surges have reportedly affected the Manila Bay coastal areas in 1960, 1964, 1965, 1971, 1972, and 1983. No details of wave heights and effects were available except for 1983 (as listed in Table I), which showed the greatest wave heights within the study area occurred in parts of Bataan, which received a substantial surge that raised water levels to an estimated height of 2 m and caused 32 deaths (PAGASA Report, 1983).

2.2.2 Habitats and Species

There are two types of mangrove forests within the bay area (Bureau of Fishery and Aquatic Resources, BFAR, 1994): natural secondary growth and mangrove reforestation. Secondary growth refers to secondary successional stage species that are smaller than old growth species and are sparsely distributed. The two types are mostly found in strips along the edges of fishponds in Bulacan, or in patches in Cavite and Bataan.

Seagrasses are found only along the coasts of Bataan and in the more protected areas of Cavite, whereas seaweeds are found in Mariveles and Orion in Bataan Province, and in Bulacan, Bacoor, Tanza, Naic, and Tarnate in Cavite Province. Seagrasses and seaweeds are sensitive to changes in sea level because of their need for sunlight.

Coral communities can still be found in the coastlines of Maragondon (Cavite), Mariveles (Bataan), and Corregidor Island toward the South China Sea. The coral reefs cover approximately 2-3 km^2, or 0.12-0.19%, of the 1,600 km^2 of the bay. Already severely stressed by pollution and harmful fishing practices, these coral reefs probably cannot keep pace with sea level rise by vertical accretion.

TABLE I

Climatological values of maximum wind direction and maximum wind speed, maximum precipitation, tidal data, tropical cyclone frequency, and storm surges in Manila Bay coastal areas

Maximum Frequency of Wind Speed/Direction [§]			Max. Recorded	Tidal Data[†]	
Annual	January	July	Rainfall[§]	Great Range[§§]	Mean Range[††]
56 $m\ s^{-1}$/W	20 $m\ s^{-1}$/ENE	36 $m\ s^{-1}$/W	472.4 mm	1.003 m	0.758 m

Tropical Cyclone Passage Statistics (1948-1992)[‡]

Metro Manila	Bataan	Bulacan	Pampanga	Cavite
36	20	19	18	16

Storm Surge Data[¥]

Associated Typhoon	Max. Observed Winds ($km\ h^{-1}$)	Areas Affected/Surge Height (m)	
Bebeng (Vera 8303) July 12-16, 1983	165-MIA	Bacon, Sorsogon/2.950[‡‡]	Orion, Bataan/0.789
		Bulabog, Sorsogon/3.354[‡‡]	Balanga, Bataan/1.955
		Legaspi City, Albay/0.650[‡‡]	Orani, Bataan/1.415
		Limay, Bataan/0.985	Manila Port Area/0.660

[§] Observed from the PAGASA weather station at the Manila International Airport (MIA, 14.5 N 121.0 E) as of 1993.
[†] From CGSD, NAMRIA (based on the latest 19 years of data).
[‡] From the Climatological and Agrometeorological Branch, PAGASA.
[¥] From PAGASA Report, 1983.
[§§] Difference between highest and lowest observed tide heights.
[††] Difference between mean highest and mean lowest observed tide height.
[‡‡] Areas outside of study site.

2.3 VULNERABILITY ANALYSIS

Among the most important sources of data for the vulnerability assessment of the impacts of accelerated sea level rise in coastal areas are topographic and bathymetric maps. For most of the developing nations, high resolution maps (less than 10 m contour intervals) are very difficult to obtain, if available at all. Traditional photogrammetric approaches are costly, both in terms of funds available and time allocated for vulnerability analysis and assessment of adaptations. Satellite imagery can also be prohibitively expensive.

A technique developed by Leatherman *et al.* (1995) called the Aerial Video-assisted Vulnerability Analysis (AVVA) was adopted for this study. In effect, this technique provides information for steps number (3) to (6) of the IPCC common methodology. The AVVA is a reconnaissance level assessment of the implications of land use in response to sea level rise. The AVVA involves (a) unrectified oblique aerial video recording of the coastal area taken from a height of 50 to 500 m with precise global positioning, (b) limited ground truth information, (c) archival research, and (d) analysis of the data in conjunction with simple land loss and response models.

2.3.1 Base Maps

The finest resolution maps used in this study have a scale of 1:10,000 with 4.0 m contour lines and spot heights. These were prepared by the Coast and Geodetic Survey Department in cooperation with Japan International Agency in 1987 and incorporate major changes from 1982 to 1986. Boundaries of regions were approximated and were verified through field surveys. Scaled land use maps were also obtained from the Coast and Geodetic Survey Department for references. Other sources of information for the land use maps were the Bureau of Lands, Bureau of Soils, Metro Manila Commission, Department of Public Works and Highways, National Economic Development Authority, and the city and municipal governments.

2.3.2 Aerial Survey

A helicopter was flown over the Manila Bay coastline. The first low level pass was taken at a 75 m altitude from Bataan to Cavite and back. The 75 m aerial survey provided the detailed view of the coast that was used for this assessment. The survey included still pictures and 8 mm video coverage. To establish locations in terms of longitude and latitude points, a Global Positioning System (GPS) was used.

2.3.3 Ground Truth Measurements

The main objective of the field survey was to delineate the specified boundary conditions for the 0.3, 1.0, and 2.0 m contours along the coast, and the major changes that have occurred there since the last updating of the topographic maps of the area. The preparatory work consisted of establishing geodetic controls (points of known geographic positions and elevations) along the coast spaced at approximately 2.5 km. Supplementary geodetic controls were established in areas where low elevations extend significantly inland.

The main activity consisted of ground profiling where spot heights were obtained. The spacing of spot heights depended largely on the nature of the terrain (i.e., closely spaced heights for steep slope). The heights and positions of these spot heights were then plotted on survey forms to obtain the 0.3 and 1.0 m contours. The 2 m contours were drawn by interpolating existing maps and validating the same through spot checking.

2.4 IDENTIFICATION OF NEEDS AND ACTIONS

Selection of the appropriate response or group of responses to accelerated sea level rise will depend on site-specific conditions (as reflected in the results of the vulnerability analysis), costs, land use, and coastal infrastructure. Some issues and concerns are the following:

1. The costs involved in the responses, including nonmonetary aspects such as risks to human lives through hazards, must be assessed.
2. If the coastal resources cannot be protected because of the high costs of protection, the main political factor to be considered is the necessity to relocate people and infrastructures likely to be affected.
3. Policies and laws should be adopted to restrict habitation and investment in vulnerable areas.
4. Public education about the coastal environment is needed to sensitize the government and the people to the issues involved, the danger, and possible solutions.
5. Any rise in sea level will increase the level of flooding both in magnitude and duration. Major development projects such as flood protection embankments or levees may be necessary to address these problems.
6. Basic research aimed at gaining a better understanding of coastal processes is needed. The rational use of coastal resources such as mangroves, sand, and coral reefs must be considered.
7. Dislocation of people and changes in livelihood are critical issues. Responsible agencies must be ready to intervene.
8. Long-term development planning in the context of integrated coastal zone management must be a priority for by all agencies concerned.

3. Preliminary Results and Discussion

3.1 AERIAL SURVEY

The aerial survey confirmed several types of coastal environment that are evident in the Manila Bay area. These are the brushland/industrial environment of Southern Bataan; the agro-aquacultural environment of Bataan-Pampanga coasts; the aquacultural and urban environment of Bulacan and Northern Metro Manila; the highly urbanized environment of Metro Manila; the tourism and agro-aquacultural environment of Cavite; and the natural environment of Southern Cavite. Initial observations revealed erosion of the Bataan coasts, of the beach areas of Cavite, and in some abandoned fishponds of Bulacan, and other shoreline changes due to reclamation for various land uses, especially in the Manila-Cavite areas. Observations also revealed that the river outlets to the bay from Pampanga and Bulacan are heavily silted with lahar deposits from the eruption of Mount Pinatubo.

The presence of squatter areas (see Figure 3) along the river deltas of Bataan, Bulacan, and Metro Manila indicate that the poorest socioeconomic sector of the population faces severe risk. This has equity implications. These resilient people may survive sea level rise, as evident by some of the houses that can be seen built right on the water. They may increase the height of the house posts to accommodate an increasing water level. However, they will be vulnerable to an increased threat of storm surges.

The agricultural lands (ricefields) of Cavite are mostly near the shore, and those of Bataan and Bulacan are landward of the fishponds.

Only two areas within the bay were identified as forested. These are at the extreme ends of Manila Bay in Bataan (Mariveles) and Cavite (Marogondon).

Fig. 3. A scene along the coast of Navotas, Metro Manila. This is one of the high density residential areas along the coast of Manila Bay.

Metro Manila shorelines have been changed greatly by the high level of urbanization. The area is almost devoid of natural vegetation, an indication of the extensive economic development. This is also the only place in the bay coastline where physical engineering protection structures (like concrete seawalls and breakwaters) are observed.

Generally, most of the shores in the northern portion of the bay are characterized by muddy substrate, and the shores toward the south have sandy substrate. There are also some short rocky stretches along the Cavite and Bataan coastlines, and various combinations of sandy, muddy, and rocky substrates.

Data from the Public Estates Development showed that a total of 17.44 km^2 for the whole coast of Manila Bay was reclaimed in the last 40 years. This figure is consists of 8.16 km^2 for fishpond development and 9.28 km^2 for the urbanized developments. Extensive land reclamation for fishponds along the coasts of Southern Bulacan is seen, and shoreline changes due to reclamation for urbanized developments are quite significant in Paranaque (residential subdivisions), Pasay (Cultural Center of the Philippines Complex), and Las Piñas-Cavite (coastal road).

3.2 VULNERABILITY TO ACCELERATED SEA LEVEL RISE SCENARIOS

Figure 4 presents a vulnerability map for parts of Cavite and Las Piñas to the accelerated sea level rise scenarios. The horizontally hatched areas are estimated to be inundated by a 0.3 m sea level rise. Even without climate change, these areas have been reported to suffer flooding during high tides. The areas covered by diagonal lines are estimated to be affected by a 1.0 m sea level rise, and the darkly shaded areas are the areas estimated to be affected by a 2.0 m sea level rise. So under a worst case scenario, the whole of Cavite City, including the Sangley Point Naval Base and Fort San Felipe, is estimated to be inundated. This area is estimated to be 11.8 km^2 and holds approximately 91,000 inhabitants (1990 estimate from the National Statistics Office, Manila). Other areas likely to be inundated include the poblacion (town proper) of Rosario and Noveleta (Cavite), including a 2.5 km stretch of the national highway. Likewise, the towns of Kawit and Bacoor will be partially submerged. The farthest distance inland to be affected by a 2.0 m sea level rise is approximately 3 km. The Manila-Cavite coastal road was not estimated to be affected by any of the accelerated sea level rise scenarios because it was elevated during construction. However, most of the roads leading to it are estimated to be under water.

Even the coastal villages of Las Piñas are likely to be inundated. Currently, this 3-km shoreline of predominantly high density residential use is one of the economically depressed areas of the town, with a population of 58,033 (MDPO, 1996). The municipality of Las Piñas has plans to expand the extent of the present Manila Bay reclamation area (MDPO, 1996). If no relocation is planned, the population in the area will be highly vulnerable to the impacts of rising sea level.

Another option is to provide coastal protection. This is a costly solution for the local government. It will cost the municipal government of Las Piñas approximately US$0.6 million per kilometer (based on current prices) to build a concrete seawall 3 m high and 1 m thick with an underwater base that is 1.5 m high and 3 m thick.

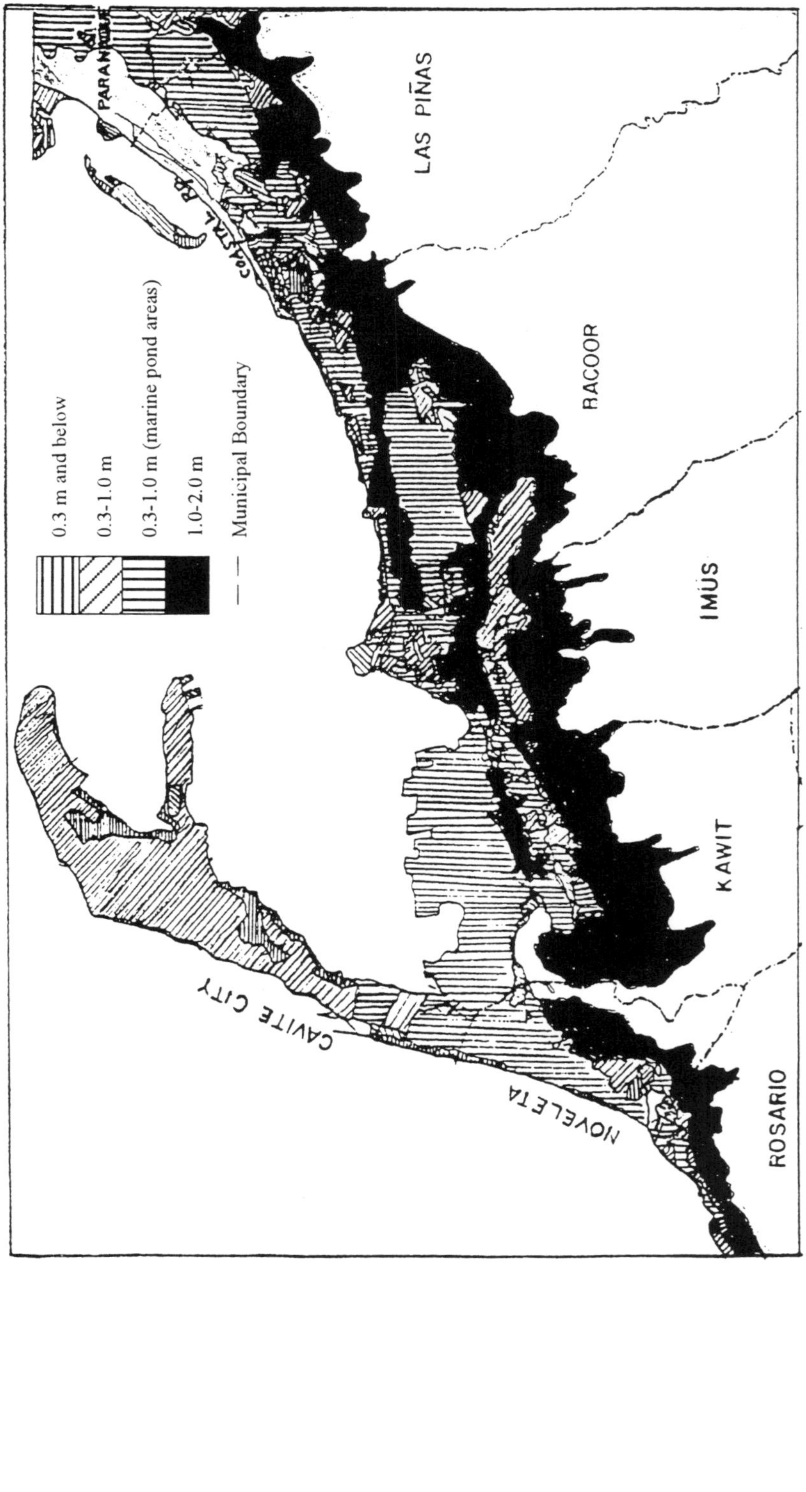

Fig. 4. Projected sea level rise in Cavite and Las Piñas.

4. Conclusion

Initial studies of developed countries regarding accelerated sea level rise show that it will occur even if the emissions of greenhouse gases that cause the global warming are arrested (IPCC, 1990). The preliminary results obtained in this study in the Manila Bay Area indicate that it is vulnerable to sea level rise both physically and socioeconomically. It must be realized that there is a long lead time required to transform a policy to reality. Take, for example, the Presidential Decree on Fishpond Leasing/Developments, which was issued to protect coastal areas from negative environmental effects. This decree is in existence but seemingly not being put into practice. Hence, it would be prudent to find a way to incorporate the information about a rise in sea level into development, land use planning, and implementation. Long-term plans for an effective coastal zone management must be developed now and put into action. It must be realized that any of the responses identified occur within a sociopolitical framework. To develop an effective response strategy requires interactions between and among several levels of the government and the local communities. Response strategies must lay heavily on government institutional actions from the national level down to the local levels. Land use planning must be examined carefully, and all laws and policies must be implemented strictly. Construction of protective structures must be carefully planned to include possible funding sources as these may require high capitalization. Disaster preparedness and mitigation at all government levels must include concern for the possible impacts of sea level rise. To be able to carry out all of these strategies, an extensive information and education campaign to include policy makers and the general public on the issues and concerns of an accelerated sea level rise must be enjoined.

Acknowledgments

This study is being supported by the U.S. Country Studies Program for Climate Change and the Philippine Government under the auspices of PAGASA. Sincere thanks go to the staff of the Natural Disaster Reduction Branch (NDRB) of PAGASA and the Coast and Geodetic Survey Department of NAMRIA. Special thanks are extended to Richard Labagnao, Alson C. Alconera, Willy Tamayo, and Nonoy About, Jr. of the NDRB. We also would like to acknowledge the cooperation extended to us by Mr. Cielito Gonzalez and Ms. Jessica Muñoz of BFAR. This study is referred to as SE-B1, or the Study Element on Vulnerability Assessment and Evaluation of Adaptations on Coastal Resources due to Accelerated Sea Level Rise, Philippine Country Study to Address Climate Change. The team leader is Commodore Renato B. Feir, Director, CGSD; the coteam leader is Dr. Rosa T. Perez, Supervising Weather Specialist, PAGASA. Other team members (research assistants) are Lt. Efren Carandang, Chief Oceanographer, CGSD; Mr. Edgardo B. Gonzalez, Sr. Weather Specialist; Engineer Martin L. Rellin, Jr., Sr. Weather Specialist; and Sharon Juliet Arruejo, Weather Specialist I, PAGASA.

References

BFAR: 1994, *Resource and Ecology Assessment of Manila Bay*, Volume 2, Bureau of Fishery and Aquatic Resources, Manila, Philippines.

Carandang E.: 1989, Preliminary Assessment of Sea Level Variations in the Philippines, Technical Report, National Mapping and Resources Information Authority, Manila, Philippines.

IPCC: 1990, Strategies for Adaptation to Sea Level Rise, Report of the Coastal Zone Management Subgroup, Intergovernmental Panel on Climate Change (IPCC) Response Strategies Working Group, Rijkswaterstaat, The Netherlands.

IPCC: 1992, Global Climate Change and the Rising Challenge of the Sea, IPCC Response Strategies Working Group, Ministry of Transport, Public Works and Water Management, The Hague, The Netherlands.

Leatherman S., Nicholls R.J., and Dennis K.C.: 1995, *Journal of Coastal Research*, Special Issue No. 14.

MDPO: 1996, Excerpts from the 1995 Socioeconomic Profile of Las Piñas, Metro Manila, Office of the Municipal Development and Planning Officer, Manila, Philippines.

National Research Council: 1987, *Responding to Changes in Sea Level — Implications*, National Academy Press, Washington, DC.

PAGASA Report: 1983, Documentation of STRIDE's Field Investigation on the Aftermath of Ty Bebeng (8302), Technical Report, Quezon City, Philippines.

SHORELINE STABILIZATION APPROACHES IN RESPONSE TO SEA LEVEL RISE: U.S. EXPERIENCE AND IMPLICATIONS FOR PACIFIC ISLAND AND ASIAN NATIONS

S.P. LEATHERMAN
Laboratory for Coastal Research, Department of Geography
University of Maryland, College Park, MD 20742, U.S.A.

Abstract. Coastal erosion is a worldwide problem; it is estimated that over 70% of the world's beaches are currently experiencing erosion (Bird, 1985), and this number may approach 90% in the United States.[1] Almost every conceivable form of shore protection has been attempted in the United States, including construction of seawalls, groins, and jetties as well as beach nourishment. The principal approach to protecting coastal property and maintaining recreational beaches in the United States today is beach nourishment. Engineering structures such as groins and seawalls have often been shown to have detrimental effects on adjacent beaches. Also, their construction and maintenance costs are quite high. Therefore, coastal communities have come to rely on a "soft" engineering solution — beach nourishment — because it is environmentally sound, aesthetically pleasing, and, so far, economically feasible. However, global warming and accelerated sea level rise will cause more rapid rates of beach loss and could make even this alternative too costly for many resort areas along the U.S. shoreline. The cost to nourish all the major recreational oceanic beaches in the United States was estimated based on various sea level rise scenarios. The beach nourishment approach involves placing enough sand on the beach to maintain stable (nonretreating) conditions in response to rising sea level. The quantity of sand required to "hold the line" is evaluated under various sea level rise scenarios from the baseline scenario to the 1-m estimate of the Intergovernmental Panel on Climate Change. Beach nourishment is not a practical alternative for most Pacific island nations because sand is a scarce resource. In fact, beaches are often mined for sand for construction, contributing to beach erosion. For mainland countries, beach fill projects are more practical. Singapore was one of the first countries to use this technology in combination with building offshore breakwaters to form artificial headlands. As Western style sunbathing and the overall popularity of beaches for recreation continue to grow, beach nourishment can be expected to become a more common way to deal with sea level rise induced coastal erosion for highly developed beach resort areas.

Key words: Asia, Pacific island nations, coastal resources, sea level rise, beach nourishment, coastal erosion

1. Introduction

A significant portion of the U.S. population lives within the coastal zone, where many buildings and facilities are located at elevations less than 3 m above sea level. These structures are currently vulnerable to storm damage, a hazard that will grow more severe as greenhouse-induced global warming causes unprecedented rates of sea level rise in the future (NRC, 1987). Despite those potential hazards, the coastal population is burgeoning. Beachfront property is some of the most valuable real estate in the country, often exceeding US$1,000 per linear meter of shoreline along the U.S. coasts.

Three general responses to accelerated sea level rise exist: (1) retreat from the shore, (2) armor the coast, or (3) nourish the beach. As part of a U.S. Environmental Protection Agency initiative to explore these options of managing erosion in response to sea level rise, this study was undertaken to assess the potential of beach nourishment as one approach to "hold the line."

1. S.P. Leatherman, testimony in U.S. Senate on impact of climate-induced sea level rise on coastal areas, 1988.

Water, Air, and Soil Pollution **92:** 149–157.
© 1996 *Kluwer Academic Publishers. Printed in the Netherlands.*

Shore stabilization measures can be divided into two categories: rigid and nonrigid (U.S. Army Corps of Engineers, 1984). The former often involves the emplacement of seawalls, bulkheads and breakwaters (shore-parallel structures), and jetties and groins (shore-perpendicular structures). Each of these structures has been shown to induce adverse effects in particular settings. For instance, the Ocean City inlet jetties have caused sand blockage along the Maryland coast for over 50 years; the result has been the downdrift erosion of northern Assateague Island (a national seashore) at a rate up to 10 m per year (Leatherman, 1986). Sea Bright, New Jersey, is protected by a massive seawall, but at the expense of the recreational beach. Elsewhere, shore-perpendicular structures, particularly groins, have been shown to cause extensive damage to downdrift beaches, the most infamous case being Westhampton Beach on Long Island, New York (Leatherman, 1991).

The nonrigid approaches of beach nourishment and dune building are generally preferred by coastal communities because the soft interface of a sandy beach is preserved for recreational pursuits, and yet storm protection can be gained if adequate sand quantities are available. This approach has the least environmental impact of any approach, but care still must be exercised. Possible problems involve both the dredging and the placement of sand. For tropical coasts, fragile ecosystems such as coral reefs must be carefully guarded to protect them during dredging operations. These carbonate beaches are maintained by sand production from the productive reefs, whose destruction would greatly accelerate beach erosion. Most U.S. cities on the beach have come to rely on beach nourishment to provide a recreational beach and serve as storm protection for very expensive beachfront developments. Over 150 beaches in the United States have been nourished, and other coastal resorts are applying for such projects (Davison *et al.*, 1992).

This research provides a preliminary estimate of the sand volumes and costs to nourish all existing and potential recreational oceanic beaches in the United States given various sea level rise scenarios. The guiding concept of beach nourishment is to place enough sand on the beach to maintain stable (nonretreating) conditions with rising sea levels. This study evaluates the quantity of sand required to "hold the line" using three sea level rise scenarios: baseline, 0.5 m, and 1.0 m. The potential costs of these nourishment projects were estimated using the best available data for existing beach nourishment projects.

The primary purpose of this paper is to describe the U.S. experience with shoreline stabilization, principally beach nourishment, for possible use by Pacific-rim countries in their assessments. The first step is to inventory existing shoreline engineering structures and past beach nourishment projects and then to project future responses and costs in response to accelerated sea level rise as has been done for U.S. beach nourishment projects in this paper.

2. Methodology

2.1 STUDY SITES

This study focused on the 21 coastal states in the United States, excluding Alaska because of its undeveloped nature. Each state's coastline was divided into three categories: (1) publicly owned, undeveloped; (2) privately owned, undeveloped; and (3) developed. Publicly owned, undeveloped areas, e.g., state parks, national seashores, and National Aeronautics and Space Administration (NASA) installations, will probably never be developed, but some areas may be nourished. In general, these locales were not considered in the nourishment assessment, unless beach filling had occurred in the past and is likely to continue.

Most of the areas in the privately owned, undeveloped coastal category are identified in the 1983 U.S. Congress Coastal Barrier Resources Act legislation, and usually are excluded from receiving federal assistance in shoreline stabilization by law. However, these areas still have the potential to be developed and maintained by private concerns, and are therefore included in the national assessment. Most such areas probably will be sparsely developed, making beach nourishment economically infeasible.

Areas in the developed category are currently urbanized or partially developed with a strong potential to be extensively developed in the future. Beaches in this category that have been nourished in the past or have undergone full-scale urbanization are the best candidates for further restoration (e.g., Miami Beach, Florida).

2.2 SEA LEVEL RISE SCENARIOS

Three eustatic sea level rise (SLR) scenarios (0.2 m or baseline, 0.5 m, and 1.0 m by 2100) were evaluated in this analysis. Local sea level rise varies on a regional basis because of variations in ground movements. Therefore, the total or relative sea level rise was projected into the future by using the IPCC (1990) eustatic sea level rise scenarios plus the local subsistence or uplift rate (back calculated from the tide gauge record).

2.3 SEAWARD LIMIT OF BEACH NOURISHMENT

Closure depth represents the seaward limit of significant sediment transport along a beach profile (Figure 1) and is the offshore extent to which beach nourishment should occur. Nourishment of the beach profile to this distance is imperative for the success and longevity of beach replenishment projects; otherwise, wave action rapidly reworks the new sediment, moving a portion of it offshore to attain profile equilibrium.

Hallermeier (1981) developed a technique using statistical wave data and local sand characteristics to determine closure depths at various coastal locations in the United States (Table I). Hallermeier's results were used in this study to determine the appropriate offshore extent of proposed beach nourishment based on average wave conditions; extreme storm events could move sand seaward of the specified closure depth. When Hallermeier's predictions were not available for a particular study area, approximate closure depths were extrapolated from the closest given locations.

U.S. Geological Survey (USGS) 7.5 minute quadrangles, supplemented by National Ocean Service nautical charts, were used to determine the approximate horizontal distance offshore to closure depth for each coastal sector. This data source was selected since USGS quadrangles

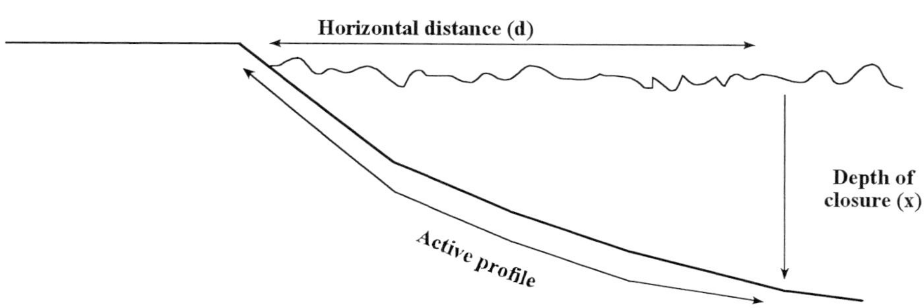

Fig. 1. Closure depth includes the active beach profile.

TABLE I
Selected closure depths for the U.S. coastline (Hallermeier, 1981)

Site	Closure Depth (m)
Atlantic Coast	
Atlantic City, NJ	7.04
Virginia Beach, VA	6.25
Assateague, MD	5.40
Nags Head, NC	7.95
Atlantic Beach, NC	5.65
Wrightsville Beach, NC	5.35
Holden Beach, NC	4.72
Bull Island, SC	4.20
Tybee Island, GA	7.40
Boca Raton, FL	4.20
Lake Worth, FL	5.04
Gulf of Mexico	
Naples, FL	2.98
Destin, FL	4.30
St. Andrews Park, FL	4.50
Crystal Beach, FL	4.80
Gilchrist, TX	4.10
Galveston, TX	3.80
Corpus Christi, TX	5.20
Pacific Coast	
Huntington Beach, CA	5.83
San Clemente, CA	7.10
Bolsa Chica, CA	6.10
Pt. Mugu, CA	5.70
Pismo Beach, CA	7.50
San Simeon, CA	6.50
Capitola Beach, CA	4.20
Stinson Beach, CA	7.30
Wrights Beach, CA	10.40
Shelter Cove, CA	7.10
Prairie Creek, CA	7.00
Umpqua, OR	7.80

(1:24,000 scale) are the most readily available and commonly utilized maps in the country, depicting both surficial features and offshore contours.

2.4 SAND VOLUME DETERMINATIONS

Volume calculations for beach and nearshore profile nourishment were made using USGS maps and knowledge of closure depth. The horizontal distance (d) from beach to offshore closure depth was measured from the appropriate maps and charts, and represents the active profile dimension. Using these data, rough estimates of sand needed for profile nourishment were obtained assuming that, for every increment (x) of sea level rise, the volume of sand required to "raise the profile" corresponds to xd per unit of shoreline length (Figure 1). This approach provides a first approximation for the purposes of this study and was adopted for its expediency given budgetary and time constraints. Although other methodologies (e.g., trend analysis and the sediment budget model) may provide better estimates of these sand quantities, they require considerably more data (NRC, 1990).

2.5 SAND RESOURCES AVAILABILITY

After estimating quantities of sand required to maintain designated beaches for various sea level rise scenarios, a determination of sand resources available to match this projected need was undertaken. Only offshore borrow sites were identified for this purpose because they are typically the preferred sources for beach nourishment projects. Many environmental and other problems exist regarding other sand sources (e.g., backbarrier lagoons and bays, mainland sand pits, ebb tidal deltas). Therefore, these sand reserves were not considered in this assessment.

Offshore sediment sources along the Atlantic coast were delineated by a series of studies undertaken in the late 1960s to mid-1970s by the U.S. Army Corps of Engineers Coastal Engineering Research Center (CERC). The Inter-Continental Shelf (ICONS) studies used high resolution seismic surveys and vibracoring techniques to perform a preliminary assessment of offshore borrow sites suitable for beach restoration (Duane and Meisburger, 1969). All of the states from New York to Florida were evaluated, and these reports were used to obtain an estimate of available offshore sand reserves. This information was updated by additional data, and information for the rest of the coast was identified (e.g., Corps district reports, USGS, and state geological surveys) for specific areas where available.

2.6 BEACH NOURISHMENT COSTS

Sand costs were based on current dredging rates per cubic meter of material as identified in the available literature or by appropriate officials. The less expensive, closer-to-shore, sand supplies are likely to be used first. Initial dredging costs (estimated nationally at US\$3/m^3) were used to determine final costs until all sediment supplies within 1.6 km of the shore were exhausted. Once these supplies are exhausted, it was assumed the costs will rise as a step function (approximately US\$0.81/m^3 per kilometer farther offshore as booster pumps are added). The "base rates" vary regionally so that the actual costs are site-specific.

3. Results

The results of this study (Tables II and III) represent a worst case scenario in that all potentially developable areas were evaluated. In reality, it is likely that only existing commercially important areas (e.g., cities on the beach) will be nourished. Therefore, nationwide summary numbers probably represent an extreme situation. On the other hand, the calculations represent a one-time nourishment at each location. It is widely recognized that most beach nourishment projects need frequent maintenance (Davison *et al.*, 1992); this study made no effort to address these requirements, and dredging costs could increase significantly in the future.

Many coastal locations or states may opt not to utilize beach nourishment at all. Some coastlines such as the rocky and/or cliffed coasts of Washington, Oregon, northern California, Maine, and other New England areas have such small sandy beaches that nourishment is not economically feasible. Other areas such as the Outer Banks of North Carolina and parts of Texas are low density development and relocation is possible. Lastly, some locations such as certain "seas islands" in the Georgia bight have higher elevations and are generally more stable

TABLE II
Quantity of sand by state for baseline and 1 m by 2100 (in 10^6 m^3)

State	Baseline (0.2 m)	1 m SLR
Alabama	2.12	20.25
California	6.87	62.37
Connecticut	13.04	121.04
Delaware	0.47	4.77
Florida-Atlantic	23.09	216.14
Florida-Gulf	28.74	274.86
Georgia	4.99	50.51
Hawaii	11.32	99.53
Louisiana	301.32	456.23
Maine	4.45	41.69
Maryland	0.73	7.51
Massachusetts	8.21	82.47
Mississippi	3.75	35.90
New Hampshire	1.57	14.12
New Jersey	9.00	99.05
New York	14.07	134.60
Oregon	4.20	29.32
Rhode Island	2.09	20.59
North Carolina	15.11	135.97
South Carolina	31.38	329.90
Texas	29.08	322.85
Virginia	2.92	31.58
Washington	9.93	69.26

than low-lying microtidal barriers. These areas would probably not require extensive beach nourishment, but they are included in the final assessment because no attempt was made to exclude historically stable areas.

Only those areas that are heavily developed and commercially important, cannot adopt relocation strategies, or have demonstrated a strong commitment to beach nourishment in the past are likely to continue these efforts. All of Florida's project expenditures are matched with equal grants from the federal government. Other states currently use or may opt for matching federal aid (e.g., Ocean City, Maryland). The potential federal expenses for beach nourishment could become staggering in coming decades (U.S. Army Corps of Engineers, 1994).

TABLE III
Nationwide estimates of sand quantities and costs required in response to sea level rise

	Sand Quantities (10^6 m^3)		Cost (US$ millions)	
Year	0.5 m	1.0 m	0.5 m	1.0 m
2000	112.03	128.28	837	958
2020	311.31	408.54	2,333	3,032
2040	576.86	821.44	4,277	6,073
2060	888.56	1,480.95	6,564	11,419
2080	1,363.51	2,116.62	10,524	15,874
2100	1,864.87	3,342.67	14,512	26,745

The quantity of sand necessary to hold the shoreline in place was assessed on a state-by-state basis for selected sea level rise scenarios (Table II). It is apparent that tremendous quantities of good quality sand will be necessary to maintain the nation's major recreational beaches. The numbers for Louisiana are suspect because the state has nourished only one beach to date (Grande Isle), and the high erosion rates (often exceeding 10 m yr^{-1}) make beach nourishment an economically unattractive option. Nationwide estimates of sand quantities required for the 0.5 and 1.0 m scenarios were also determined (Table III). These costs are based on the current expenses of offshore dredging and onshore pumping of locally derived material. Obviously, the costs will increase with inflation, but more important the expenses could be greatly underestimated if sand must be acquired from areas at considerable distances from the beach requiring nourishment.

The offshore sand is not evenly scattered along the U.S. coastline, so some areas will run out of local (the least expensive) sand in a few decades. The state of Florida recently (1995) sent a delegation to the Bahamas to see about buying offshore shoal sand to be barged and then pumped onto South Florida beaches at considerable expense (perhaps as high as US\$8/m^3).

4. Pacific/Asian Experience

The implementation of response options to accelerated sea level rise and coastal erosion is often limited by financial considerations in developing and transitional countries. Protection strategies will almost always involve hard engineering structures, which are costly to both construct and maintain. For example, the present cost of protection in the Maldives is approximately US\$800/m of shore (IPCC, 1996). The annual cost, defined as a percentage of GNP spent on protection, is highest for the Marshall Islands at 7% (Holthus *et al.*, 1992) and Nevis in the Caribbean at 2.7% (Cambers, 1994), emphasizing the vulnerability of small island states. In Kiribati, for instance, implementation of any major protection measure will require external assistance (Abete, 1993). Even without climate change and associated sea level rise, all of these small island states will continue to experience increasing coastal vulnerability due to high population growth rates as well as continued uncontrolled and unrestricted exploitation of coastal resources (e.g., sand mining, over-fishing, tourism development, and water pollution).

Although beach nourishment is generally impossible economically or environmentally (e.g., no source of suitable sand) for most small island states in the Pacific, mainland countries are likely to select this response option to protect tourist infrastructure and the sandy beach itself. Singapore was one of the first Asian countries to experiment with large-scale beach nourishment in combination with structural controls through the construction of large offshore headlands (Silvester, 1974). Also, the amount of sand required to maintain a beach in the face of accelerated, long-term, sea level rise is uncertain (Stive *et al.*, 1991), and the costs of beach nourishment could dominate basic response costs if countries depend heavily on such a response strategy (Nicholls and Leatherman, 1995).

More recently, Repulse Bay in Hong Kong has been nourished by pumping sea floor sediments onshore. This beach had been depleted during World War II for sand bags, so the project was actually restorative. This nourishment of a relatively small pocket beach has been quite successful in that little erosion has been experienced, but it has not yet been tested by direct impact of a typhoon (W. Yim, 1996, personal communication).

The Japanese have used beach nourishment more extensively than elsewhere in Asia, but largely as a consequence of harbor development. Because sand filling of an eroding beach must be done over and over through time, the Japanese government has usually elected to employ hard shoreline stabilization technology such as rubble-mound seawalls.

5. Conclusions

This study represents the first estimation of sand requirements necessary to stabilize the U.S. coastline in response to accelerated sea level rise. The volume of sand and associated costs to nourish the beach profile in response to sea level rise have been assessed. A number of assumptions were incorporated into making these calculations so that the numbers can be refined as more data become available.

The cost to stabilize the U.S. coast through the "soft" engineering approach of sand filling ranges from approximately US$14.5 billion to US$26.7 billion by 2100 (Table III). Considering the enormous value of coastal property (e.g., estimated at US$2 trillion for U.S. beachfront development), it is safe to assume that densely developed areas will be nourished and maintained. It is unclear at which point moderate-density areas will be forced by economic considerations to choose another approach (e.g., retreat from the eroding beach).

The next stage in these analyses is to compare potential stabilization costs with the value of the affected coastal property to determine which beaches qualify for beach nourishment under existing conditions, and in the future in response to accelerated sea level rise.

This kind of analysis should be conducted for Asian and Pacific countries. For small island nations there will probably be a continuing reliance on seawalls for coastal protection because beach nourishment is not possible environmentally (e.g., no source of suitable sand or problems with disturbance of living coral reefs) and economically (e.g., the cost would be prohibitively high concerning the economies of most such areas). Also, it is different to defend against the sea when there is so much shoreline per area of land mass, as on island atolls, and flooding and erosion can be a problem on both the seaward and lagoonal sides (e.g., Tarawa atoll, Kiribati).

Repulse Bay has become a popular urban beach in Hong Kong and was first nourished in the late 1980s by pumping sand onshore. This is a relatively small, pocket beach, and the benefits clearly outweigh the costs under present conditions. Elsewhere, there have been few nourishment projects, with most reliance placed on hard shoreline engineering, mainly seawalls and groins. In fact, over 50% of the Japanese coast is reported to be protected by hard coastal engineering structures.

Acknowledgment

This study was supported by the U.S. Environmental Protection Agency, Office of Policy, Planning, and Evaluation, Washington, DC, Mr. Jim Titus, Project Officer.

References

Abete T.: 1993, The Kiribati preliminary assessment to accelerated sea level rise, in: *Vulnerability Assessment to Sea Level Rise and Coastal Zone Management*, Proceedings of IPCC WCC93 Workshop, Tsukuba, Japan, pp. 91-98.

Bird E.C.F.: 1985, *Coastline Changes: A Global Review*, Wiley Interscience, Chichester, England.

Cambers G.: 1994, Assessment of the vulnerability of coastal areas in Antigua and Nevis to sea level rise, in: *Global Climate Change and the Rising Challenge of the Sea*, Proceedings of the IPCC CZMS Workshop, Margarita Island, Venezuela, pp. 11-27.

Davison A.T., Nicholls R.J., and Leatherman S.P.: 1992, *Journal of Coastal Research* **8**, 984.

Duane D.B. and Meisburger E.P.: 1969, Geomorphology and Sediments of the Nearshore Continental Shelf, Miami to Palm Beach, Florida, TM No. 29, U.S. Army Corps of Engineers, Vicksburg, Mississippi.

Hallermeier R.J.: 1981, *Coastal Engineering* **4**, 253.

Holtus P., Crawford M., Makroro C., and Sullivan S.: 1992, Vulnerability Assessment for Accelerated Sea Level Rise Case Studies, Majuro Atoll, Republic of the Marshall Islands, SPREP Report No. 60, Apia, Western Samoa.

IPCC (Intergovernmental Panel on Climate Change): 1990, *Strategies for Adaptation to Sea Level Rise*, Rijkswaterstaat, The Netherlands.

IPCC (Intergovernmental Panel on Climate Change): 1996, *Climate Change 1995: The IPCC Second Assessment Report, Volume 2: Scientific-Technical Analyses of Impacts, Adaptations, and Mitigation of Climate Change*, Watson R.T., Zinyowera M.C., and Moss R.H. (eds), Cambridge University Press, Cambridge and New York.

Leatherman S.P.: 1986, Shoreline response to sea level rise: Ocean City, Maryland, in: *Proceedings of Icelandic Conference on Coasts and Rivers*, Reykjavik, Iceland, pp. 267-276.

Leatherman S.P.: 1991, Coasts and beaches, in: *Heritage of Engineering Geology: The First Hundred Years*, Geological Society of America, Boulder, Colorado, pp. 183-200.

National Research Council: 1987, *Responding to Changes in Sea Level: Engineering Implications*, National Academy Press, Washington, DC.

National Research Council: 1990, *Managing Coastal Erosion*, National Academy Press, Washington, DC.

Nicholls R.J. and Leatherman S.P.: 1995, *Journal of Coastal Research*, Special Issue No. 14, 323.

Silvester R.: 1974, *Coastal Engineering*, Elsevier, Amsterdam.

Stive M.J.F., Nicholls R.J., and DeVriend H.J.: 1991, *Coastal Engineering* **16**, 147.

U.S. Army Corps of Engineers: 1984, *Shore Protection Manual*, 3 volumes, U.S. Army Corps of Engineers, Vicksburg, Mississippi.

U.S. Army Corps of Engineers: 1994, *Shoreline Protection and Beach Erosion Control Study*, Institute for Water Resources, Alexandria, Virginia.

TROPICAL CYCLONES AND COASTAL INUNDATION UNDER ENHANCED GREENHOUSE CONDITIONS

A.B. PITTOCK, K. WALSH, and K. MCINNES
CSIRO Division of Atmospheric Research
PMB 1, Aspendale 3195, Australia

Abstract. The behavior of tropical cyclones under enhanced greenhouse conditions has been the subject of considerable speculation. Typical spatial scales of these cyclones are on the order of tens to hundreds of kilometers. Therefore they cannot be simulated in global climate models with resolutions of several hundred kilometers. Thus speculation has been largely based on extrapolation from their present observed distribution, and on simple parametric relationships. However, the conditions under which tropical cyclones form, the intensities they reach, and their usual paths depend on a number of dynamic and thermodynamic factors that may change in complex ways with changing climate. Recent studies using finer resolution global and regional climate models, sensitivity studies that model individual cyclones, and parametric studies have been reviewed. These suggest that the future behavior of tropical cyclones remains an open question, with changes of either sign possible in numbers and intensities. The paper also describes the combined effect on coastal inundation of mean sea level rise and changes in storm surges due to tropical and extratropical cyclones. Impact studies highlight the importance of taking both these factors into account and the highly site-specific nature of the problem.

Key words: Australia, coastal resources, limited area models, storm surges, tropical cyclones

1. Introduction

Tropical cyclones are destructive. Since 1970, hundreds of thousands of people have been killed by them (Anthes, 1982), and in August 1992, Hurricane Andrew caused US$20-25 billion in property damage to Florida (Mayfield *et al.*, 1994). It is therefore important to obtain an estimate of any likely changes in their behavior as a result of the enhanced greenhouse effect. Any substantial changes in the climatology of tropical cyclones, either in the numbers of such storms or in their geographical distribution, could cause increases (or decreases) in the amount of damage caused in the regions of the globe where they occur, or could cause them to affect new regions. Any changes in tropical cyclone climatology need to be assessed in the context of demographic changes in coastal regions. This paper provides an overview of current research efforts on tropical cyclones under enhanced greenhouse conditions and a detailed discussion of a study on tropical-cyclone-like vortices in a limited area model.

2. Current Research Efforts

An estimate of changes in tropical cyclone climatology might be derived from a climate model. Climate models are the main tools used for the development of climate change scenarios. They are useful, but they have limitations, particularly for modeling tropical cyclones. One of these limitations is their low horizontal resolution. Typically, a climate model has a resolution of a few hundred kilometers. Some entire cyclones are smaller than this (e.g., Hurricane Andrew; Mayfield *et al.*, 1994), and all cyclones have important features such as the eye which are only about 10-100 km wide (Anthes, 1982). Therefore these cyclones are not reliably simulated in coarse resolution global climate models. The impact of climate change on tropical cyclones (hereafter called cyclones) thus has generally been assessed by using various indirect techniques

Water, Air, and Soil Pollution **92:** 159–169.
© 1996 *Kluwer Academic Publishers. Printed in the Netherlands.*

rather than direct simulation of the storms themselves. In general, the effect of climate change on these cyclones remains controversial (Lighthill *et al.*, 1994; Emanuel, 1995; Broccoli *et al.*, 1995).

One of these techniques is based on the assumption that it is reasonable to examine the variation of numbers and intensities of cyclones with sea surface temperature (SST) in the current climate and extrapolate the results to those temperature changes expected as a result of an enhanced greenhouse effect. However, Nicholls (1989) found little evidence from observations of any relationship between underlying monthly average SST and cyclone numbers. For intensities, Evans (1993) showed that there was little direct relationship between SST and the intensity of cyclones in most of the regions in which they form, with the exception of intense storms in the North Atlantic. In the North Atlantic formation region, DeMaria and Kaplan (1994) found an empirical exponential relationship between SST and the maximum potential intensity (MPI) of cyclones, i.e., the strongest storms likely to be encountered over a long period of time in any one location.

Another technique assumes that present-day cyclone numbers or intensities can be related by theoretical arguments to large-scale fields (e.g., temperature and wind fields). Geographical and vertical variations of these fields can then be related to the formation and strength of cyclones. Gray (1975) showed that it was possible to predict realistic numbers of cyclones from such fields as low-level vorticity, midtropospheric relative humidity, and SST when they were combined in a seasonal genesis parameter (SGP). Ryan *et al.* (1992) showed that when forced by the output of a general circulation model, the SGP could give a good representation of the observed cyclone climatology, but that changes in the SGP under enhanced greenhouse conditions were dominated by changes in SST. Given that there is little evidence from observations of a close relationship between numbers of cyclones and SSTs, this suggests that the SGP may need modification for use in studies of climate change. Watterson *et al.* (1995) evaluated the ability of the SGP, when forced both by observations and by climate model output, to simulate the observed interannual variations in cyclone numbers and geographical distribution. While the SGP demonstrated some skill in simulating the observed year-to-year variations, this skill was present only in certain regions.

Cyclone intensities and numbers can be predicted from large-scale fields. The work of Emanuel (1988, 1991) and Holland (in press) suggests that the MPI of cyclones may be related to aspects of the mean state of the atmosphere and ocean. MPI may also be related to regions of cyclone formation, since if the MPI is less than the required threshold intensity of a cyclone, a cyclone could not form in that location. A preliminary investigation of the ability of a climate model to predict the observed MPI using Holland's MPI formulation was that of Henderson-Sellers *et al.* (1995). The model they used had a reasonable ability to predict the observed MPI in the northwest Pacific. When the MPI formulation was applied to the results of an enhanced greenhouse simulation, there was a small increase in MPI in this region but no significant change in areas of cyclone formation, despite simulated increases in SSTs. The lack of a direct association between MPI and SST alone in this study, as well as the results of observational studies, suggests that the dynamical factors governing MPI are more complex than a simple association with SST. Additionally, one drawback of these studies is that the results depend on the quality of the climate model. Models are continually evolving and improving, but they have yet to simulate regional climate accurately. Numbers and geographical distributions of cyclones are also highly correlated with the El Niño-Southern Oscillation (ENSO) phenomenon (Philander, 1990), whose characteristics in a changed climate are quite uncertain at this time despite some preliminary experiments (Meehl *et al.*, 1993; Knutson and Manabe, 1994; Tett, 1995; Tokioka *et al.*, 1995).

Low-resolution climate models are unable to simulate cyclones reliably, but finer-resolution models have recently shown some potential to simulate features of cyclones, including their regions of origin, tracks, and aspects of their structure. Bengtsson *et al.* (1995) analyzed the simulation of a fine-resolution (about 120 km) coupled global climate model and found that it could generate numbers, intensities, and a geographical distribution of cyclones that were similar to reality. A preliminary enhanced greenhouse experiment (Bengtsson *et al.*, 1994) suggested that although the simulated cyclone-like lows were generated in roughly the same locations, their numbers were reduced in the Southern Hemisphere by more than a factor of two. This response seems unusually strong, and suggests that it would be premature to base a scenario on this simulation. There is also some uncertainty surrounding the climate response in the Southern Hemisphere in recent coupled model simulations caused by the poor representation in current ocean models of the uptake of heat in the Southern Ocean (England, 1995). Recently, limited area models (LAMs; Giorgi *et al.*, 1994; McGregor and Walsh, 1994; Walsh and McGregor, 1995) have been used to simulate the climate of specific regions at resolutions finer than those that can be employed economically by global climate models. It is of interest to determine whether these models can also simulate aspects of the structure and climatology of cyclones.

3. Discussion

3.1 TROPICAL-CYCLONE-LIKE VORTICES IN A LIMITED AREA MODEL

Analysis is in progress of the day-to-day variability of the LAM simulation of Walsh and McGregor (1995). The simulation is of 20 months (10 separate Januarys and 10 Julys), and was performed at a horizontal resolution of 125 km. The results suggest that, like the global model used by Bengtsson *et al.* (1995), a LAM is capable of generating cyclone-like vortices. Figure 1 shows the simulated occurrence of such vortices in the Australian region. The vortices have similar regions of origin to those observed in this region, although a small fraction of midlatitude disturbances are detected falsely as cyclone-like vortices. The tropical vortices also have many of the structural features that are observed in real cyclones: warm cores and low-level wind maxima. Although the structure and climatology of these vortices differ from reality in some respects, the simulation suggests that, with finer horizontal resolution and a better model representation of important atmospheric processes such as precipitation, an even more realistic simulation may be achieved. A 60 km resolution, seasonally varying, DARLAM simulation is currently being performed at the Commonwealth Scientific and Industrial Research Organization (CSIRO). It is hoped that such simulations will be used for climate change experiments to simulate directly the effects of climate change on cyclones. However, it should be noted that such experiments retain inherent limitations: DARLAM is forced by the output of a global model, and both models are not perfect representations of the current climate.

3.2 INUNDATION AND COASTAL IMPACTS

Even if the climatology of cyclones were unaffected by the enhanced greenhouse effect, there would likely be coastal impacts associated with the predicted rise in sea level (Warrick and Oerlemans, 1990; Wigley, 1995). Inundation of low-lying coastal terrain due to rising sea level is a predicted outcome of the enhanced greenhouse effect. However, additional consideration needs to be given to the temporary sea level fluctuations that occur as a result of severe

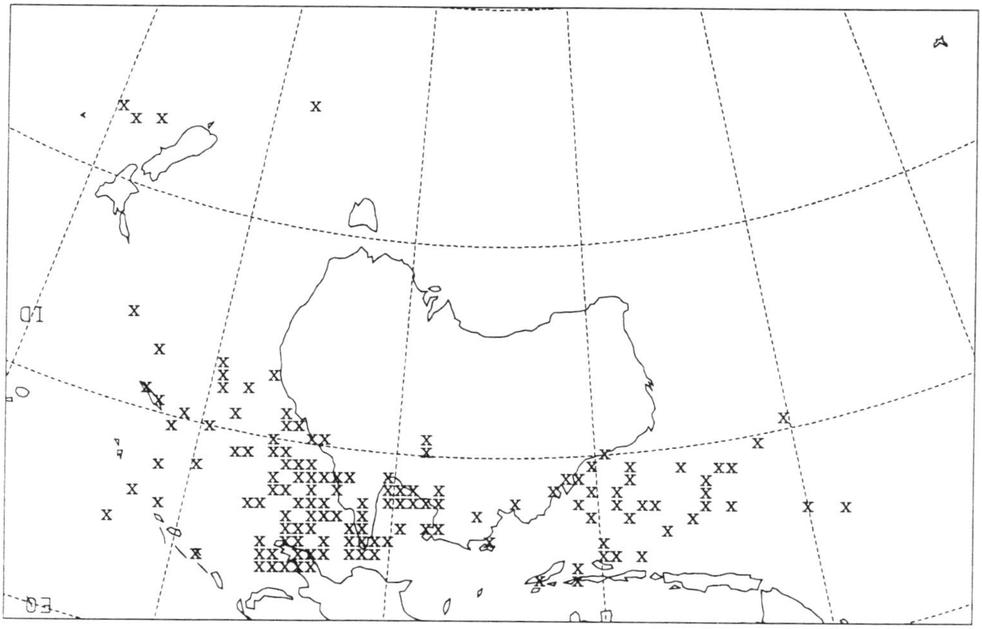

Fig. 1. Occurrence of tropical-cyclone-like vortices in the LAM simulation of Walsh and McGregor (1995) for the Australian region. Simulation consisted of 10 Januarys at a horizontal resolution of 125 km. Crosses represent positions of vortices.

atmospheric storms. Studies have shown that the return period of cyclone storm surge events reaching a given height above the present mean sea level is greatly reduced (i.e., the frequency of surge events is increased) by an increased sea level typical of that predicted for the end of the next century (Love, 1988). In coastal regions, these storms can produce not only destructive winds and extreme rainfall amounts, but also oceanic responses that are potentially hazardous to both natural and artificial coastal structures. A combination of severe winds and rapid atmospheric pressure reductions over the ocean surface can lead to the generation of storm surges, which cause most of the damage associated with the passage of cyclones (Anthes, 1982). Storm surges can cause flooding of low-lying coastal regions, and can enable the destructive forces of wind-generated waves to penetrate farther inland. In addition, elevated sea level can combine with high rainfall amounts to reduce river drainage rates and prolong floods upstream.

The issue of coastal inundation, both on long and short time scales, is highly site specific. In terms of mean sea level rise, significant variations in local sea level may occur as a result of changes in ocean circulation and local atmospheric pressure associated with global warming and climate change. In terms of day-to-day variations in sea level due to storm surges, the site-specific nature of the problem is caused in part by the meteorology affecting a particular region and in part by the local coastal geomorphology, which must be examined in the context of the prevailing meteorological forcing to determine its vulnerability in the event of climate change. Important factors include the frequency, intensity, and movement of severe storms. In addition, changing currents and wave climates may shift sand and sediments, thus changing beach alignments.

To illustrate some of these issues, some key results of a recent study (McInnes and Hubbert, 1995) on the potential impacts of climate change on Port Phillip Bay in Australia are presented (Figure 2). Port Phillip Bay is located on the southeast of the Australian continent; the central

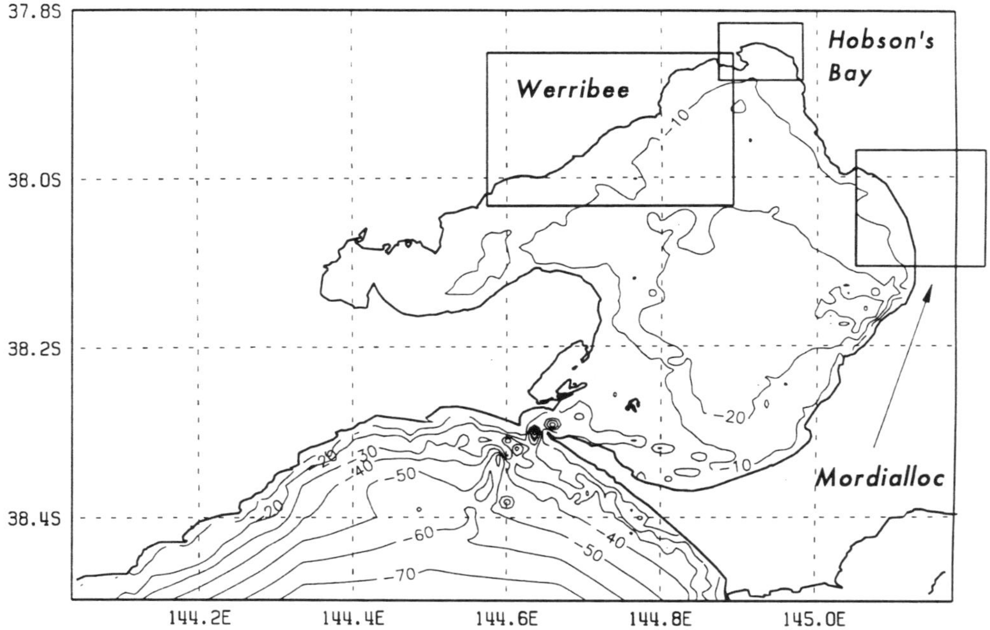

Fig. 2. The Port Phillip Bay domain showing the three subdomains of Hobson's Bay, Werribee, and Mordialloc.

business district of the city of Melbourne is located near Hobson's Bay at the northern extremity of Port Phillip Bay. Much of the coastline surrounding the bay is made up of residential areas, which extend to the beachfront. The meteorology that most commonly gives rise to storm surges along the southern coastline is characterized by westerly winds associated with the passage of cold fronts.

McInnes and Hubbert (1995) modeled two present-day, midlatitude, severe storms on the entire Port Phillip Bay domain and on the three smaller domains shown in Figure 2. One storm from May 1994 and one storm from November 1994 were selected for the study because they were the two largest storm surge events to occur during 1994. A sophisticated storm surge model capable of simulating the inundation associated with such events was used (Hubbert *et al.*, 1990). The model is driven by atmospheric surface winds and pressure at the ocean-atmosphere interface, and (on the larger domain) by the observed sea level signal on the open southern lateral boundary. The lateral boundaries of the three smaller domains are nested within the modeled sea surface heights and currents of the larger domain. However, no existing or planned flood mitigation structures (such as levee banks) were included in the simulations.

Figure 3 shows the modeled sea surface height compared with observed sea surface height at a location within the Hobson's Bay domain for the two events. The simulated sea surface heights show close agreement with the observations. The modeled tidal signal is also shown and indicates that the storm surge component of this event exceeds 1 m in both cases. A control experiment (Exp. 1) as well as two sensitivity experiments (Exp. 2 and Exp. 3) were performed on both storms; since the results of each sensitivity experiment were similar for both storms, only the inundation patterns for the November 1994 storm are presented (Figure 4). The first sensitivity experiment (Exp. 2) involved running the model on all domains with an imposed background sea level rise of 80 cm. This value is used because it is the worst case scenario for

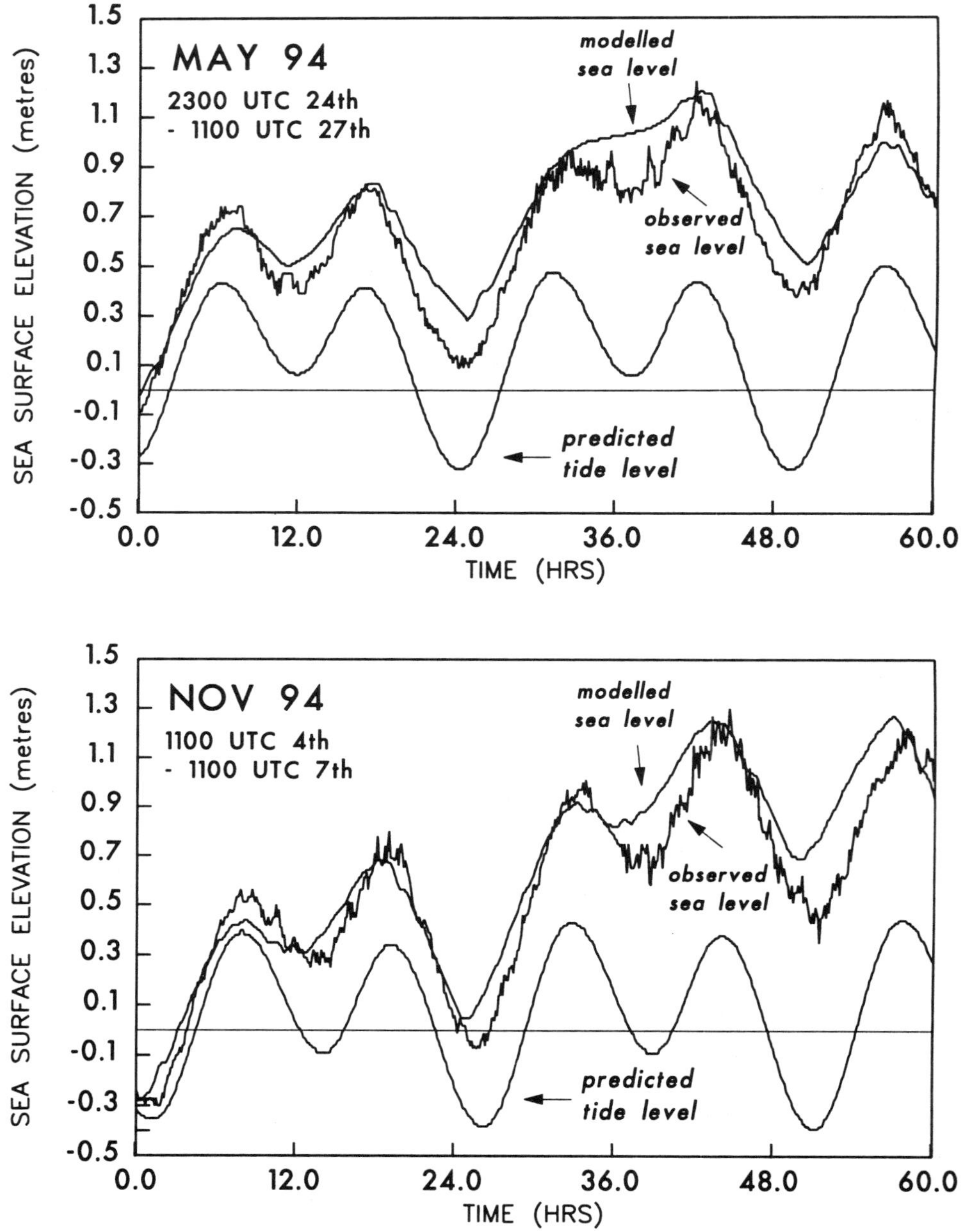

Fig. 3. The observed and modeled sea levels at St. Kilda on the Hobson's Bay domain over a 60 h interval for the two modeled events. The predicted tide level is also shown.

TABLE I

Summary of the ratio of area inundated in Exp. 2 and Exp. 3 to the area of inundation in Exp. 1 for the same event and model domain

	May 1994						November 1994					
	Werribee		Hobson's Bay		Mordialloc		Werribee		Hobson's Bay		Mordialloc	
	Ratio	Total Area Inundated (km^2)	Ratio	Total Area Inundated (km^2)	Ratio	Total Area Inundated (km^2)	Ratio	Total Area Inundated (km^2)	Ratio	Total Area Inundated (km^2)	Ratio	Total Area Inundated (km^2)
Exp. 1 Baseline	1.0	(11.2)	1.0	(0.9)	1.0	(0.7)	1.0	(15.8)	1.0	(0.9)	1.0	(0.7)
Exp. 2 0.8 m SLR	2.7	(30.4)	1.9	(1.7)	9.3	(6.5)	2.2	(34.4)	2.0	(1.8)	8.9	(6.2)
Exp. 3 0.8 m SLR 10% Increase in Wind Forcing	3.1	(34.3)	2.6	(2.3)	11.0	(7.7)	2.6	(41.0)	3.2	(2.9)	12.6	(8.8)

mean sea level rise for 2070 (Whetton *et al.*, in press). It should be noted that the actual sea level rise at this time is likely to be considerably less. In the second sensitivity experiment (Exp. 3), a 10% increase in the wind forcing was imposed in addition to the 80 cm sea level rise.

The areas of inundation on each of the three grids for the three experiments are summarized in Table I. In the control experiment (Exp. 1), the area of inundation is approximately 1 km^2 on the Hobson's Bay and Mordialloc grids, whereas it exceeds 10 km^2 on the Werribee grid. This is because the Werribee grid covers considerably more area. On the Werribee and Hobson's Bay domains, the effect of increasing the mean sea level for the same two storm events is an increase in the area of inundation by a factor of two. In contrast, on the Mordialloc domain, the increase is ninefold compared with the control simulation. This is due to the location of the domain relative to the storm, as well as the structure of the coastal terrain. Located on the eastern side of the bay, this region receives the maximum effect of the westerly winds and hence the greatest storm surge heights. The Mordialloc domain also contains low-lying wetlands behind the foreshore area with rivers flowing into the bay from the north and south (Figure 4a). In Exp. 2 (Figure 4b), water is backed up through the rivers into the low-lying area behind the foreshore. A 10% increase in wind strength in Exp. 3 (Figure 4c), in combination with the sea level rise, increases the area of inundation still further on all grids, again with the most pronounced impact on the Mordialloc domain.

These results highlight two important conclusions. First, estimation of climate impact in coastal regions is a highly site-specific problem that requires consideration of both the detailed topography of the region as well as the meteorology that leads to extreme sea levels. Second, although mean sea level may pose a threat to low-lying coastal areas, the occurrence of severe storms and storm surges can cause fluctuations in sea level that are far greater than the worst case estimates of mean sea level rise.

(a) Exp. 1

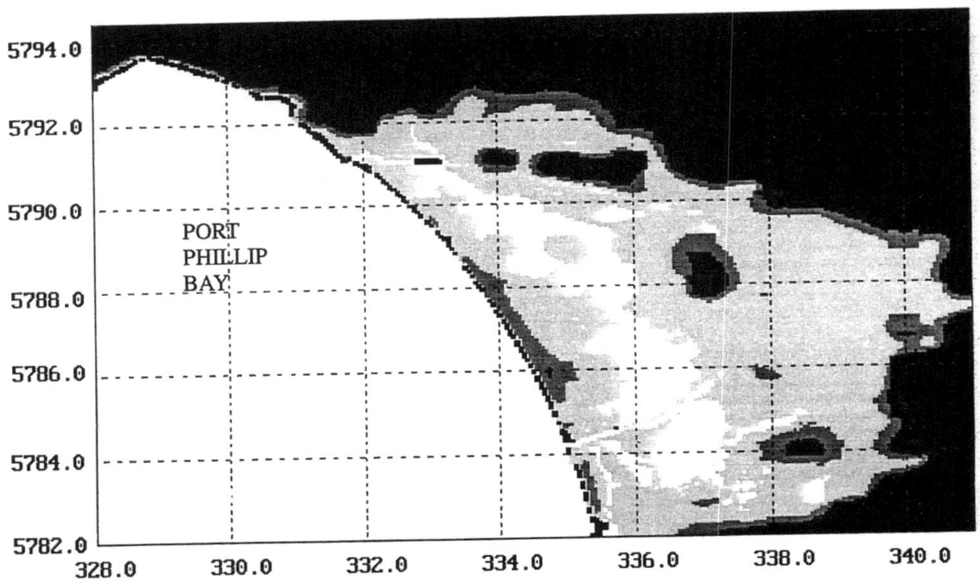

(b) Exp. 2

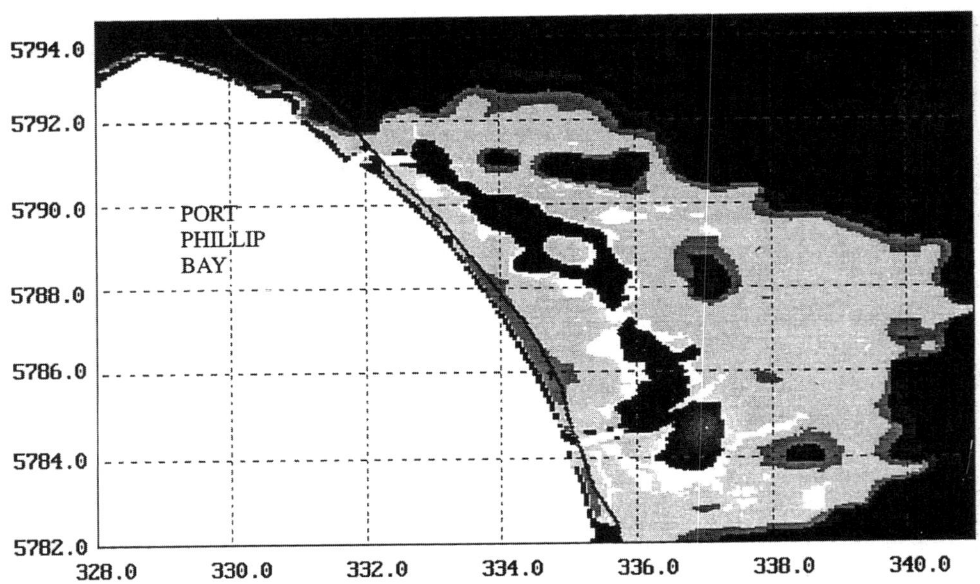

Fig. 4(a) and (b). See caption on next page.

(c) Exp. 3

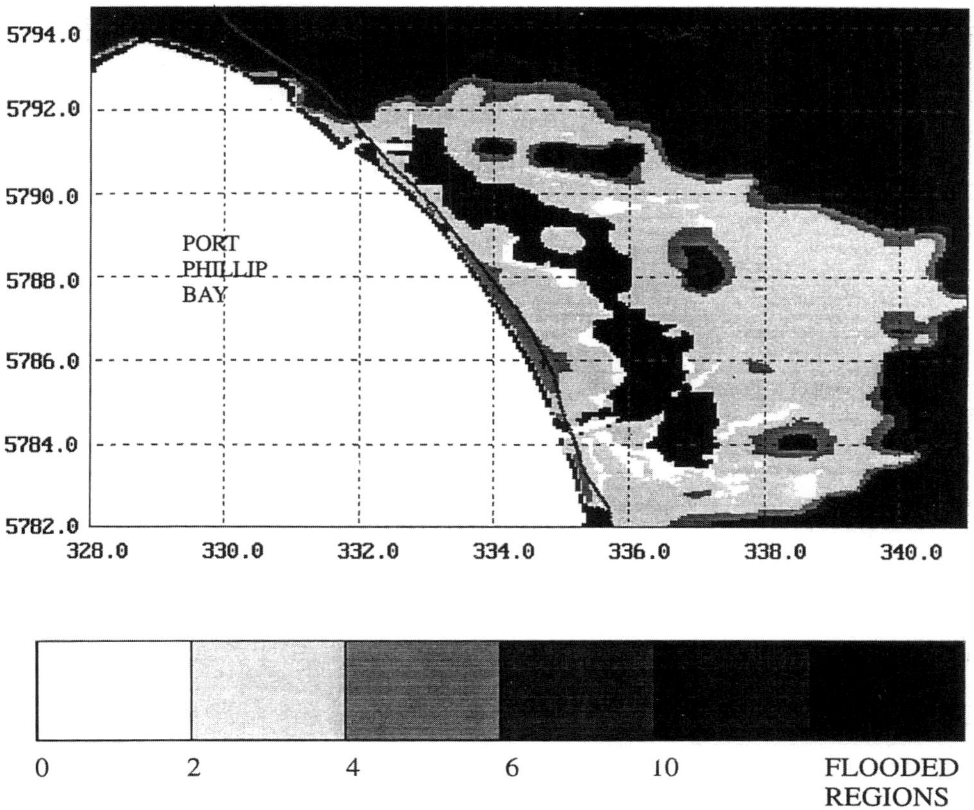

Fig. 4. Diagram showing the modeled inundation on the Mordialloc domain for (a) Exp. 1, (b) Exp. 2, and (c) Exp. 3. Black shading indicates areas that are inundated. Lightest grey indicates the lowest lying areas, and darkest grey the highest in elevation.

3.3 ADAPTATION TO COASTAL IMPACTS

The responsibility for successful mitigation of and adaptation to the consequences of sea level rise rests largely with governmental authorities. The institutional arrangements made by these authorities vary widely even within Australia. The state government of New South Wales has a vigorous and coordinated approach to the problem of urban flooding; the government of Queensland has no coordinated strategy at the state level, and instead devolves the responsibility to the local government authorities (Smith, 1995). Thus in Queensland, which is frequently affected by cyclones, the potential for successful adaptation varies widely among locations. The possibility of damage by storm surge in Queensland in the absence of comprehensive planning has been described as "alarming" (Smith, 1995). One approach to adaptation requires floodplain studies to determine areas at risk, and coordinated policies at the state level (including financial and technical assistance to local governments) for floodplain and surge planning and management. Risk from storm surges and other coastal inundation events is a product of the physical event, the local topography, and the exposure measured in terms of the detailed location and magnitude of population and infrastructure. Planning, zoning regulations, and warning and evacuation procedures can all be used to minimize the risk.

4. Conclusion

The effect of the enhanced greenhouse effect on the climatology of tropical cyclones is currently unknown. Current climate models lack the resolution to simulate tropical cyclones accurately, but the development of fine resolution global and limited area climate models holds the promise of increased knowledge of future tropical cyclone behavior. Simulations have shown that even with no change in tropical cyclone climatology, the predicted rise in mean sea level will cause significant increases in coastal inundation associated with storm surges. The magnitude of such increases depends heavily on the specific characteristics of the site, including both meteorology and coastal geomorphology. This requires site-specific inundation modeling studies. The effectiveness of adaptation to the consequences of sea level rise depends on comprehensive planning at all levels of government.

Acknowledgments

We would like to thank the Climate Modeling Group of the CSIRO Division of Atmospheric Research for developing the CSIRO climate models. We thank Dr. Graeme Hubbert, who developed the storm surge/inundation model. We acknowledge the generous support of the Victorian Environmental Protection Agency, Melbourne Water, the governments of Queensland, the Northern Territory, and Western Australia, the federal Department of the Environment, Sport and Territories, and CSIRO.

References

Anthes R.A.: 1982, Tropical Cyclones: Their Evolution, Structure, and Effects, *Amer. Meteor. Soc. Meteorological Monographs* No. 41, Boston, Massachusetts.

Bengtsson L., Botzet M., and Esch M.: 1994, Will Greenhouse Gas-Induced Warming over the next 50 Years Lead to Higher Frequency and Greater Intensity of Hurricanes? Max-Planck-Institut für Meteorologie Report No. 139.

Bengtsson L., Botzet M., and Esch M.: 1995, *Tellus* **47**A, 175.

Broccoli A.J., Manabe S., Mitchell J.F.B., and Bengtsson L.: 1995, *Bull. Amer. Meteor. Soc.* **76**, 2,243.

DeMaria M. and Kaplan J.: 1994, *J. Climate* **7**, 1,324.

Emanuel K.A.: 1988, *J. Atmos. Sci.* **45**, 1,141.

Emanuel K.A.: 1991, *Annu. Rev. Fluid. Mech.* **23**, 179.

Emanuel K.A.: 1995, *Bull. Amer. Meteor. Soc.* **76**, 2,241.

England M.H.: 1995, *Geophys. Res. Letters* **22**, 3,051.

Evans J.L.: 1993, *J. Climate* **6**, 1,133.

Giorgi F., Shields Brodeur C., and Bates G.T.: 1994, *J. Climate* **7**, 375.

Gray W.M.: 1975, *Tropical Cyclone Genesis*, Dept. of Atm. Sci. Paper No. 234, Colorado State University, Ft. Collins, Colorado.

Henderson-Sellers A., Holland G.J., McGuffie K., Tonkin H., and Li S.: 1995, Implications for anthropogenic climate change for tropical cyclone intensity, in: *Proceedings of 21st Conference on Hurricanes and Tropical Meteorology*, April 24-28, American Meteorological Society, pp. 354-356.

Holland G.J.: In press, *J. Atmos. Sci.*

Hubbert G.D., Leslie L.M., and Manton M.J.: 1990, *Q. J. R. Meteor. Soc.* **116**, 1,005.

Knutson T.R. and Manabe S.: 1994, *Geophys. Res. Letters* **21**, 2,295.

Lighthill J., Holland G., Gray W., Landsea C., Craig G., Evans J., Kurihara Y., and Guard C.: 1994, *Bull. Amer. Meteor. Soc.* **75**, 2,147.

Love G.: 1988, Cyclone storm surges post greenhouse, in: *Greenhouse: Planning for Climate Change,* Pearman G.I. (ed), Brill E.J. Leiden, The Netherlands, pp. 202-215.

Mayfield M., Avila L., and Rappaport E.N.: 1994, *Mon. Wea. Rev.* **122**, 517.

McGregor J.L. and Walsh K.: 1994, *J. Geophys. Res.* **99**, 20,889.

McInnes K.L. and Hubbert G.D.: 1995, *Extreme Events and the Impact of Climate Change on Victoria's Coastline,* Victorian Environment Protection Authority, Melbourne, Victoria, Australia.

Meehl G.A., Branstator G.W., and Washington W.M.: 1993, *J. Climate* **6**, 42.

Nicholls N.: 1989, Global warming, tropical cyclones and ENSO, in: Proceedings of *Workshop on Responding to the Threat of Global Warming. Options for the Pacific and Asia,* Argonne National Laboratory, Argonne, Illinois, pp. 2.19-2.36.

Philander G.: 1990, *El Niño, La Niña and the Southern Oscillation,* Academic Press, New York.

Ryan B.F., Watterson I.G., and Evans J.L.: 1992, *Geophys. Res. Letters* **19**, 1,831.

Smith D.I.: 1995, The effectiveness of planning for urban flood and surge in Australia, in: *Proceedings of Workshop on Atmospheric Hazards: Process, Awareness and Response,* University of Queensland. St. Lucia, Queensland, Australia.

Tett S.: 1995, *J. Climate* **8**, 1,473.

Tokioka T., Noda A., Kitoh A., Nikaidou Y., Nakagawa S., Motoi T., Yukimoto S., and Takata K.: 1995, *J. Meteor. Soc. Japan* **74**(4), 817.

Walsh K. and McGregor J.L.: 1995, *J. Climate* **8**, 2,387.

Warrick R.A. and Oerlemans H.: 1990, Sea level rise, in: *Climate Change. The IPCC Scientific Assessment,* Houghton J.T., Jenkins G.J., and Ephramus J.J. (eds), Cambridge University Press, Cambridge, England.

Watterson I.G., Evans J.L., and Ryan B.F.: 1995, *J. Climate* **8**, 3,052.

Whetton P.H., Mullan A.B., and Pittock A.B.: In press, Climate change scenarios for Australia and New Zealand, in: *Proceedings of Greenhouse 94,* Pearman G.I. and Manning M. (eds), Wellington, New Zealand.

Wigley T.M.L.: 1995, *Geophys. Res. Letters* **22**, 45.

VULNERABILITY OF BANGLADESH TO CLIMATE CHANGE AND SEA LEVEL RISE THROUGH TROPICAL CYCLONES AND STORM SURGES

A. ALI

Bangladesh Space Research and Remote Sensing Organization (SPARRSO)
Agargaon, Sher-e-Bangla Nagar, Dhaka, Bangladesh

Abstract. Bangladesh is frequently visited by natural disasters such as tropical cyclones, storm surges, floods, droughts, tornadoes, and "norwesters." Of these, tropical cyclones originating in the Bay of Bengal and associated storm surges are the most disastrous. There are various reasons for the disastrous effects of cyclones and storm surges in Bangladesh. Superimposed on these disastrous effects, climate change and any consequent sea level rise are likely to add fuel to the fire. A rise in temperature is likely to change cyclone activity: cyclone intensity, if not cyclone frequency, may increase. As a result, storm surges may also increase substantially. Sea level rise, an increase in cyclone intensity, and consequent increases in storm surge heights will have disastrous effects on a deltaic country like Bangladesh, which is not much above the mean sea level. This paper examines the climatology of cyclones in the Bay of Bengal for the last 110 years and trends in cyclone frequency and intensity. The phenomenon of storm surges in the Bay of Bengal is examined along with the primary reasons for the severity of storm surges in Bangladesh. The paper discusses both qualitatively and quantitatively the impacts of rises in temperature on tropical cyclone intensity in Bangladesh. With the use of a mathematical model developed for the simulation of storm surges along the Bangladesh coast, various scenarios of storm surges are developed. Using lower and upper bounds of sea surface temperature rise of 2 and 4°C and of sea level rise of 0.3 and 1.0 m (according to the Intergovernmental Panel on Climate Change standard), the model simulates the maximum possible surges that are likely to occur under these conditions.

Key words: Bangladesh, coastal resources, storm surges, tropical cyclones

1. Introduction

Bangladesh is situated at the interface of two quite different settings. To the north of the country lie the Himalayas and the Khasi-Jaintia hills, and to the south are the Bay of Bengal and the Indian Ocean. Both settings control, modify, and regulate the climate of the country and the region. Embedded in any effects of global climate changes are local perturbations due to the Himalayan range and the bay.

The geographical location of Bangladesh and its geomorphic conditions have made the country easily vulnerable to natural disasters such as tropical cyclones and accompanying storm surges, floods, tornadoes, norwesters, droughts, and river bank and coastal erosion. The havoc caused by the disastrous cyclones of 1970 and 1991, when about 500,000 and 138,000 people, respectively, perished, is still vivid in the memory of the people (Ali, 1980; Choudhury, 1989; Haider *et al.*, 1991; Chowdhury, 1995). There were also disastrous floods in 1984, 1987, and 1988 (Miah, 1988; Choudhury, 1989); in 1988, about two-thirds of the country was flooded. The country is often overwhelmed by these disasters, which cause significant loss of life and property. Being one of the most densely populated and poorest countries in the world, Bangladesh can hardly withstand such disasters.

Climate change and sea level rise are likely to worsen the effects of the naturally occurring disasters. This paper discusses the vulnerability of Bangladesh to climate change and sea level rise, concentrating on the effects on tropical cyclones and storm surges.

Water, Air, and Soil Pollution **92:** 171–179.
© 1996 *Kluwer Academic Publishers. Printed in the Netherlands.*

2. Temperature and Sea Level Rise

This increasing concentration of greenhouse gases in the atmosphere is causing an increase in the atmospheric temperature. One of the direct and major consequences of an increase in temperature is a rise in sea level. In the South Asian region, the current rise in sea level is reported to be about 1.0 mm per year (in India and Pakistan). Two estimates of potential future sea level rise for Bangladesh are 0.30-1.5 m and 0.3-0.5 m for 2050 (DOE, 1993). Whatever the magnitude, the rate of increase of sea level rise is not a linear function of time. Several other factors such as geological subsidence and sedimentation may influence this rate. The Bangladesh Task Force (1990) estimates that any sea level change would be 90% due to sea level rise and 10% due to subsidence. River sedimentation, though, is considered an influential counterfactor for Bangladesh. It has been thought that some 4,000-5,000 years ago, the northern boundary of the Bay of Bengal was farther north (at or near Rajmahal), and that the coast migrated southward through deposition carried by river systems such as the Ganges, the Brahmaputra, and the Meghna, one of the largest in the world, and by their tributaries (Fergusson, 1863). This has been contradicted by Chowdhury (1959), who said that the present northern boundary of the Bay of Bengal was farther south, that the present northern bay was dry land during the late Pleistocene, and that the region was submerged because of a rise in sea level.

Ali and Ahmad (1992) estimated potential coastal and inland inundation in Bangladesh due to different sea level rise scenarios. These are given in Table I. These estimates show that a rise of 1.0 and 1.5 m would inundate 10% and nearly 16% of the country, respectively.

TABLE I
Estimated inundation in Bangladesh under different sea level rise scenarios

Sea Level Rise (m)	Inundation (km^2)	% of Total Area
1.0	14,000	10.0
1.5	22,320	15.5

3. Tropical Cyclones

The Bay of Bengal is favorable to cyclogenesis. Gray (1968) estimated that 10% of the world's tropical cyclones form in this bay; more recently, however, Gray (1985) estimated that only 7% of tropical cyclones form in the North Indian Ocean, which includes both the Bay of Bengal and the Arabian Sea. Since cyclones form in the Bay of Bengal about five to six times more frequently than in the Arabian Sea, the bay's share of the world total would be about 6%. An analysis of all the cyclones that formed in the Bay of Bengal between 1881 and 1990 (110 years) shows that about 14% of them hit Bangladesh, and about 66% hit India. If the bay's share is 10% of the world cyclones, the percentage of the world total hitting Bangladesh is 1.4%, and the percentage hitting India is about 6.6%. This percentage is lower if the above-mentioned 6% for the bay is considered.

In terms of death tolls due to tropical cyclones alone, Bangladesh suffers the most. A look at tropical cyclone disasters in which death tolls exceeded 5,000 shows that 15 out of 21 major cyclone disasters occurred in Bangladesh and India, and 9 of them occurred in Bangladesh.

About 49% of the world total of deaths due to these cyclones occurred in Bangladesh alone, and about 22% occurred in India (Frank and Husain, 1971; Ali, 1980).

If climate change and sea level rise affect global cyclone activity, Bangladesh may suffer even more. The two major aspects of cyclones that are most likely to be affected by climate change are cyclone frequency and cyclone intensity.

3.1 CYCLONE FREQUENCY

One of the necessary, but not sufficient, conditions for tropical cyclone formation is that the sea surface temperature should have a threshold value of 26-27°C. This temperature dependence has led to the expectation that any rise in sea surface temperature will be accompanied by an increase in cyclone frequency.

The relationship between sea surface temperature and cyclone formation has been well established: almost all tropical cyclones form in warm water. The area just west of Central America has the highest frequency of tropical cyclone genesis per unit area in the world (Frank, 1985). This region has an average sea surface temperature above 29°C. The highest number of tropical cyclones (33% of the world total) form in the Western North Pacific, which is an extremely large area of very warm waters, some above 30°C (Frank, 1985). Positive correlations between North Atlantic cyclone activity and sea surface temperature immediately west of Africa have been found, as well as correlations between cyclone occurrence and sea surface temperature in the oceans surrounding Australia (McBride, 1995).

Gray's (1985) analysis of tropical cyclones forming in different ocean basins showed no long-term trends in cyclone frequency for 1958-1984. The author of this paper also analyzed the cyclone frequency in the Bay of Bengal for 1881-1990. Ten-year plots of cyclones were made, and one plot was made for all types of cyclones — depressions, cyclonic storms, and severe cyclonic storms. Depressions are cyclones with maximum wind speed less than 34 knots, cyclonic storms have wind speeds between 34 and 47 knots, and severe cyclonic storms have wind speeds over 47 knots. No increasing or decreasing tendency was observed in cyclone numbers between 1881 and 1990. Another plot was made for just cyclonic storms and severe cyclonic storms, and there seems to be a period of oscillation of some 30-40 years for these storms. The reasons for such an oscillatory period are yet unknown. The Bay of Bengal is currently going through a decreasing trend in number of storms. The decreasing trend started in about 1970. One may crudely infer from the trend that from about the beginning of the next century, the number of storms will start increasing. A more active cyclonic period may be expected in the coming decades; however, it cannot be said whether this will have any correlation with global warming.

The monthly distribution of cyclone frequency in the Bay of Bengal shows a maxima in about August. This maxima is dominated by the depressions that form in the north Bay of Bengal during the southwest monsoon. During the southwest monsoon, strong winds blow over the north Indian Ocean from the southwest direction toward land in the north, and bring huge amounts of moisture that turn into heavy precipitation. The southwest monsoon period extends from June to September. The cyclonic storms and severe cyclonic storms show two maxima, one in April-May and the other in October-November, i.e., during the pre-monsoon (before the southwest monsoon) and post-monsoon periods, which correspond to two sea surface temperature maxima (Chowdhury, 1995). This shows that the monthly distribution of cyclones in the Bay is strongly related to sea surface temperature.

Most of the storms that hit Bangladesh originate in the south Bay of Bengal. Sometimes they are the remnants of typhoons in the Pacific. These storms generally move in their initial stages to the north/northwest, and then curve northeastward toward Bangladesh. The reasons for this recurvature are not clear. One may be the influence of steering currents that prevail in the region. It is also likely that this recurvature has something to do with the warm western boundary current in the Bay of Bengal. The warm western boundary current is a warm zone of water that has a characteristic sea surface temperature of greater than 25°C; it moves northward along the west coast of India, is deflected toward the east approximately along 19°N, and becomes diffuse. This current has been observed on satellite imagery by Huh *et al.* (1985) and Legeckis (1987). Johns and Ali (1991) simulated the warm western boundary current with a numerical model. However, the influence (if any) of the current in channeling the cyclone track toward Bangladesh needs to be investigated. If the current has any influence on the cyclone track, then an increase in sea surface temperature might affect the current and hence affect the track of cyclones in the region. An increase in sea surface temperature may further extend the excursion of the warm western boundary current northward toward Bangladesh because more warm waters will then be available in the north for its extension. Such a situation may increase the likelihood that Bangladesh will be hit.

3.2 CYCLONE INTENSITY

Although it is not clear whether global warming and sea level rise will have any effect on cyclone frequency, there are speculations that cyclone intensity might be affected. Miller (1958) showed that cyclone intensity increases with an increase in sea surface temperature. Merrill (1985) diagrammed the relationship between sea surface temperature and the maximum intensity of North Atlantic cyclones. Emanuel (1987) showed the relationship between sea surface temperature, minimum sustainable central pressure, and maximum wind speed in a cyclone. This relationship, presented in Table II shows that a normally occurring cyclone may have a higher intensity as a result of warmer seas due to global warming. Table II also gives the ratio of maximum wind speed (V_m) at different temperatures to maximum wind speed at 27°C (V_{27}).

Assuming a lower bound of sea surface temperature rise of 2°C and an upper bound of 4°C (according to the IPCC limits), the corresponding increases in maximum cyclone intensity would be 10 and 22%, respectively, using the threshold temperature of 27°C. The maximum wind speed of the April 1991 cyclone that hit Bangladesh, killing about 138,000 people, was

TABLE II
Relationship of maximum wind speed (V_m) in cyclones to sea surface temperature

Sea Surface Temperature (°C)	V_m (m s^{-1})	V_m/V_{27} [§]
27	72	1.00
28	75	1.04
29	79	1.10
30	83	1.15
31	88	1.22
32	93	1.29
33	99	1.38
34	106	1.47

[§] V_m/V_{27} is the ratio of maximum wind speed at different temperatures to maximum wind speed at 27°C.
Source: Emanuel, 1987.

about 225 km h^{-1}. If this cyclone had occurred with sea surface temperatures that were 2 and 4°C higher, the wind speeds would have been about 248 and 275 km h^{-1}, respectively. The coastal inundation caused by such increases is discussed in the section on storm surges.

It should be pointed out that there is significant uncertainty concerning the effect of change in sea surface temperature on cyclone wind speed, partly because sea surface temperature is not the lone factor affecting wind speed. The results given in Table II should not be taken as predictive but rather indicative. They are, however, useful to examine the sensitivity of storm surges to changes in cyclone intensity.

4. Storm Surges

Most of the casualties from cyclones in Bangladesh, as in other parts of the world, are caused by storm surges. Storm surges are generated by cyclonic winds and the atmospheric pressure drop associated with a cyclone. The major contribution comes from the winds, which exert a stress on water proportional to the square of wind speed. Consequently, any increase in wind speed (cyclone intensity) due to climate change will increase the height of storm surges, leading to greater coastal flooding.

A mathematical model of storm surges for the Bay of Bengal has been used to generate storm surge scenarios for Bangladesh under increased cyclone intensity and sea level rise.

4.1 MODEL GENERATION OF STORM SURGE SCENARIOS

A hydrodynamic numerical model, described by Ali and Hoque (1994), was used to generate storm surge scenarios for Bangladesh under potential sea level rise induced by climate change. The model equations are:

$$\frac{\delta h}{\delta t} + \frac{\delta(u,d)}{\delta x} + \frac{\delta(v,d)}{\delta y} = 0 \tag{1}$$

$$\frac{\delta u}{\delta t} - fv = g\frac{\delta h}{\delta x} + \frac{1}{\rho(h+d)}(F_{sx} - G_{bx}) \tag{2}$$

$$\frac{\delta v}{\delta t} - fu = g\frac{\delta h}{\delta y} + \frac{1}{\rho(h+d)}(F_{sy} - G_{by}). \tag{3}$$

Equation (1) is the continuity equation, and equations (2) and (3) are the horizonal components of the momentum equation. In these equations, x and y denote the east-west and north-south Cartesian coordinates, respectively; u and v are the depth-averaged components of water velocity in the x and y directions, respectively; g is the acceleration due to gravity; ρ is the density of water; f is the Coriolis parameter; F_{sx} and F_{sy} represent surface wind stresses in the x and y directions, respectively; G_{bx} and G_{by} represent bottom stresses in the x and y directions, respectively; h is the height of water (storm surge or tide) above the mean sea level; d is the depth of water below mean sea level; and t is time.

The model area was the Bay of Bengal area north of 20°N. The equations were solved numerically using a finite-difference scheme. The grid spacings were 16 and 15 km in the east-west and north-south directions, respectively. The stress terms were formulated using quadratic stress laws. For simplicity, the details of the model are not given here.

A lower bound of sea surface temperature rise of 2°C and an upper bound of 4°C were assumed. Similarly, the lower and upper bounds of sea level rise of 0.3 and 1.0 m were used. The April 1991 cyclone that hit Bangladesh was used as the model case. This is a typical case. The results of the model calculation are given in Table III for a grid point corresponding to a place near Chittagong. Chittagong is the largest seaport of Bangladesh, and the area in and around Chittagong is one of the areas most vulnerable to cyclonic casualties. As shown in Table II, a 2°C and 4°C rise in sea surface temperature will increase cyclone wind speed by 10 and 22%, respectively, from maximum wind speeds at the threshold temperature of 27°C. Three wind speeds were used as the forcing function: 225 km h^{-1} (corresponding to current temperature), 248 km h^{-1} (corresponding to 2°C rise), and 275 km h^{-1} (corresponding to 4°C rise). Surge heights under different conditions are given in the last three rows in Table III.

It can be seen that rises in sea level tend to reduce storm surges if wind speed remains constant. On the other hand, surge height increases with increased wind speed. It should be mentioned here that the model had a fixed boundary and, as a result, could not simulate the situation pertaining to a moving shoreline. This is the limitation of the model.

The percentage increase or decrease in storm surge heights from the present surge height of 7.6 m is given in parentheses. The maximum increase of about 49% occurs with present sea level and wind speed corresponding to a 4°C rise in temperature.

4.2 INLAND FLOODING BY SURGE WATERS

Storm surges that hit the coast of Bangladesh move inland, leading to flooding. These waters moving over land cause most of the disasters in the country. Surges traveling along rivers propagate for long distances. This has been demonstrated by numerical modeling for Bangladesh (Johns and Ali, 1980). The movement of surge waters over coastal land is very complex. It is extremely difficult to estimate the distance the surge water will move inland. The movement is governed by a number of factors. Numerical hydrodynamic models developed for a limited number of coasts (Flather and Heaps, 1975) to simulate the penetration of storm surges

TABLE III

Storm surge heights (m) under different sea surface temperatures and sea level rise scenarios (wind speed of 225 km h^{-1} corresponds to that of the April 1991 cyclone)

	Current Temp. (27°C)	2°C Increase	4°C Increase
Wind (km h^{-1})	225	248	275
No Sea Level Rise Surge Height (% change)	7.6 (0%)	9.2 (21%)	11.3 (49%)
Sea Level Rise = 0.3 m Surge Height (% change)	7.4 (-3%)	9.1 (20%)	11.1 (46%)
Sea Level Rise = 1.0 m Surge Height (% change)	7.1 (-7%)	8.6 (13%)	10.6 (40%)

TABLE IV
Intrusion distance of surge water inland under different surge heights and friction

Surge Height (m)	Intrusion Distance (km)					
	c[§] = 0.01	c = 0.02	c = 0.03	c = 0.04	c = 0.05	c = 0.10
7.6	15.42	9.91	7.30	5.78	4.78	2.57
9.2	17.48	11.24	8.28	6.56	5.43	2.91
11.3	20.21	12.99	9.57	7.58	6.27	3.37

[§] c = friction factor.

over dry land have been used to examine the movement of the shoreline with the rise and fall of surge heights.

In the absence of any such model for the Bay of Bengal, the Multipurpose Cyclone Shelter Programme (MCSP, 1992) produced an empirical formula following that given by Freeman and Mehaute (1964). The formula used was:

$$L = \frac{4(d + 1.5h)^2}{3(d + h)(s + c/8)}, \tag{4}$$

where L is the maximum distance traveled by the surges, d is the depth of water approximately at the point where h is determined, s is the bed slope, and c is the friction factor. In line with the argument given by MCSP (1992), d is taken as 10 m. This formula has been shown to give a reasonable estimate of inland travel of water for the April 1991 cyclone.

Equation (4) is used here to calculate the values of L for generating scenarios under different climate conditions. Table IV presents the results for present sea surface temperature and 2°C and 4°C rises of temperature (corresponding to wind speeds of 225, 248, and 275 km h^{-1} and thus storm surges of 7.6, 9.2, and 11.3 m, as shown in Table III). A value of s = 0.001 has been assumed, as per MCSP (1992). This value gives a reasonable simulation of inland surge penetration along the flat land around the Meghna river estuary.

Table IV gives the maximum surge intrusion values for different friction factors (c) as given in MCSP (1992). The increase of surge height from 7.6 to 9.2 m increases the distance of inland inundation by about 13%, and an increase in surge height from 7.6 to 11.3 m increases inundation by about 31%. These results are only indicative because of several assumptions and simplifications used in their derivation, detailed discussions about which are given in MCSP (1992).

5. Adaptation

Although efforts are being made worldwide to minimize or stop greenhouse emissions and arrest global warming, we should take all precautionary measures to safeguard ourselves against any impending dangers likely to arise from climate change and sea level rise. As far as cyclones and the accompanying storm surges are concerned, we cannot stop them, even if we stop greenhouse emissions. These natural disasters have been occurring for centuries; only their ferocity might increase with the climate change and sea level rise. Since we cannot stop them, we should learn how to better live with them. We discuss here a few important adaptation strategies with respect to Bangladesh.

There are three kinds of adaptive measures: retreat, accommodation, and protection. Considering the high population density, future population projections, and the area of the country, retreat is not a feasible option for Bangladesh. Therefore, Bangladesh should pursue the other two options. Two protection options — construction of embankments and creation of a coastal green belt — are discussed below.

As discussed in MCSP (1992) and as seen in Table IV, intrusion distance of surge water is sensitive to resistance (with small slope). One of the major sources of surface resistance is forests. The western coastal region of Bangladesh has a large mangrove forest. It has been found that a cyclone hitting this coastal area causes less disasters than one hitting nonforested areas like the Meghna estuary and the surrounding region. Thus surface resistance can be increased by afforestation in the coastal area. In fact, massive afforestation programs are going on in Bangladesh to create a green belt along the coast. The afforestation will also help stabilize land and help increase accretion process in the coastal area. A huge amount of sediment (about 2.6×10^9 t per year) is carried by the rivers in Bangladesh into the bay, and it is lost to the bay and the Indian Ocean. If these sediments could somehow be trapped, it would increase the land area and also raise the topographic height, which would help circumvent the effect of sea level rise.

Another important protection option is the construction of embankments and polders. Though these may have other adverse effects, they would substantially dissipate the energy associated with storm surges. The existing embankments are not sufficient for storm surge protection, because they were designed for protection against salinity intrusion due to normal tides. The design of the embankments should consider these environmental aspects.

Protective measures are expensive for Bangladesh. Nonetheless, they should be pursued within the limited resources available in the country. Assistance from and experience of the technologically advanced countries may also be sought for undertaking protective measures.

Accommodation of changing situations has been one of the traditional methods adopted by the people of Bangladesh. These kinds of adaptation processes have been quite extensively used in cases of natural disasters such as cyclones, storm surges, floods, droughts, and erosion, which are important components of climate. More research and development should be carried out on conventional adaptation processes by involving people at the grassroots level.

The Integrated Coastal Zone Management program is considered to be one of the most effective adaptive measures against the possible impacts of climate changes. In implementing such programs, a marriage of modern technology and the people's adaptive measures is considered very effective. This will improve implementation of adaptation policies.

6. Conclusions

A brief examination of the frequency of tropical cyclones that formed in the Bay of Bengal during 1881-1990 was made. A 30-40 year period of oscillation was observed in the cyclone frequency, but no trend as far as the relationship to global warming was observed. Using the temperature and wind speed relation given by Emanuel (1987), the likely wind speeds of the April 1991 cyclone that hit Bangladesh under two sea surface temperature scenarios of 2 and 4 °C have been estimated. These wind speeds were used to generate storm surge scenarios in Bangladesh for a sea surface temperature rise of 2 and 4 °C, and sea level rise of 30 and 100 cm. Storm surge scenarios were developed using a numerical model. The inland penetration of storm surges under the present and above-mentioned sea surface temperatures were also estimated.

The results, however, should not be treated as conclusive but indicative, because there are many uncertainties in their derivation. For example, there is significant uncertainty concerning the effect of sea surface temperature on cyclone wind speed, which has been pointed out earlier.

Nevertheless, the results indicate the gravity of the situation likely to occur in Bangladesh in the event of climate change and sea level rise. Finally, some adaptation options have been suggested. However, more detailed and in-depth research on vulnerability and adaptation processes needs to be done so that better adaptation and protective measures may be undertaken against possible climate change and sea level rise.

References

Ali A.: 1980, Storm Surges in the Bay of Bengal and Their Numerical Modeling, SARC Report No. 125/80, Atomic Energy Commission, Dhaka, Bangladesh.

Ali A. and Ahmad A.A.Z.: 1992, Impact of Sea Level Rise on Other Disasters in Bangladesh, presented at an IOC/UNEP Workshop on Impact of Sea Level Rise Due to Global Warming for the South Asian Region, 16-19 December, Dhaka, Bangladesh.

Ali A. and Hoque M.A.: 1994, *The Journal of NOAMI*, Bangladesh **II(1)**, 27.

Bangladesh Task Force: 1990, *Report of the Task Force on Bangladesh Development Strategies for 1990s*, Vol. 4, University Press, Dhaka, Bangladesh.

Choudhury A.M.: 1989, Forecasting and Warning System of Disaster, a SPARRSO (Bangladesh Space Research and Remote Sensing Organization) Report, Dhaka, Bangladesh.

Chowdhury M.H.K.: 1995, Critical Assessment of Tropical Cyclones in the Bay of Bengal: Bangladesh Perspective, presented at the Workshop on Global Change and Tropical Cyclones, 18-21 December, Dhaka, Bangladesh.

Chowdhury M.I.: 1959, The Morphological Analysis of the Bengal Basin, MSc thesis, University of Cambridge.

DOE: 1993, Assessment of the Vulnerability of Coastal Areas to Sea Level Rise and Other Effects of Global Climate Change, Pilot Study Bangladesh, report prepared by Department of Environment, Govt. of Bangladesh, Dhaka.

Emanuel K.A.: 1987, *Nature* **326**, 483.

Fergusson J.: 1863, *J. Geol. Soc.* **XIX/I**, 321.

Flather R.A. and Heaps N.S.: 1975, *Geophys. J.R. Astr. Soc.* **42**, 489.

Frank N.L. and Husain S.A.: 1971: *Bull. Am. Meteor. Soc.* **52**, 438.

Frank W.M.: 1985, Tropical cyclone formation, in: *A Global View of Tropical Cyclones*, R.L. Elsberry (ed), based on the International Workshop on Tropical Cyclones, 25 November-5 December, Bangkok, Thailand.

Freeman J.C. and Mehaute B.Le.: 1964, *ASCE J. of the Hydraulic Division* **90**, 187.

Gray W.M.: 1968, *Mon. Wea. Rev.* **96**, 669.

Gray W.M.: 1985, Tropical Cyclone Global Climatology, WMO Technical Document WMO/TD No. 72, Vol. I, WMO, Geneva, Switzerland, pp. 3-19.

Haider R., Rahman A.A., and Huq S.: 1991, *Cyclone '91, an Environmental and Perceptional Study*, Bangladesh Centre for Advanced Studies, Dhaka.

Huh O.R., Ali A., and Quadir D.A.: 1985, Observations on the Surface Features in the Bay of Bengal with NOAA Satellite AVHRR Imagery, Technical Report, Baton Rouge, Coastal Studies Institute, Louisiana State University.

Johns B. and Ali A.: 1980, *Quart. J. R. Met. Soc.* **106**, 1.

Johns B. and Ali A.: 1991, *J. Marine Systems* **3**, 267.

Legeckis R.: 1987, *J. Geophys. Res.* **92**, 12974.

McBride J.L.: 1995, Tropical cyclone formation, in: *Global Perspectives on Tropical Cyclone*, WMO Technical Document WMO/TD-No. 693, Report No. TCP-38, pp. 63-105 (reprinted 1996).

MCSP: 1992, Storm Surge Analysis, Vol. IV, Draft Final Report, Multipurpose Cyclone Shelter Programme, UNDP/World Bank/GOB Project BGD/91/025, Dhaka, Bangladesh.

Merrill R.T.: 1985, Environmental Influences on Hurricane Intensification, Dept. of Atmos. Sci. Paper No. 394, Colorado State University, Fort Collins.

Miah M.M.: 1988, *Flood in Bangladesh,* Academic Publishers, Dhaka, Bangladesh.

Miller B.I.: 1958, *J. Meteor.* **15**, 184.

VULNERABILITY AND ADAPTATION OF HONG KONG TO HAZARDS UNDER CLIMATIC CHANGE CONDITIONS

WYSS W.-S. YIM
Department of Earth Sciences, The University of Hong Kong
Pokfulam Road, Hong Kong

Abstract. Given the evidence from past experience, the coastal city of Hong Kong would be vulnerable to a variety of hazards that could be exacerbated by climatic change, including potential increases in typhoons, landslides, floods (both storm surges and rainstorms), and droughts, as well as the threat of future sea level rise. The greatest death tolls in the past have been inflicted during typhoon-induced storm surges. During two unnamed typhoons, in 1906 and 1937, the death tolls were 10,000 and 11,000, respectively. In contrast, death tolls associated with landslides and other hazards were relatively small. Under a climatic change scenario of an increase in frequency and severity of typhoons and rainstorms, and rising sea level, the vulnerable areas of Hong Kong can be identified. The shift in development from hillslope areas to coastal land reclamations over the last 25 years is a matter of concern. Although the hillslopes are susceptible to landslides during rainstorms, virtually all landslides can be traced to some important anthropogenic causes and are preventable with appropriate measures. In the coming decades, the low-lying areas created through coastal land reclamations could be susceptible to flooding during storm surges and rainstorms and as the sea level rises. These areas are also where there is high density urban and industrial development, and the vulnerability is the greatest. High seawalls to protect the reclaimed areas from storm surge flooding and sea level rise may increase the risk of rainstorm flooding. To reduce vulnerability, a combination of better monitoring, planning, protection, maintenance, and warning is proposed for adaptation. Monitoring is needed not only to detect future sea level changes but also to identify areas with the greatest risk.

Key words: Hong Kong, coastal resources, sea level rise, floods, landslides, storm surges, typhoons

1. Introduction

There has been a marked global trend of rural depopulation through migration to coastal urban centers in the twentieth century. This has led to a growing number of coastal megacities, which are defined as having populations exceeding five million, throughout the world. Because of the high population densities resulting from rapid population growth, and local circumstances such as overcrowding and poverty, these megacities are particularly vulnerable to a variety of weather-induced hazards, including potential increases in typhoons, landslides, floods (both storm surges and rainstorms), and droughts. These hazards may increase in intensity or frequency under climatic change conditions. It is therefore important to assess the vulnerability and adaptation of these megacities to the effects of climatic change.

In this study, Hong Kong, a coastal megacity in subtropical southern China, was selected for a vulnerability and adaptation assessment. The objectives of the study are:

- to examine the instrumental temperature and precipitation records since 1884 to identify possible future trends
- to examine the frequency of typhoons since 1946 to identify possible future trends
- to review the evidence for future sea level changes
- to learn from the historical record of damage caused by typhoons, landslides, floods, and droughts
- based on the above, to give recommendations for adaptation to reduce vulnerability to climatic change induced hazards.

Water, Air, and Soil Pollution **92:** 181–190.
© 1996 *Kluwer Academic Publishers. Printed in the Netherlands.*

The present approach differs from studies based on future climatic scenarios (i.e., general circulation model scenarios) in that damage caused by variability of weather in the past is used to provide a better indication of future vulnerability.

2. History and Background

Hong Kong lies immediately south of the Tropic of Cancer on the northern coastline of the South China Sea. As a megacity, Hong Kong has a surprisingly short history. At the end of the nineteenth century, the population was only about 0.25 million compared to today's estimated 6.5 million. The current climate is subtropical with a prominent monsoonal influence. The wet southwest monsoon in the summer and autumn is associated with typhoons, and the northeast monsoon in winter and spring is usually appreciably drier.

The Royal Observatory's Nathan Road station (Figure 1), which was established in Hong Kong in 1883, is responsible for monitoring all meteorological parameters, as well as sea levels. Temperature and rainfall records are available for the Nathan Road station since 1884, with the exception of a seven-year gap that includes the Second World War. Tide gauge stations in operation in 1990 are also depicted in Figure 1. However, the earliest records of tide gauge measurements date back to only 1954.

3. Methods

The methodology adopted is aimed at assessing the vulnerability and adaptation of Hong Kong to weather-induced hazards under climatic change conditions based mainly on local evidence and past experience. This is carried out initially by an examination of the instrumental temperature and rainfall records of the Nathan Road station from 1884 to 1990. The number of typhoons within the area of 0°N-45°N and 100°E-160°E from 1946 to 1990 is then analyzed (pre-1946 records are less reliable). A review is then made of past hazards, including typhoons, landslides, sea-level rise, floods, and droughts to provide some insight on the extreme events that are likely to be increasingly common in the event of climatic change. Finally, the significance of the results is examined for the recommendation of policy decisions.

3.1 TEMPERATURE RECORD

At the Royal Observatory's Nathan Road station (Figure 1), the minimum, mean, and maximum temperatures have risen by 2.79, 0.78, and 0.93°C, respectively, for the 1891-1906 and 1970-1985 periods (Koo, personal communication). The rate of minimum temperature rise before the Second World War was 0.010°C per year; for the post-war years, the rate was 0.025°C per year (Koo, 1990). However,when compared to the Macau station, located about 40 km west of Hong Kong, only a weak upward trend can be observed. Only the minimum and maximum temperatures have risen, by 0.28 and 0.02°C, respectively, for 1906-1916 and 1971-1985, and the mean temperature fell 0.23°C between the two periods. This difference can be accounted for by the topographical difference between the Hong Kong and Macau stations. The Nathan Road station is located on a small hill surrounded by mountains within the Victoria Harbour basin in Kowloon Peninsula. Because the station is surrounded by higher ground and there is a concentration of developed areas within the basin, heat dissipation by wind is

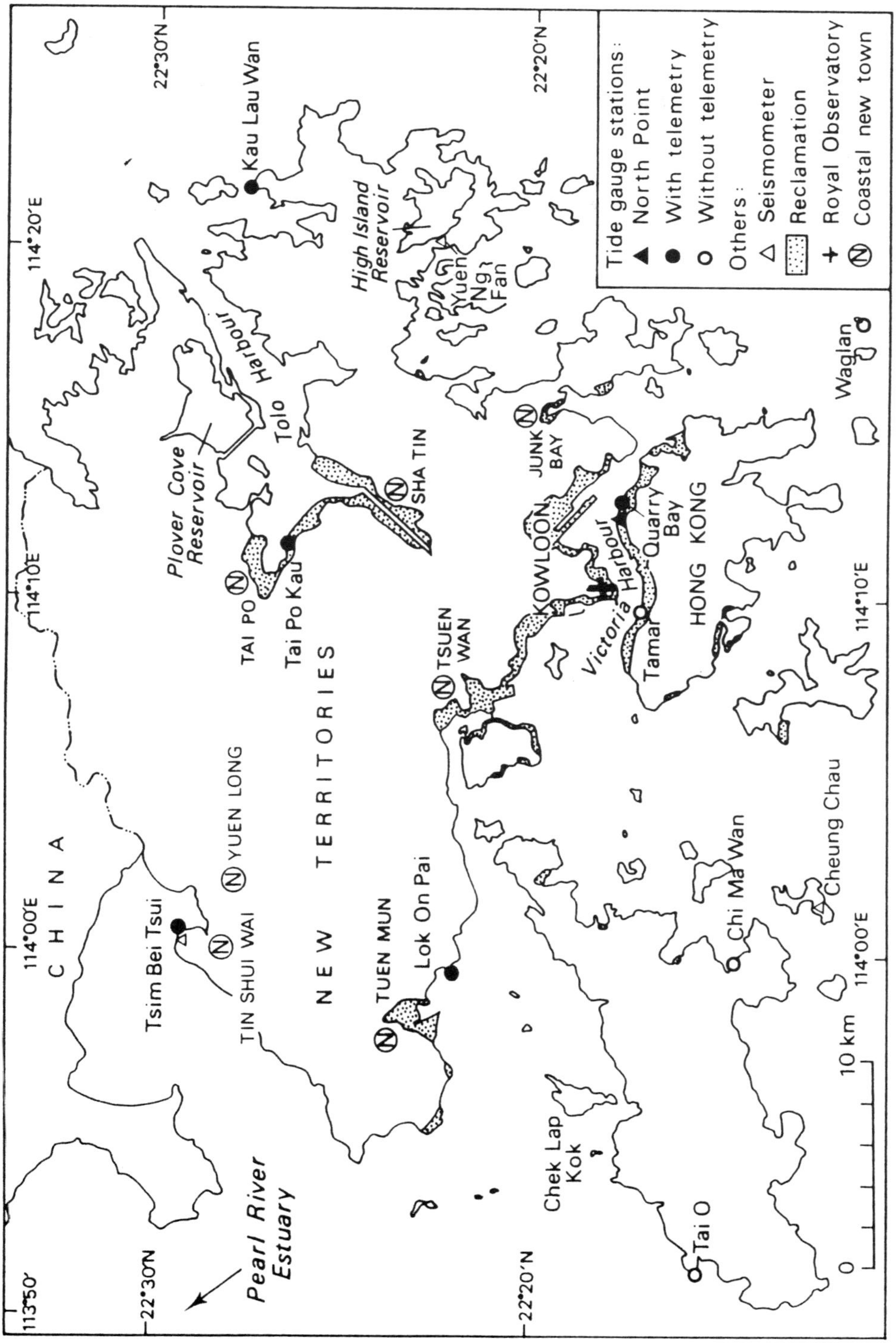

Fig. 1. Map of Hong Kong showing the coastal land reclamation areas, coastal new towns, tide gauge stations in 1990, and the Royal Observatory's Nathan Road station.

inefficient. The Macau station is located at the top of a small hill, but is not surrounded by higher ground. Heat dissipation is therefore more efficient. Consequently, Victoria Harbour has a stronger urban-heat island effect. This conclusion is supported by Koo (1990), who found that the plot of the three temperatures for Waglan Island, a small rural oceanic station southeast of Hong Kong (Figure 2), failed to show any rising trend. Based on this regional experience, the global warming of 0.5-0.7 °C in the past century found by Hansen and Lebedeff (1987) is at least in part attributable to the urban heat-island effect since the majority of temperature gauging stations with the longest record are located within cities.

3.2 RAINFALL RECORD

There have been five main periods when precipitation consistently exceeded the annual mean value of 2,183 mm at the Royal Observatory's Nathan Road station from 1884 to 1990. They are 1884-1895, 1920-1928, 1958-1961, 1971-1979, and 1981-1987. It is also seen that the post-1946 period was a time of greater variability, including both the driest and wettest years, in 1963 and 1982, respectively.

Tables I and II show the 10 years with the highest and lowest precipitation, respectively, expressed as a percentage of average at the Nathan Road station between 1884 and 1990. Between the driest and wettest year, precipitation varied from a low of 41.3% of average and a high of 148.8% of average. The post-Second World War years included 7 of the 10 years with the highest precipitation, but only 4 of the 10 years with the lowest precipitation. This suggests a wetter post-war period with greater rainfall variability. A closer examination of the years with the highest annual precipitation supports a link with El Niño/Southern Oscillation (ENSO) episodes. For example, the strongest ENSO episode in the instrumental record is marked by very high precipitation in 1982 and 1983, with 1982 as the wettest year ever.

Tables III and IV show the 10 highest hourly and 24-hour precipitation values, respectively, recorded at the Nathan Road station between 1884 and 1990. Eight and six of the highest recorded values for hourly and 24-hour precipitation, respectively, are found during the post-war period. This is indicative of a period associated with greater rainfall intensity, possibly

TABLE I
Ten years with the highest precipitation expressed as a percentage of average at the Royal Observatory's Nathan Road station between 1884 and 1990

Year	Precipitation (mm)	% of Average
1885	2769.1	126.8
1889	3041.8	139.3
1891	2974.5	136.3
1957	2950.3	135.1
1959	2797.4	128.1
1972	2807.2	128.6
1973	3100.4	142.0
1975	3028.7	138.7
1982	3247.5	148.8
1983	2893.8	132.6

TABLE II
Ten years with the lowest precipitation expressed as a percentage of average at the Royal Observatory's Nathan Road station between 1884 and 1990

Year	Precipitation (mm)	% of Average
1895	1164.2	53.3
1898	1447.7	66.3
1901	1418.8	65.0
1912	1625.2	74.4
1933	1585.2	72.6
1938	1405.8	64.4
1954	1367.0	62.6
1956	1649.3	75.5
1963	901.1	41.3
1967	1570.6	71.9

TABLE III
Ten highest hourly precipitation recorded at the Royal Observatory's Nathan Road station between 1884 and 1990

Date	Precipitation (mm)
12 June 1966	108.2
2 May 1989	104.8
19 July 1926	100.7
13 June 1968	100.0
18 June 1972	98.7
30 April 1975	92.4
15 July 1986	88.4
30 May 1889	86.4
1 July 1947	85.0
22 September 1948	84.0

TABLE IV
Ten highest 24-hour precipitation recorded at the Royal Observatory's Nathan Road station between 1884 and 1990

Date	Precipitation (mm)
30 May 1889	697.1
19 July 1926	552.2
25 August 1976	416.2
31 October 1920	408.8
12 June 1966	401.2
29 May 1982	394.3
21 May 1989	387.8
17 October 1978	379.8
9 June 1960	368.4
15 July 1886	363.8

reflecting climatic change. Hazards increasingly common with this change in Hong Kong are therefore likely to include rainstorm flooding and landslides.

3.3 TYPHOON RECORD

The occurrences of typhoons within the area of 0°N-45°N and 100°E-160°E in the western Pacific from 1946 to 1990 were examined by Yim (1995). From 1960 to 1969, there is a peak in the five-year running mean of typhoon occurrence up to 38 typhoons per year, while the maximum number of typhoons was 45 in 1961. Since 1969, the five-year running mean did not exceed 31 typhoons per year. Therefore, there is no supporting evidence for an increase in frequency of typhoons based on the 45-year record examined. On the other hand, the lower frequency of typhoons since 1969 may be linked to the increase in frequency of ENSO episodes: these occurred in 1969-1970, 1972-1973, 1976-1977, and 1982-1983. Projecting this trend, Hong Kong appears to be less vulnerable to typhoon hazards, at least in terms of their frequency. However, it is important to note that the reduced frequency does not imply that the typhoons to hit Hong Kong would be less ferocious.

3.4 HISTORICAL RECORD OF THE IMPACTS FROM WEATHER-INDUCED HAZARDS

Because Hong Kong is already confronted with a variety of weather-induced hazards through extreme events, it is useful to learn from the historical record of damage to assist the present assessment. The hazards examined include typhoons, landslides, sea level rise, floods, and droughts.

In terms of death tolls, casualties, and damage to property, typhoons are the most severe by far. However, since typhoons are normally accompanied by storm surges and heavy rainfall, it may be necessary to distinguish typhoon-induced floods from those resulting solely from rainstorms in a manner similar to Peart and Yim (1992). The main damage caused by typhoons is from wind, storm surges, flooding, and landslides.

Wind damage is always associated with typhoons, but its contribution is often hidden in typhoon damage statistics. On average, between four and five typhoons with tracks falling within 800 km of Hong Kong and wind strength reaching 41 km h^{-1} affect Hong Kong each year (K.P. Wong, personal communication), all resulting in storm surges. Table V provides a summary of death tolls, casualties, and damage to ships and boats caused by selected typhoons in Hong Kong since 1906. The two worst disasters by far were the unnamed typhoons in 1906 and 1937, with estimated death tolls of 10,000 and 11,000, respectively. In relation to the population of Hong Kong at the time, the death tolls amounted to about 3% and 1% of the total population, respectively. During both events, the majority of deaths were attributable to drowning of boat dwellers because of the large number of boats capsizing, and the absence of a typhoon warning system, which is currently available through radio and television. As a consequence of these catastrophes, numerous typhoon shelters were built throughout Hong Kong to provide shelter for boats during typhoons.

The difference in maximum sea level at Tai Po Kau and North Point (Figure 1) during the unnamed typhoon in 1937 is, according to Yim (1993a), explainable by the difference in coastal configuration. Seiching, in the landlocked Tolo Harbour, is responsible for the higher maximum sea level at Tai Po Kau.

Flooding resulting from both storm surge and rainstorm related precipitation is not considered to be severe because of the temporary nature of those events. However, because of the steep hillslope gradients in Hong Kong, the flood peak from basin headwaters is often very short. Peart and Jayawardena (1993) carried out an analysis on the Shek Kong drainage basin in the New Territories, which is only 11.7 km^2. For 36 storms, the average time it took flood waters to peak was 2.82 h. Consequently, monitoring water levels would give at best only about two hours warning. The coastal land reclamations, which are naturally low lying and are covered by impermeable concrete, were identified by Peart and Yim (1992) to be prone to flooding. This is made worse by the "trough effect," where there is a trough created by the reclamations at the old coastline, which are located between the steep hillslopes, and the newer reclamations, which are reclaimed at a higher level. The trough areas are particularly prone to flooding because of surface runoff from both the hillslopes and the newer reclamations.

After typhoons, landslides are historically the second worst hazard in terms of damage. Table VI lists the locations and death tolls of the 10 landslides, including boulder falls, with the highest death tolls in Hong Kong since 1917. Table VII shows the 10 years with the highest death toll caused by landslides and boulder falls in Hong Kong since 1917. Landslides tend to

TABLE V

Summary of death toll, casualties, and damage of selected typhoons in Hong Kong since 1906. Based mainly on data from the Royal Observatory.

Tropical Cyclone	Year	Death Toll	Persons Injured or Missing	Oceangoing Vessels in Trouble	Small Craft Sunk or Damaged
Unnamed	1906	10,000§	Unavailable	Unavailable	Unavailable
Unnamed	1937	11,000§	Unavailable	28	1,855
Mary	1960	45	138	6	814
Wanda	1962	130	Unavailable	36	2,053
Ruby	1964	38	306	20	314
Dot	1964	26	95	2	90
Rose	1971	110	291	34	>303
Hope	1979	12	260	29	374
Ellen	1983	10	345	44	360

§ Estimated.

be associated with very heavy downpours during the wet season (Lumb, 1975), even though the damage caused by individual landslides would depend on other, nonweather variables, including the degree of human disturbance of the slope and geological conditions. Two of the three most disastrous landslides occurred in 1972, with death tolls of 71 and 67 at Sau Mau Ping and the Mid-levels, respectively. Although 1972 is the eighth wettest year on record, this conceals the fact that the Sau Mau Ping landslide was judged by an international panel of geotechnical experts to be caused by a combination of heavy rainfall and an inadequately compacted fill slope created by humans.

The analysis of data from tide gauge stations in Hong Kong by Yim (1991a, 1993a, 1993b) has failed to reveal a rising sea level trend. Even in the event of future sea level rise, some degree of protection is already available in the seawalls designed to protect the reclamations from storm surges (Yim, 1995). Consequently, storm surges are identified to be a greater problem than any long-term, gradual future sea level rise. However, because reliable data for long-term rates of ground settlement in the different coastal land reclamations are lacking, it is necessary to carry out further monitoring to detect relative sea level changes in respect to the level of the coastal land reclamations. Between 1954 and 1983, a total settlement of 16.3 cm, a rate of 5.4 mm per year, was indicated by check surveying of the North Point tide gauge located in Figure 1 (Yim, 1991b). The uncorrected and corrected five-year running mean of annual sea level at the longest running North Point tide gauge from 1954 to 1988 failed to reveal a rising sea level trend (Figure 2). Ground settlement is therefore potentially a much more serious problem.

Droughts may be defined as periods with precipitation of up to 75% of normal. Based on this definition, 9 of the 10 years with the lowest precipitation shown in Table II are drought years. In an analysis of droughts in Hong Kong by Peart and Jayawardena (in press), an 11-year moving average plot of precipitation revealed two dry spells between 1884 and 1990. The first occurred at the turn of the century and the second occurred in the 1930s. Droughts are more frequent in the six month dry season than in the wet summer season, and there is a high probability that a dry spell will have a duration of three or more years. Since 1963, the worst drought year on record, Hong Kong appears to be going through a relatively wet phase associated with ENSO episodes. Hong Kong is a highly urbanized coastal city with moderately high rainfall. Groundwater usage is minimal, and agriculture is relatively small scale. There is

TABLE VI
Ten landslides, including boulder falls, with the highest death toll in Hong Kong since 1917. Based on K. Ho (personal communication).

Year	Locality	Death Toll
1925	Po Hing Fong	73
1972	Sau Mau Ping	71
1972	Mid-Levels	67
1978	Sau Mau Ping	18
1968	Ma Shan Village	16
1966	La Salle Road	8
1917	St. Joseph College	6
1968	Brown Street	6
1926	Pokfulam	5
1994	Kwun Lung Lau	5

TABLE VII
Ten years with the highest death toll caused by landslides and boulder falls in Hong Kong since 1917. Based on K. Ho (personal communication).

Year	Death Toll
1972	149
1925	73
1966	38
1976	29
1982	27
1968	22
1994	6
1917	6
1971	5
1926	5

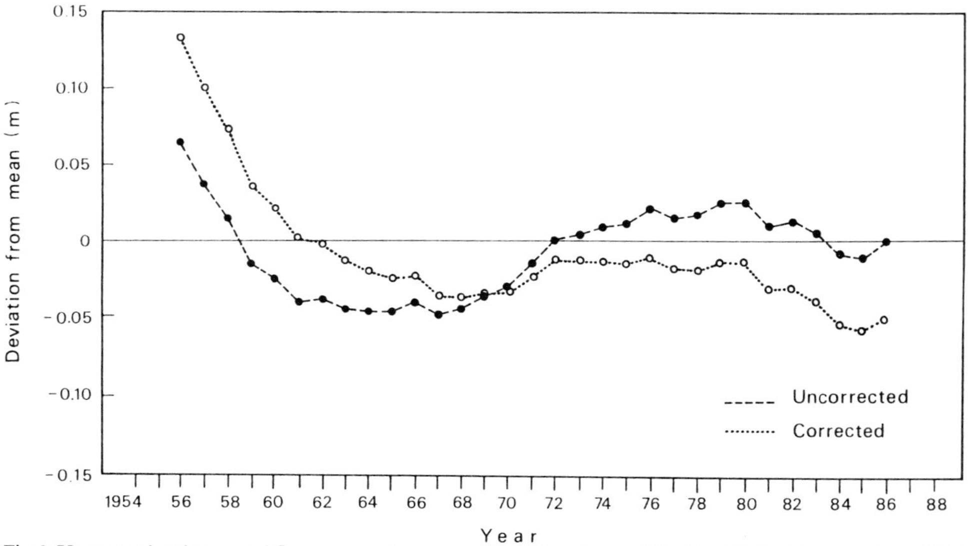

Fig. 2. Uncorrected and corrected five-year running mean of annual sea level at the North Point tide gauge from 1954 to 1988. Note that data since 1986 are based on the Quarry Bay tide gauge. Reproduced from Yim (1995) after R. Wardlaw (personal communication).

also an agreement with China for the importation of water for consumption. Therefore droughts are not a major concern. However, the increase in frequency or intensity of droughts in southern China would inevitably lead to increases in water supply costs. In addition, hillfires are likely to be increasingly disastrous.

4. Adaptation

The lesson learned from the historical record of typhoons is that death tolls, casualties, and damage due to storm surges are minimized through better warning and protection provided by typhoon shelters. Today, satellite tracking of typhoons permits regular updates of the latest information for broadcast on radio and television. Breakwaters with armoring may be used to dissipate the storm-wave energy to provide adequate protection in typhoon shelters. Furthermore, the death tolls of the unnamed typhoons in 1906 and 1937 will not reoccur because of the sharp decline in the population of boat dwellers.

Measures to protect the coastal land reclamations in Victoria Harbour were examined in a study by Yim (1995). The best response will differ for existing urban areas and for new coastal land reclamations. For existing urban areas, there is no possibility of retreat or accommodation (IPCC, 1990). However, during urban renewal the formation level of the ground could be raised. Protection of the existing low-lying coastal land reclamations is the most likely option in the absence of urban renewal. Based on the 46 km length of the Victoria Harbour coastline, the cost of raising the seawalls from the existing 3.7 m P.D. (principal datum is the reference datum used to measure ground elevation) to 4.2 m P.D. is about US$35 million at 1993 costs. The cost rises to US$75 million for a 4.7 m P.D. seawall, and US$175 million for a 5.5 m P.D. seawall. Before decisions to increase protection measures are made, it is necessary to identify the rate of ground settlement (Yim, 1995). Unstable reclamations showing clear signs of ground settlement, and stable reclamations showing no sign of ground settlement should be dealt with separately. Their identification should be based on regular surveying against bedrock benchmarks and the use of satellite altimetry (Yim, 1991b). Yim (1995) suggested a combination of 13 responses that may be applied after obtaining estimates on the magnitude of the ground settlement problem. However, high seawalls to protect reclamations from storm surge flooding and future sea level rise could increase the risk of rainstorm flooding. Based on the limited information, no local evidence has been found for future sea level rise; however, continuous monitoring to provide warning for future sea level changes should be given high priority.

Landslides in Hong Kong are initiated by anthropogenic deforestation, including hillfires, followed by soil and rock excavation. Heavy rainfall changes the state of the soil cohesion through seepage, eventually leading to slope failure. All landslides are preventable, however, with appropriate measures. Therefore, input from individuals conducting climatic change vulnerability assessments is needed to assist geotechnical engineers in determining the level of protection required and to adopt the necessary preventive measures.

To reduce vulnerability to the potential effects of climatic change, a combination of better monitoring, planning, protection, maintenance, and warning is proposed for adaptation. Possible methods include:

- monitoring temperature, precipitation, typhoons, and sea level elevations
- monitoring soil microfabrics in all cutslopes
- monitoring ground settlement in coastal land reclamations
- providing adequate protection of all cutslopes
- reducing infiltration into cutslopes through appropriate drainage measures
- constructing adequate stormwater drains and installing flood pumping stations
- landscaping reclamations to encourage runoff
- planning land use of flood-prone areas to minimize flood damage
- using improved construction methods to ensure that the long-term ground settlement is tolerable
- promoting research work on the mitigation of climatic change hazards.

The above methods will go some way to minimize the impact of the hazards that may be increased by climatic change. Floods and landslides are likely to be of the greatest concern in Hong Kong because of the creation of low-lying coastal land reclamations for urbanization and continuous urbanization of the steep hillslope areas under a trend toward an increasingly wetter climate associated with higher intensity precipitation.

5. Conclusions

Hong Kong is vulnerable to weather-related hazards that could be exacerbated by the effects of climatic change, including potential increases in typhoons, landslides, and floods caused by both storm surges and rainstorms. Although storm surges have the greatest historical record of damage, they are unlikely to cause as much damage or loss of life as in the past because of the change in circumstances, including a sharp decline in the population of boat dwellers and the establishment of a better warning system through radio and television. Floods and landslides are identified as the greatest future concern because of the trend toward an increasingly wetter climate associated with higher intensity precipitation. Both are made worse by the action of humans: the former through the creation of low-lying coastal land reclamations, and the latter mainly through the cutting into the steep hillslopes. For adaptation, a combination of better monitoring, planning, protection, maintenance, and warning is proposed to mitigate the effects of climatic change-induced hazards.

Acknowledgments

This study is funded by the Hong Kong Research Grants Council, the Office of Policy Analysis of the U.S. Environmental Protection Agency, and the Croucher Foundation. The U.S. Country Studies Program is thanked for funding my attendance at the Regional Workshop on Climate Change Vulnerability and Adaptation in Asia and the Pacific in Manila, where this paper was presented orally. I am grateful to the Director of the Civil Engineering Department and the Director of the Royal Observatory, both of the Hong Kong Government, for the provision of relevant information to assist this study. I would also like to thank I.J. Ayson, M.L. Chalmers, Elaine Koo, Dr. S.L. Leatherman, Dr. M.R. Peart, Dr. R.J. Nicholls, K.H. Tam, Dr. R. Wardlaw, and K.P. Wong for their assistance. This paper is a contribution to the Commission on Quaternary Shorelines of the International Union for Quaternary Research and the IUGS-UNESCO project on Climates of the Past.

References

Hansen J. and Lebedeff S.: 1987, *Journal of Geophysical Research* **92**, 13345.

IPCC: 1990, Strategies for adaptation to sea level rise, in: *Report of the Coastal Zone Management Sub-Group*, IPCC Response Strategies Working Group, Rijkswaterstaat, The Netherlands.

Koo E.H.: 1990, Climate change — secular trends in urban temperature, in: *Proceedings of the World Congress on Climate and Development*, Trinidade S.C., Karpe H.-J., and Otten D. (eds), Hamburg, Springer-Verlag, Berlin, Germany, pp. 285-292.

Lumb P.: 1975, *Quaternary Journal of Engineering Geology* **8**, 31.

Peart M.R. and Jayawardena A.W.: 1993, Flood forecasting and warning in Hong Kong, in: *Proceedings of the 25th International Association for Hydraulics Research Congress*, Tokyo, pp. 241-248.

Peart M.R. and Jayawardena A.W.: In press, Drought in Hong Kong, in: *Proceedings of the 4th International Conference on the Evolution of the East Asian Environment*, Hong Kong.

Peart M.R. and Yim W.W.-S.: 1992, Flood hazard in the coastal land reclamations of Hong Kong, in: *Proceedings of the 2nd U.S.-Asia Conference on Engineering for Mitigating Natural Hazards Damage*, Chiu A.N.L. and Danuatmodjo A.S. (eds), Yogyakarta, pp. F14/1-F14/8.

Yim W.W.-S.: 1991a, *Hong Kong Meteorological Society Bulletin* **1**, 16.

Yim W.W.-S.: 1991b, Relative sea-level change and long term ground settlement in coastal land reclamations — their assessment and future monitoring, in: *Reclamation — Important Current Issues*, Blacker P. (ed), Hong Kong Institution of Engineers, Hong Kong, pp. 139-151.

Yim W.W.-S.: 1993a, Future sea level rise in Hong Kong and possible environmental effects, in: *Climate and Sea level Change: Observations, Projections and Implications*, Warrick R.A., Barrow E.M., and Wigley T.M.L. (eds), Cambridge University Press, Cambridge, United Kingdom, pp. 349-376.

Yim W.W.-S.: 1993b, Recent sea-level changes in Hong Kong and their implications, in: *Proceedings International Workshop on Sea Level Changes and their Consequences for Hydrology and Water Management*, Session 4: Socioeconomic Impacts, Noordwijkerhout, pp. 89-98.

Yim W.W.-S.: 1995, *Journal of Coastal Research* **14,** 167.

VULNERABILITY ASSESSMENT OF ANGAT WATER RESERVOIR TO CLIMATE CHANGE

A.M. JOSE,[1] L.M. SOSA,[2] and N.A. CRUZ[1]
[1] *Philippine Atmospheric, Geophysical and Astronomical Services Administration (PAGASA) 1424 Quezon Avenue, Quezon City, Philippines*
[2] *National Water Resources Board (NWRB) NIA Compound, EDSA, Quezon City, Philippines*

Abstract. Global warming due to an anticipated doubling of carbon dioxide concentration in the atmosphere is expected to alter the earth's climate system within the next century. The potential changes in the climate system could affect hydrological cycles and processes. Possible impacts of climate change on water resources should be assessed to evaluate probable adaptation measures. In the Philippines, a preliminary assessment of the vulnerability of water resources to climate change and variability was undertaken. For this particular study, the Angat Reservoir was chosen as the study area. Because of its socioeconomic importance, it is useful to assess its vulnerability to climate change. A rainfall-runoff simulation model, WATBAL, was used to determine the effect of temperature and rainfall changes, based on CO_2 doubling, on inflow to the reservoir. Climate change scenarios developed from results from three general circulation models and incremental changes were used. The results showed that changes in temperature and rainfall could affect runoff either positively or negatively. Using the temperature and rainfall changes from the Geophysical Fluid Dynamics Laboratory model there was a 32% increase in runoff, and with the Canadian Climate Centre Model, there was a 15% decrease in runoff. Under a climate scenario generated by the United Kingdom Meteorological Office model, runoff is estimated to increase by 5%. The use of incremental scenarios revealed the strong sensitivity of runoff to changes in rainfall as compared with changes in temperature.

Key words: Philippines, water resources, CCCM, GFDL, UKMO, WATBAL

1. Introduction

This paper presents a preliminary assessment of the vulnerability of the Angat water reservoir to impending climate change using two climate change scenarios. One of these scenarios is based on the simulation results of three general circulation models (GCMs): the Canadian Climate Centre Model (CCCM), the United Kingdom Meteorological Office model (UKMO), and the Geophysical Fluid Dynamics Laboratory model (GFDL). The other climate change scenario is based on incremental changes of rainfall and precipitation. WATBAL (Yates *et al.*, 1994), a lumped integral model, was calibrated and validated to simulate future values of runoff based on temperature and rainfall under a doubled carbon dioxide concentration ($2\times CO_2$).

The mean average annual atmosphere temperature is projected to increase by 2.0°C in the next century (IPCC, 1996) as a consequence of the expected $2\times CO_2$ concentration in the atmosphere. Regional temperature increases may alter hydrologic cycles and processes in many parts of the world. These alterations could ultimately affect water availability and management.

Because of the potential impacts of climate variability and climate change on water resources, efforts are under way to investigate and assess such impacts to major reservoirs in the Philippines. During the past decades, extreme events such as droughts and floods have affected the Philippine economy. Several studies have investigated long-term rainfall and temperature trends in the Philippines (Jose, 1989; Jose *et al.*, 1993). The results of these studies revealed a general increasing temperature trend in most areas. Hydrologic parameters such as inflow in some multipurpose dams were also examined, and they exhibited a decreasing trend.

Water, Air, and Soil Pollution **92:** 191–201.
© 1996 *Kluwer Academic Publishers. Printed in the Netherlands.*

2. Methods

2.1 STUDY AREA SELECTED

The Angat multipurpose reservoir and its hydrologic basin (Figure 1) were selected because of its socioeconomic importance to Luzon, the principal island of the Philippines. The reservoir is the main municipal water source for Metro Manila. It also supplies irrigation water to thousands of farms planted with rice and vegetables, and serves as a flood control for areas downstream. As one of several sources of hydropower on Luzon, it contributes to Luzon's power requirement with an annual average of 500 GWh generating capacity. Its inflow is almost totally dependent on rainfall. Past records show that annual and seasonal rainfall variability has affected the dam's operation, with attendant water management problems.

Figure 2 shows the monthly average rainfall, runoff, and evapotranspiration measured at the reservoir from 1968 to 1993. As shown in the figure, the rainfall and runoff distributions exhibit a bimodal characteristic. Two maxima occur during a year: one in July and the other in October. The observed maximum values in July are associated with the activity of the southwest monsoon. The highest monthly average for both rainfall and runoff in October, is brought about during passage of tropical cyclones within or near the basin. Actual evapotranspiration averages 150 mm per month during the dry season (January-May) and decreases to an average of 60 mm per month during the rainy season (June-October). The mean annual temperature ranges from 24 to 28 °C, with January as the coldest month and May the warmest (PAGASA, 1995).

Rainfall data were gathered from three telemetering stations and weather stations, located within and around the basin. Average annual precipitation is about 3,500 mm, and the reservoir has an annual average runoff of 1.7×10^9 m^3. Runoff data were gathered from the National Power

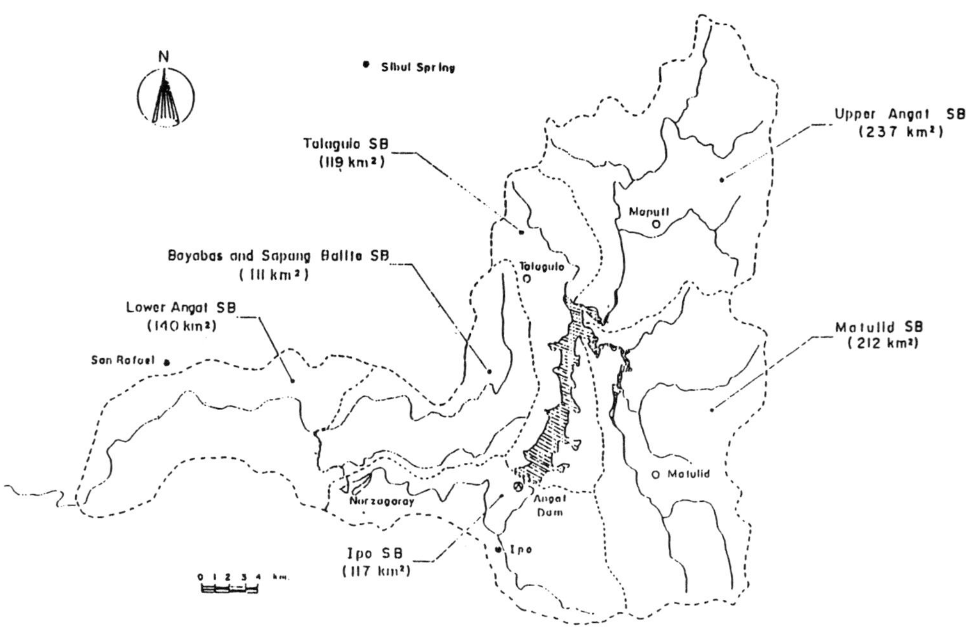

Fig. 1. Location map of Angat River Basin.

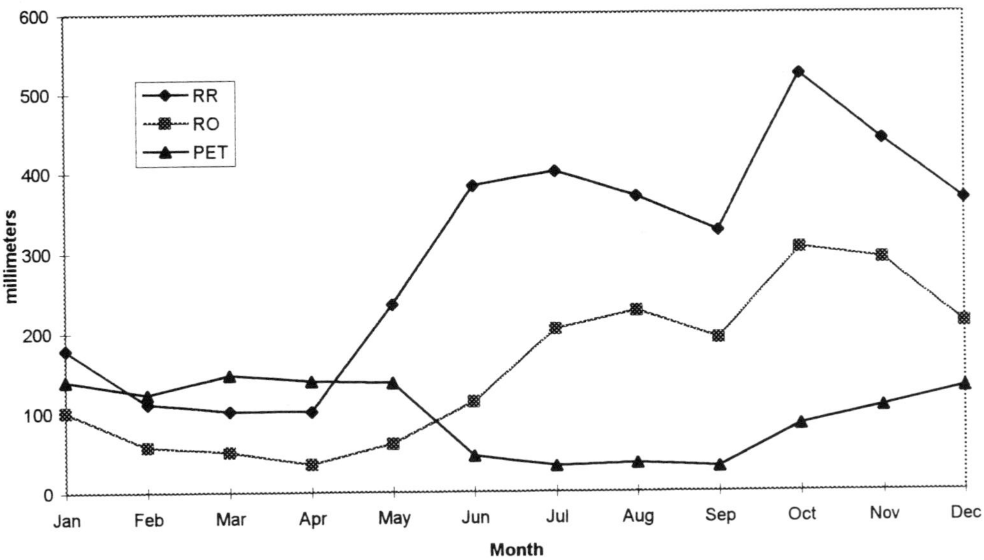

Fig. 2. Mean monthly rainfall (RR), runoff (RO), and evapotranspiration (PET) at Angat Water Reservoir (1968-1993). Source: Report of the National Power Corporation to the Water Resources Study Group.

Corporation, and precipitation data were taken from Philippine Atmospheric, Geophysical and Astronomical Services Administration (PAGASA) records. Figure 3 shows the comparison of annual rainfall and runoff for 1968 to 1993. The graph clearly shows the strong relationship between rainfall and runoff. Years with low runoff such as 1969, 1976, and 1982 are associated

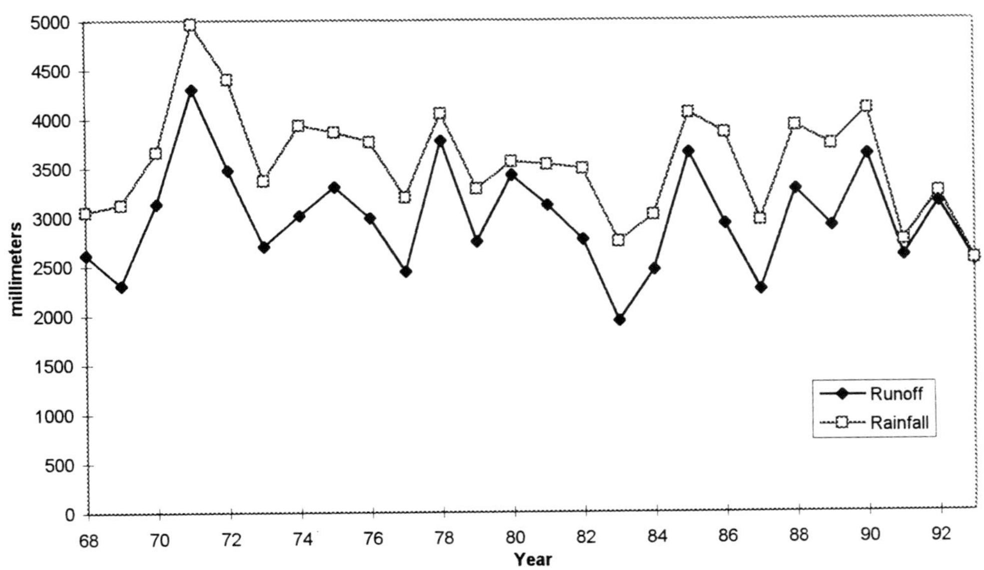

Fig. 3. Comparison of annual precipitation and runoff at Angat water reservoir (1968-1993). Source: PAGASA

with low rainfall values that are thought to be a manifestation of episodes of El Niño-Southern Oscillation (ENSO).

2.1.1 Utilization of Angat Reservoir

Irrigation. The Angat Maasim River Irrigation System (AMRIS) of the National Irrigation Administration takes water mainly from Angat Dam, with approved water rights of 36 $m^3 s^{-1}$. Based on the latest statistics from the National Irrigation Administration (NIA), out of the 3.4×10^4 ha potential service area of AMRIS, its current service area is 31,485 ha. Dry season operation usually starts on 16 June and ends on 15 November. The average irrigated area during the dry season is 2.66×10^4 ha. In the wet season, from 16 November to 30 April, the average irrigated area is about 2.21×10^4 ha. Occurrences of extreme events generally associated with ENSO greatly affect the operation of AMRIS. Thousands of hectares of croplands can be affected, with significant reductions in crop yield due to water shortages (taken from Report of the NIA to the water resources study group).

Domestic Supply. The Metropolitan Waterworks and Sewerage System (MWSS), which is responsible for Metro Manila's domestic water requirement, gets 90% of its feed water from the Angat reservoir. The remaining 10% is from underground water accessed through artesian wells. MWSS has a water right of 22 $m^3 s^{-1}$ from the Angat reservoir; however, with the ever-increasing demand for water, MWSS has had to seek an additional water right of 15 $m^3 s^{-1}$. Figure 4 shows the annual population of MWSS service areas (made up of Metro Manila and adjoining towns) and water consumption. Faced with increasing demand in the future, MWSS is looking for other sources of fresh water as well as increasing the efficiency of its current water distribution system (taken from Report of the MWSS to the water resources study group).

Power Generation. The National Power Corporation has operated and maintained Angat Dam since its construction in 1968. In line with the government policy for optimum water resource utilization, the water releases for irrigation and domestic uses are first routed through the

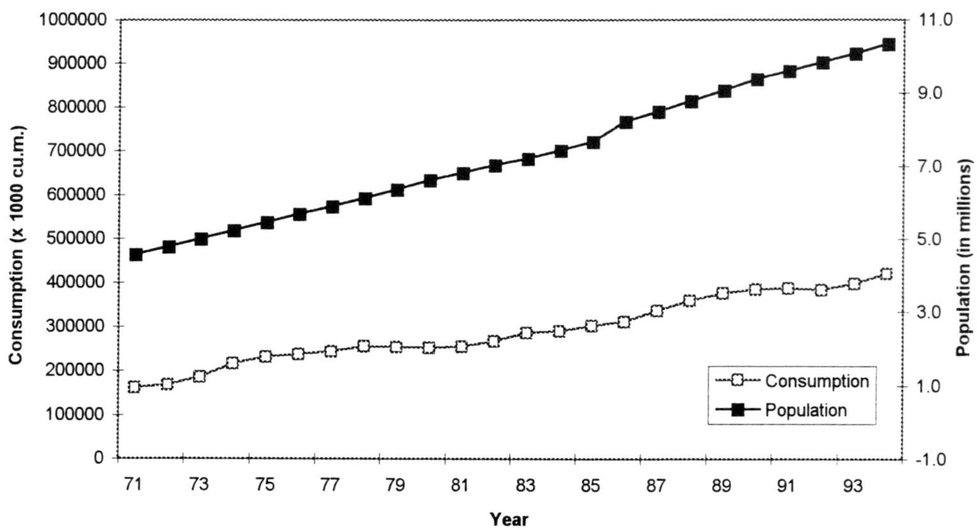

Fig. 4. Annual population of MWSS services areas and total water consumption (1971-1994). Source: Report of the MWSS to the Water Resources Study Group.

reservoir's hydroelectric turbine generators. Turbine inflow is bypassed only when the reservoir's water elevation is above the operation rule curve and flooding is imminent. For the period from 1968 to 1993, Angat has generated an annual average of 500 GWh, which is about 5% of Luzon's power demand. The reservoir can only generate power when the water level is 180 m above mean sea level, which is considered the minimum operating level; the maximum operating level is 217 m. The normal high water level of the reservoir is 212 m from 1 December to 30 April; the flood season high water level is 210 m (1 May to 30 November) (taken from the Report of the National Power Corporation to the water resources study group).

2.1.2 Hydrologic and Climatological Characteristics of the Angat River Basin

The reservoir, located about 60 km northeast of Metro Manila, has an effective capacity of 6.4×10^8 m^3 and regulates an average annual inflow of 1.7×10^9 m^3. The Angat River Basin has a total drainage area of 936 km^2, and the upper basin, which covers the area from the farthest watershed divide down to Angat Dam, is 568 km^2 (Rosacay, 1989). Figure 1 shows the location of the river basin. It is bounded on the northeast by the Upper Pampanga River Basin, on the southwest by Metro Manila, and on the east by the Sierra Madre range. Situated along the slopes of the Sierra Madre, the basin is exposed to prevailing southwesterlies during the summer and northeasterlies during the winter monsoon.

2.2 GENERAL CIRCULATION MODELS

Predictions from GCMs based on doubled CO_2 emissions were used in this study as the basis for climate change scenarios. Three GCMs were chosen to represent potential future values of rainfall and precipitation under a $2{\times}CO_2$ scenario: the CCCM (Boer *et al.*, 1992), the UKMO model (Wilson and Mitchell, 1987), and the GFDL model (Mitchell *et al.*, 1990).

In selecting the most appropriate GCMs, estimates of precipitation and temperature under $1{\times}CO_2$ (current conditions) from these models were compared with existing climatic normals (In-na *et al.*, 1991). The comparison was based on large-scale temporal and spatial distributions of precipitation and temperature. Available long-term averages of precipitation and temperature (1951-1975) covering the Southeast Asia region and the western Pacific (an area bounded by 20°S-40°N and 80°E-160°E) were used in the comparison (Asuncion *et al.*, 1980; ASEAN Climate Atlas, 1982). Maximum and minimum rainfall and temperature values associated with large-scale, quasi-permanent weather systems, such as the Intertropical Convergence Zone and subtropical high pressure areas were compared with the GCM results.

Table I shows the temperature changes and rainfall ratios between $1{\times}CO_2$ and $2{\times}CO_2$ generated by the three models. The temperature changes from the three models show an increase, which is consistent with the expected warming. The GFDL model and the UKMO model estimate an increase in annual temperature of 3.1°C and the CCCM predicts an increase of 2.0°C annually. The three models also predicted varying changes in rainfall. The UKMO and GFDL models both predict an annual increase of rainfall and the CCCM gives a slight decrease in annual rainfall.

Rainfall and temperature scenarios for present conditions from the selected GCMs were then compared with the monthly observed average temperature and rainfall (PAGASA, 1995 and taken from the Report of the National Power Corporation to the water resources study group) of Angat. Figure 5 shows the mean monthly rainfall for Angat as compared with the results from the three GCMs for the $1{\times}CO_2$ runs. It can be seen that none of the three models has the bimodal distribution that is seen in the current climate data, which demonstrates that the models

TABLE I
Rainfall ratio and temperature increases between current ($1\times CO_2$) and estimated ($2\times CO_2$) as generated by three GCMs

	CCCM		UKMO		GFDL	
Month	Rainfall Ratio	Temp. °C Change (+)	Rainfall Ratio	Temp. °C Change (+)	Rainfall Ratio	Temp. °C Change (+)
January	0.69	2.0	0.62	4.0	1.11	3.1
February	1.11	2.1	2.00	3.7	1.35	2.4
March	1.19	1.6	1.14	2.5	0.93	2.3
April	0.75	1.9	0.53	3.5	1.09	2.3
May	1.13	1.9	1.42	2.8	1.33	2.3
June	0.87	2.0	0.86	2.7	0.77	2.3
July	0.90	2.0	1.06	2.7	0.93	2.2
August	1.03	2.2	1.26	2.4	1.08	2.0
September	1.13	2.2	1.45	2.6	1.80	2.4
October	1.03	2.3	0.92	2.4	0.88	2.6
November	0.70	2.2	0.69	3.4	1.35	2.4
December	0.80	2.0	0.46	4.3	1.18	2.8

are not capable of simulating individual cyclonic events that contribute significantly to the reservoir's performance. A comparison of the mean air temperatures for the region in which the Angat Reservoir is located with those of the GCMs is shown in Figure 6. The GFDL model tends to underestimate the temperature significantly, and the CCCM has some overestimation, particularly for the second half of the year. The UKMO model shows the same temperature distribution as that of the mean except for overestimated April and May values.

2.3 INCREMENTAL SCENARIOS

Incremental changes in rainfall and temperature were also used to determine the impact of climate change on runoff. Incremental scenarios were applied because they offer the advantage

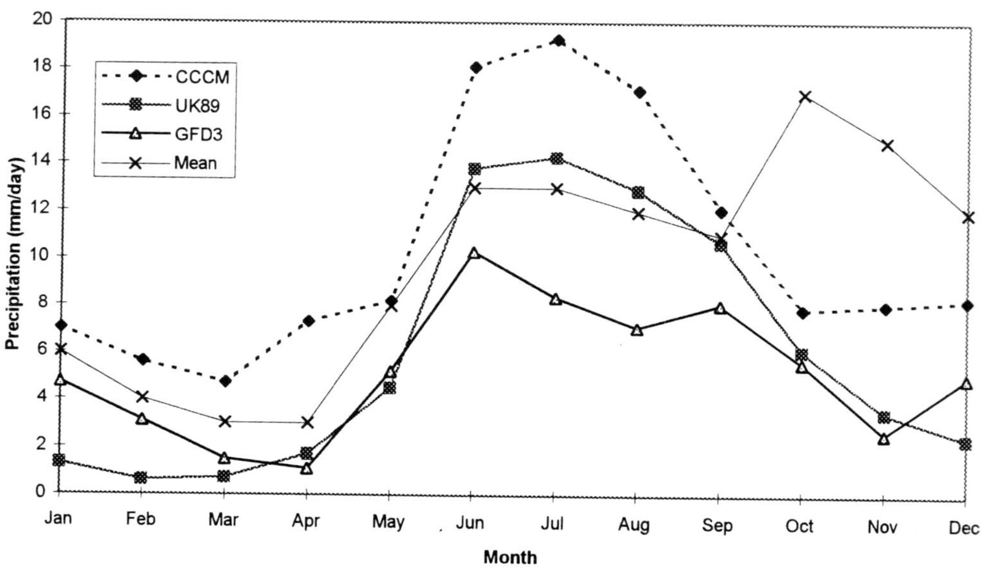

Fig. 5. Comparison of Angat's mean rainfall with estimated rainfall values at present condition ($1\times CO_2$) from three selected GCMs.

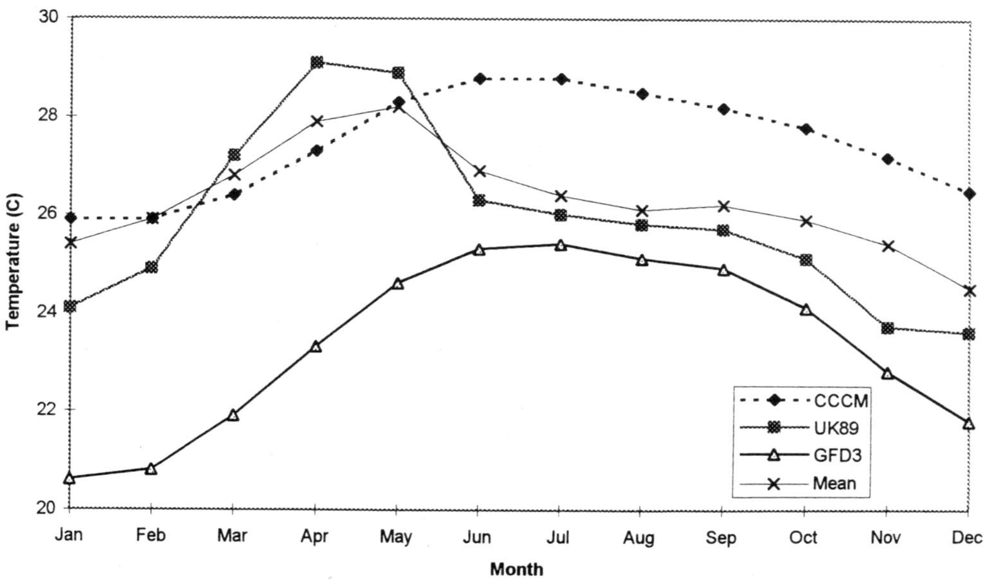

Fig. 6. Comparison of Angat's mean temperature with estimated temperature values at present condition ($1\times CO_2$) from three selected GCMs.

TABLE II
Incremental climate change scenarios where T = temperature (°C) and P = precipitation (%)

T+0, P+0%	T+0, P+10%	T+0, P+20%	T+0, P-10%	T+0, P-20%
T+2, P+0%	T+2, P+10%	T+2, P+20%	T+2, P-10%	T+2, P-20%
T+4, P+0%	T+4, P+10%	T+4, P+20%	T+4, P-10%	T+4, P-20%

of capturing a wide range of climatic variability. The scenarios chosen (Table II) represent uniform annual changes in temperature and precipitation. Two and four degree increases in temperature coupled with ±10% and ±20% in precipitation were employed. Monthly observed average values of Angat temperature and rainfall were used as the base temperature and rainfall in conjunction with the data in Table II to develop incremental scenarios.

2.4 HYDROLOGIC MODEL

To assess the reservoir's vulnerability, WATBAL (Yates, 1994), a lumped conceptual hydrologic model, was used to determine rainfall-temperature-runoff relationship. Although there are other existing hydrologic models that can be used for runoff prediction, WATBAL was suggested by the U.S. Country Studies Program because of its fewer data requirements and its simplicity. There are two main modeling components within WATBAL. One is the water balance component, which uses continuous functions to describe water movement into and out of a conceptualized basin, and the other is the calculation of potential evapotranspiration using the Priestly-Taylor radiation method. The method uses an approximation to the Penman Equation, which gives an estimate of reference crop evapotranspiration. Twenty-six years of climatological data such as rainfall, temperature, and relative humidity from the adjoining weather stations were provided by PAGASA, and National Power Corporation (NPC) provided

TABLE III
Estimated water demand (10^6 m^3 yr^{-1}) of Metropolitan Waterworks and Sewerage System (MWSS) and National Irrigation Administration (NIA)

	MWSS				NIA
Year	Industrial	Commercial	Domestic	Total	Irrigation
1995	108	269	534	911	872
2000	121	348	805	1,274	872
2025	210	1,133	1,219	2,563	872
2050	339	2,600	1,630	4,567	872

the set of runoff data. The model was calibrated using 18 years of data (1968-1985) and validated based on 8 years of data (1986-1993).

2.5 FUTURE WATER DEMAND

The NIA, the MWSS and the NPC provided projected water demand up to 2050. Table III presents the projected water demand of the MWSS and irrigation requirements of the NIA. Future requirements for domestic (which includes commercial demand) and industrial purposes show remarkable increases. The domestic requirement has the highest growth rate, attributable to the rapid migration of people into Metro Manila and its suburbs, which is expected to persist through several decades. Another factor is the projected industrialization of the country that the government hopes to attain in the next century. Therefore, industrial demand may grow more rapidly than projected. The irrigation requirement is projected to remain the same until 2050. However, for the last 10 years, the conversion of existing agricultural areas to commercial land is projected to accelerate, which could decrease the water required for agriculture. But the decrease in farmland is then offset by the conversion of still idle lands for agricultural purposes. The projected demand for power is expected to reach 5,934,269 GWh in 2050, which could be provided by other hydropower plants on the island as well as power generating plants that rely on oil or coal. Thus the domestic and industrial needs in the future would be the main problem facing the water management agencies. It should be noted, however, that the projections made by the three agencies did not consider any effects of the expected global warming, which could lead to even greater demand.

3. Results and Discussion

3.1 ESTIMATED CHANGES IN RUNOFF BASED ON GCM SCENARIOS

Table IV shows the changes in precipitation and temperature and the corresponding changes in runoff based on the climate change scenarios generated by the three GCMs. The UKMO and GFDL models showed an increase in runoff by 5 and 32%, respectively. The increase in runoff is attributable to an increase in precipitation predicted by the two models. GFDL has the highest percent change in runoff; it has a 15% increase in precipitation coupled with a 2.4 °C increase in temperature. The UKMO model predicted only a small increase in precipitation but predicted a higher increase in temperature than the other models. Under the CCCM scenario a 12%

TABLE IV
Changes in precipitation, temperature, and runoff for Angat water reservoir from three GCMs (based on $2 \times CO_2$)

GCM	Change in Precipitation	Change in Temperature (°C)	Change in Runoff
CCCM	-6%	+2.0	-12%
UKMO	3%	+3.1	5%
GFDL	15%	+2.4	32%

annual runoff reduction is estimated based on the scenario's estimated decrease in annual rainfall and increase in temperature.

Based on historical data, the Angat water reservoir has in the past experienced extreme changes in runoff. The 1971 annual runoff, for example, was 2.44×10^9 m^3 which is a 44% increase in the average runoff. This extreme event resulted in flooding of downstream areas, damaging crops worth millions of pesos and causing loss of lives. The GFDL model results suggest that greater frequency of flooding could be expected. Again, looking into historical records, the lowest annual runoff recorded was 1.1×10^9 m^3 in 1983, a 35% reduction in the average runoff. The drought of 1983, associated with the El Niño phenomenon, affected much of the regions' agricultural crops. If the CCCM scenario predominates in the future, the expected decrease in runoff would mean more serious problems in water availability, perhaps through more frequent episodes of drought. It should be noted that the monthly distribution of rainfall is far more critical than the annual distribution because of the varying monthly demand by the MWSS and the NIA.

3.2 ESTIMATED CHANGES IN RUNOFF BASED ON INCREMENTAL SCENARIOS

The runoff results for the 14 incremental scenarios are shown in Table V. The table indicates that runoff is more sensitive to precipitation variability than to changes in temperature. Increases of 10% and 20% in precipitation are estimated to result in increases in runoff of 8 and 18%, respectively. Alternatively, precipitation decreases of 10 and 20% are estimated to result in decreases in runoff of 15 and 25%, respectively. This suggests that a decrease in precipitation has a greater impact on runoff than does an increase in precipitation. On the other hand, the effect of temperature on runoff is not as significant as the effect of precipitation. A 1 °C increase would lead to only less than a 1% reduction in runoff. Scenarios that include a change in both temperature and precipitation, such as a 2 °C increase in temperature and a corresponding increase and decrease of 10% in precipitation, showed an 8% increase and 15% decrease in runoff, respectively. These preliminary results do not support the assumption that higher temperatures would mean higher evaporation and thereby reduce runoff significantly.

Further studies should be done to investigate the effect of temperature increase as well as the effect of other meteorological variables on the river basin's runoff since this vulnerability assessment assumes that with the change in temperature and rainfall, other meteorological parameters such as relative humidity and wind, among others, remain constant.

TABLE V
Percent change in Angat water reservoir runoff based on incremental changes in precipitation (P) and temperature (T)

	ΔP 0%	ΔP 10%	ΔP 20%	ΔP -10%	ΔP -20%
ΔT 0°C	0%	8%	18%	-15%	-25%
ΔT 2°C	-1%	8%	17%	-15%	-26%
ΔT 4°C	-1%	7%	17%	-16%	-26%

3.3 ENSO-RELATED DROUGHT EVENT OF 1992

ENSO episodes during past years have had significant effects on the reservoir's performance. Based on historical records, inflows have been quite low during ENSO events, particularly in 1969, 1973, 1977, 1983, 1987, and 1992. The 1992 ENSO episode demonstrates the adverse impacts of extreme rainfall and runoff variability on the country's socioeconomic conditions in general and on the reservoir's performance in particular (Jose, 1992). Total inflow during the first half of the year (January to June) at the reservoir was only 1.27×10^8 m^3, a 69% reduction from the average inflow for the six-month period. With a critical decrease in inflow, there was a curtailment in the releases to meet the domestic and irrigation requirements of the Metropolitan Water Works and Sewerage System and the National Irrigation Administration. As a result, the Metropolitan Water Works and Sewerage System reported an equivalent of a 20% reduction in its domestic water production, which created the need for water rationing in many of the areas it serviced. The National Irrigation Administration reported that 1,575 ha of land planted with rice were affected. The start of the cropping season for the year was also delayed because of very low releases in June and July, which herald the start of the wet season crop. The reduction in irrigated areas affected the country's rice production in 1992. For power generation, the National Power Corporation reported a 37% reduction in electricity production (202 GWh). During that particular year, the lowest water level ever in Angat reservoir was recorded at 150 m, far below the minimum operating level of 180 m.

4. Conclusions

As mentioned previously, runoff to the Angat Reservoir depends almost totally on rainfall. The climate change scenarios employed in this assessment suggest large differences compared to today's rainfall and runoff. Such variability has been experienced in the past, as shown by historical data. Even 20-30% decreases or increases in rainfall have been observed in the past. These are greater than the values used in the climate change scenarios. Based on runoff simulation results, the expected rise in temperature in the future will not be a significant factor in runoff variability, although a more comprehensive study on this aspect is needed. Additionally, the impacts of temperature increase could be manifested in water demand. We can thus conclude that the Angat Reservoir, which is already exposed to extreme rainfall variability and its adverse consequences, as in 1992, faces more threats from increased climatic variability as suggested by results of the selected GCMs. Results from the GCMs also show great variability in rainfall with respect to time. CCCM, for example, gives very high rainfall estimates during the rainy season (May to September) and lower rainfall estimates during the dry season, compared with actual conditions. We can possibly attribute high rainfall to more typhoon passages, or heavy thunderstorm activities, as suggested by other studies.

The threat to Angat Reservoir from climate change could be aggravated by the increasing population, which translates to more demand for water, and degradation of the physical environment, which could affect to some extent the hydrology of the area. For this reason, policy makers and planners from water related agencies should start formulating possible adaptation measures to lessen potential impacts of climate change to the country's water resources.

Acknowledgments

The authors wish to express their gratitude for the support of the U.S. Country Studies Program for Climate Change and the PAGASA. The authors are also thankful to Susan Espinueva, Florentino David, Jorge Estioko, Alberto Guanzon, and Jose Dimatulac, members of the study group, for their support and cooperation.

References

The ASEAN Climatic Atlas: 1982, ASEAN Subcommittee on Climatology, ASEAN Subcommittee on Science and Technology, ASEAN Secretariat, Jakarta, Indonesia.

Asuncion J. and Parong E.M.: 1980, Mean Streamlines and Isotachs for Selected Pressure Levels for Southeast Asia and Adjacent Areas and an Attempt to Forecast the 500 and 300 Millibar Wind Flow, PAGASA Research Paper, Quezon City, Philippines.

Boer G.J., McFarlane N., and Lazare M.: 1992, *Journal of Climate* **5**, 1045.

In-na N. and Hemantha E.J.: 1991, Simulating the Effects of Global Warming on Evapotranspiration and Rainfall, Report, Thammas at University, Rangit Campus, Thailand.

IPCC (Intergovernmental Panel on Climate Change): 1992, *Scientific Assessment of Climate Change*, Houghton J.T., Callander B.A., and Varney S.K. (eds), WMO/UNEP, Cambridge University Press, Cambridge, Massachusetts.

IPCC (Intergovernmental Panel on Climate Change): 1996, *Climate Change 1995: The IPCC Second Assessment Report, Volume 2: Scientific-Technical Analyses of Impacts, Adaptations, and Mitigation of Climate Change*, Watson R.T., Zinyowera M.C., and Moss R.H. (eds), Cambridge University Press, Cambridge and New York.

Jose A.M.: 1989, Seasonal Rainfall Prediction in the Philippines, Doctoral Thesis, University of the Philippines, Manila, 172 pp.

Jose A.M.: 1992, Preliminary Assessment of the 1991-1992 ENSO-Related Drought Event in the Philippines, PAGASA Research Paper, Quezon City, Philippines.

Jose A.M., Francisco R.V., and Cruz N.A.: 1993, A Preliminary Study on the Impact of Climate Variability/Change on Water Resources in the Philippines, PAGASA Research Paper, Quezon City, Philippines.

Mitchell J.F.B, Manabe S., Tokioka T., and Meleshko V.: 1990, Equilibrium change, in: *Climate Change: The IPCC Scientific Assessment,* Houghton J.T., Jenkins G.J., and Ephramus J.J. (eds), Cambridge University Press, New York.

PAGASA: 1995, *Monthly Mean Annual Climatic Data (1961-1994)*, PAGASA Climate Data Section, Manila, The Philippines.

Rosacay D.A.: 1989, Angat Dam Level Management and Hydroforecasting, Technical paper presented at the First Workshop on Operation and Control Issues: ASEAN Cooperation Project No. 3, Philippines.

Yates D.: 1994, WATBAL — An Integrated Water Balance Model for Climate Impact Assessment of River Basin Runoff, IIASA Working Paper WP-94-64, Laxenburg, Austria.

Yates D. and Strezepek K.: 1994, Comparison of Models for Climate Change Assessment of River Basin Runoff, IIASA Working Paper WP-94-45, Laxenburg, Austria.

VULNERABILITY OF FRESHWATER RESOURCES TO CLIMATE CHANGE IN THE TROPICAL PACIFIC REGION

G.A. MEEHL
National Center for Atmospheric Research
P.O. Box 3000
Boulder, Colorado 80307-3000, U.S.A.

Abstract. El Niño events and associated droughts adversely affect freshwater resources on islands in the tropical Pacific region. Particularly vulnerable are low-lying atolls because rainwater collection is the main freshwater source on such islands. During El Niño-related droughts, water can be drawn only from the limited freshwater lenses beneath the islands. If drought conditions such as these intensify, the depletion of freshwater resources could affect the habitability of atolls. Average climate change in the Pacific region from increased anthropogenic carbon dioxide in a global coupled climate model resembles present-day El Niño conditions as well as the decadal time-scale sea surface temperature and precipitation anomalies observed during the 1980s and early 1990s. These anomalies are a consequence of greater warming of sea surface temperatures in the eastern equatorial Pacific than over the western Pacific warm pool with increased carbon dioxide in the climate model. Attendant increases in precipitation in the central equatorial Pacific are also accompanied by precipitation decreases in the northern and southern tropical Pacific (roughly 5°N to 15°N and 5°S to 15°S), as well as in the Australasian and eastern Indian Ocean regions. Associated effects in the midlatitude North Pacific also resemble El Niño conditions and the decadal time-scale signals from the 1980s. Future possible increases of drought conditions in certain tropical Pacific regions, as indicated by the climate model results, could limit the sustainability of atoll populations in those regions, causing migration and increased urbanization, with all the attendant problems, on larger high islands with more stable water supplies.

Key words: Australasia, Pacific island nations, water resources, El Niño, drought

1. Introduction

Island countries with inhabited atolls are particularly vulnerable to precipitation fluctuations. In particular, droughts associated with El Niño events that typically occur every two to six years disrupt freshwater supplies. Such droughts involve the virtual absence of precipitation for periods of several months to a year in certain regions of the Pacific. If these drought conditions were to intensify, as suggested by some climate model results, there could be implications for habitability of certain islands in the future. The purpose of this paper is (1) to draw attention to the vulnerability of freshwater resources on atolls to such precipitation deficits using the example of Kapingamarangi Atoll; (2) to show, through the presentation of climate model results, that increased carbon dioxide in the model reduces precipitation over considerable areas of the tropical Pacific and Australasia similar to precipitation anomalies associated with present-day El Niño events; and (3) to suggest, using the example of Tau Island in American Samoa, that if habitability of an island is threatened by ecological perturbations (the impacts on Tau from tropical cyclones are used as an analog to the ecological perturbation of precipitation deficits and freshwater depletion on atolls), migrations to main or high islands could increase, thus further contributing to urbanization and associated economic and societal problems on those islands.

Water, Air, and Soil Pollution **92:** 203–213.
© 1996 *Kluwer Academic Publishers. Printed in the Netherlands.*

2. Potential Vulnerability of Kapingamarangi Atoll to Climate Change

Kapingamarangi Atoll lies just north of the equator near 154°E longitude. It is one of several inhabited atolls that are part of Pohnpei, one of the Federated States of Micronesia. Among the 20 or so small low-lying islets of Kapingamarangi, two are inhabited and connected by a causeway (Figure 1). The roughly 500 people who live on Kapingamarangi depend on rainfall collected in cisterns (Figure 2) for their freshwater supply. During El Niño events, the region around Kapingamarangi typically experiences drought conditions. When water collected in the cisterns runs out (typically in about a month), the islanders must rely on water from shallow wells (Figure 3) that draw from the freshwater lens just beneath the surface (Rappa *et al.*, 1995). At the time the photograph in Figure 3 was taken, in February 1992, drought associated with the El Niño event of 1991-1992 had resulted in islanders drawing somewhat brackish water from this well for several months. The islanders had to boil the water before using.

The islanders are usually able to subsist on the freshwater lens for a period of months, depending on the hydrology of the lens and the particular details of the islet geography. Yet, during the El Niño event of 1918, severe drought conditions on Kapingamarangi resulted in a partial evacuation of islanders to the main high volcanic island of Ponape (now Pohnpei) (Stanley, 1985). These early ecological refugees not only added to the population of Pohnpei Island, but also required services and jobs, and played a part in further urbanization of the main town of Kolonia.

Of course, in 1918, issues involved with urbanization were virtually nonexistent in the island Pacific due to sparse population and residents widely distributed on islands throughout the various island groups. However, after 1945, rapid population growth coupled with economic migrations from outer islands to main population centers accentuated the problems associated with urbanization (Rappa *et al.*, 1995). If drought conditions become particularly widespread or persistent, the one small instance of islander migration from Kapingamarangi to a main island with a more stable water supply could be multiplied throughout the island Pacific, resulting in more severe problems associated with urbanization in existing population centers.

3. Pacific Region Climate Change Background

It has been noted that decadal time-scale climate fluctuations in the Pacific region have been characterized by a relatively greater warming of the surface waters of the central and eastern tropical Pacific Ocean compared to the western tropical Pacific during the 1980s and early 1990s (Nitta and Yamada, 1989; Graham, 1994; Nitta and Kachi, 1994; Trenberth and Hurrell, 1994; Meehl and Washington, in press). Climate model experiments have shown that these persistent warm ocean surface temperatures have probably contributed to observed global warming during this period (Graham, 1995; Kumar *et al.*, 1994; Salinger *et al.*, 1995). Those studies could not distinguish whether the warming trends were themselves a product of CO_2-related global warming, a manifestation of increased frequency of El Niño events in that region, or some other fluctuation of the climate system.

The balance between two processes, the super greenhouse effect (from increased atmospheric moisture — as SSTs warm, evaporation and deep convection increase and pump even more moisture higher in the troposphere and enhance the trapping of infrared radiation due to the greenhouse effect, thus the term "super greenhouse effect") and cloud albedo feedback (from increased clouds reflecting more solar radiation), has been shown to be important for sea surface

Fig. 1. Kapingamarangi Atoll; bridge at left connects the two inhabited islets in the atoll. Typical of Pacific atolls, the highest point above sea level is only several meters.

Fig. 2. The main source of freshwater on Pacific atolls is rainwater collected in cisterns such as this one on Kapingamarangi.

Fig. 3. When droughts occur, often associated with El Niño events, freshwater collected in cisterns is exhausted in about a month or two. Islanders then rely on water from shallow wells such as this one on Kapingamarangi that draw from the freshwater lens just beneath the surface.

temperature (SST) response in the warm tropical oceans (Meehl and Washington, 1995). The results of those studies, and the suggestions of other studies (Ramanathan and Collins, 1992), allude to the importance of cloud-albedo feedback for climate change in the Pacific region. That is, as SSTs warm, a combination of dynamical and radiative feedbacks leads to increased deep convection and high albedo cirrus cloud anvils that reflect incoming solar radiation and inhibit further warming of SSTs. Thus, if there is a warming of the tropical Pacific Ocean surface, the western Pacific warm-pool SSTs might increase at a slower rate than the SSTs in the tropical eastern Pacific, where this effect is not as strong. If this hypothesis is correct, a mean or decadal time-scale warming in the Pacific region would be manifested by a relaxation of the SST gradient across the Pacific, not unlike what occurs in a present-day El Niño event. And, like what is observed during El Niño events (Ropelewski and Halpert, 1987; Kiladis and van Loon, 1988), attendant precipitation anomalies would feature enhanced precipitation in the central equatorial Pacific with precipitation deficits to the north and south in the tropical Pacific and over Australasia.

Because of the vulnerability of freshwater resources on atolls such as Kapingamarangi (as noted above), such precipitation deficits could have adverse impacts on atoll populations.

4. Model, Methodology, and Limitations

To test this hypothesis, I analyzed results from an experiment with a global coupled ocean-atmosphere general circulation model (GCM) with increased atmospheric CO_2, and I attempted to gain insights into the recent observational results, with implications for freshwater resources, by studying model-simulated phenomena. The changes in mean climate due to increased CO_2 described here could also affect phenomena associated with El Niño-like variability in the model (Meehl *et al.*, 1993).

A second-generation global coupled general circulation climate model (Meehl and Washington, 1995) developed at the National Center for Atmospheric Research was integrated for 75 years of atmospheric CO_2 increasing at a rate of 1% per year compounded. The last 20 years of this experiment are analyzed (near the time of CO_2 doubling at about year 70) and compared to a control integration with present-day amounts of CO_2.

The atmospheric model has an approximate horizontal resolution of 4.5° latitude and 7.5° longitude with nine vertical levels. The ocean and sea-ice components have 1° latitude-longitude resolution with 20 levels in the ocean. Sea ice includes a three-layer thermodynamic scheme along with dynamic sea ice. No flux adjustments are used at the air-sea interface in the coupled model (see discussion of flux adjustment in Meehl, 1995). Of particular relevance to this experiment is the inclusion of a simple cloud-albedo feedback parameterization and a mass flux convective scheme in the atmospheric model. The former parameterization accounts for the observed relationship between very warm SSTs, deep convection, and bright clouds (Ramanathan and Collins, 1992; Washington and Meehl, 1993). The latter represents the super greenhouse effect in the model (Ramanathan and Collins, 1992; Meehl and Washington, 1995).

Because of the uncertainty involved with some cloud processes, cloud feedbacks can be tuned to produce a wide variety of sensitivities (Senior and Mitchell, 1993). Even though the cloud effects in the present model are fairly simply accounted for, their inclusion compares favorably to a model with a more sophisticated feedback scheme for cloud optical properties (Boer, 1993), and also reasonably reproduces observed measures of cloud albedo feedback and the super greenhouse effect (Ramanathan and Collins, 1992; Meehl and Washington, 1995).

A sensitivity experiment performed with this version of the coupled model, where the cloud-albedo feedback was strengthened, showed that (1) there was a large-scale response of the climate system combining radiative and dynamic feedbacks, and (2) the maximum values of tropical SSTs were a function of the strength of the cloud-albedo feedback as represented by the cloud-albedo feedback parameterization in the model (Meehl and Washington, 1995).

Thus, even though there are limitations and simplified parameterizations in the coupled model, important processes affecting the sensitivity of SSTs in the warm tropical oceans have been documented, compared favorably to observations, and analyzed in a sensitivity experiment (Meehl and Washington, 1995).

5. Geographical Patterns of Climate Change

The SST anomalies in the tropical Pacific region, with increased CO_2 minus control, for the December-January-February (DJF) season (DJF is shown here because observational studies have focused on this season; other seasons in the model show similar results) show least warming (less than 2°C) where mean SSTs are greatest in the model (Figure 4). These areas include the tropical Pacific near 20°N in the region of the Intertropical Convergence Zone (ITCZ), in the South Pacific Convergence Zone (SPCZ) region southeast of Papua New Guinea in the tropical southwestern Pacific, and over Australasia and the eastern Indian Ocean (Meehl and Washington, in press). These are not only the regions of warmest SSTs and greatest mean rainfall in the Pacific region, but also where the cloud-albedo feedback effects are greatest in the model (Meehl and Washington, 1995). Meanwhile in the equatorial eastern Pacific, there is relatively greater warming of the ocean surface (2-4°C). Therefore, the CO_2-related surface warming is not uniform at the ocean surface across the tropical Pacific (Meehl and Washington, in press). There is a reduction of the meridional SST gradient not unlike what is seen during a present-day El Niño event (Meehl, 1987). This is not due to a first order change of frequency of El Niño-like events in the model since preliminary results from an analysis of El Niño frequency do not show significant changes between control and increased CO_2 experiments.

Observations from the Pacific region (Bottomly *et al.*, 1990) show that such differential warming occurred during the 1980s. During this time period, when global temperatures also increased, the mean warming in the tropical western Pacific (about 0.15°C) was roughly less than one half that in the tropical eastern Pacific (about 0.35°C). This relatively greater warming of mean SSTs in the eastern tropical Pacific compared to the western tropical Pacific during the 1980s has been noted in other studies as well (Houghton *et al.*, 1992; Kumar *et al.*, 1994; Salinger *et al.*, 1995). This could be interpreted as increased incidence of El Niño events in relation to the mean pre-1980 climate, but more likely represents an increase of the mean SSTs after 1980 (Houghton *et al.*, 1992; Kumar *et al.*, 1994; Salinger *et al.*, 1995) such that warm and cold oscillations continue to occur in relation to the warmer post-1980 mean (Wang, 1995).

Precipitation anomalies from the coupled model, increased CO_2 minus control (Meehl and Washington, in press), show that precipitation increased in the central equatorial Pacific and decreased in the warm-water regime regions of the ITCZ (5°N to 15°N), SPCZ (5°S to 15°S), Australasia, and the eastern Indian Ocean (Figure 5a). These mean climate-change patterns in the coupled model due to increased CO_2 resemble those associated with present-day El Niño events in the tropical Pacific region (Figure 5b), as well as similar decadal time-scale changes observed during the 1980s (Kumar *et al.*, 1994; Salinger *et al.*, 1995).

Sea level pressure (SLP) anomaly patterns from the coupled model (Figure 6a) also show the El Niño-like feature seen in the observations of a deepened Aleutian low-pressure center in the north Pacific (van Loon and Madden, 1981; Meehl and Washington, in press), as well as the changes in SLP in the model in the North Pacific due to increased CO_2 and those seen in the similar decadal time-scale SLP anomalies in that region (Trenberth and Hurrell, 1994; Nitta and Kachi, 1994; Trenberth, 1990; Chen *et al.*, 1992) (Figure 6b). This similarity between the decade of the 1980s in the observations is associated with greater SST warming in the central and eastern Pacific compared to the western Pacific warm-pool region in both model and observations.

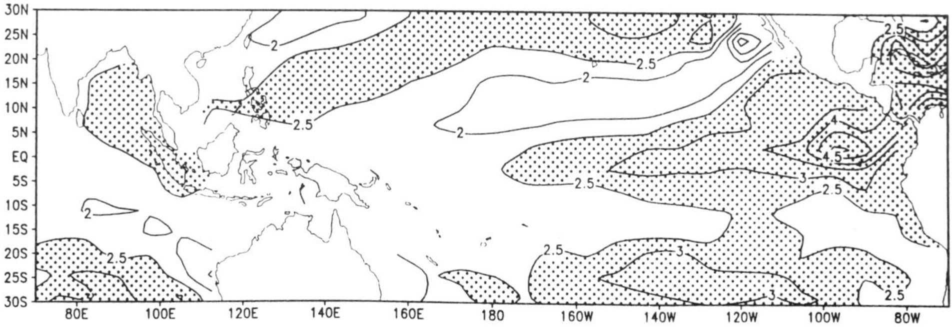

Fig. 4. SST differences(°C), transient CO_2 increase experiment minus control, for DJF, the last 20 years of 75-year transient and control integrations with the coupled model (CO_2 has doubled in the atmosphere at about year 70 of the transient experiment).

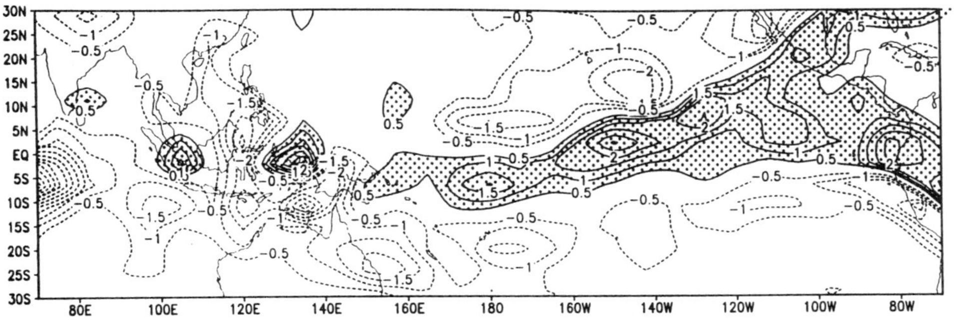

Fig. 5a. Precipitation differences (mm/day), transient CO_2 increase experiment minus control, for DJF, the last 20 years of 75-year transient and control integrations with the coupled model (CO_2 has doubled in the atmosphere at about year 70 of the transient experiment), areas of precipitation increase are gray-shaded.

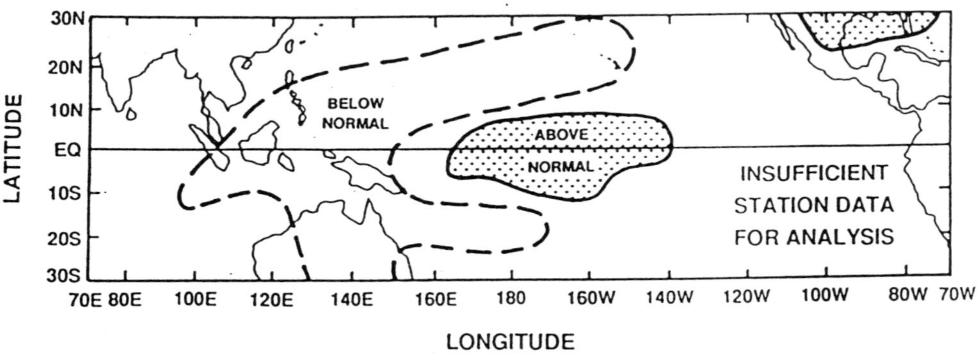

Fig. 5b. Schematic of composite areas in observed ENSO events where there is a consistent precipitation signal during northern winter based on station data (Ropelewski and Halpert, 1987). Stippled areas indicate precipitation increase.

(a) (b)

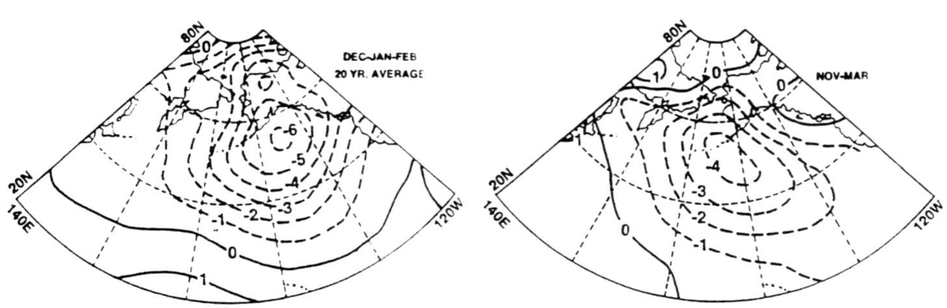

Fig. 6(a). Sea level pressure (SLP) differences (mb) for the north Pacific region, transient CO_2 increase experiment minus control, for DJF, the last 20 years of 75-transient and control integrations with the coupled model (CO_2 has doubled in the atmosphere at about year 70 of the transient experiment); (b) observed SLP differences (mb), 1977 to 1988 minus 1924 to 1976, for November to March season. Dashed contours indicate negative differences.

6. Vulnerability

As noted for the example of Kapingamarangi, low-lying atolls in Pacific island countries are particularly vulnerable to precipitation fluctuations that result in persistent droughts. Islanders can adapt to such drought conditions for periods of months by subsisting on water from the freshwater lenses lying beneath the surface of the islets in the atoll. However, severe or prolonged drought conditions can deplete the freshwater lenses. In the Kapingamarangi example, a portion of the population was evacuated to Pohnpei during the 1918 El Niño event. That much larger high volcanic island had more stable water resources that were able to sustain the population.

Since World War II, there has been a general trend toward lower outer island populations as islanders migrate to urban centers on main islands (and in some cases to countries outside of the Pacific), mainly for economic reasons (Shankman, 1993). This has resulted in increasing problems associated with urbanization, including crime, unemployment, and alcoholism (Stanley, 1985). There are also recent cases of migration due to ecological perturbations that could provide analogs to what some atoll populations may experience in the future. For example, on the island of Tau in American Samoa, a succession of three hurricanes in 1987 (Tusi), 1990 (Ofa), and 1991 (Val) caused extensive damage to crops and property. Instead of rebuilding on Tau, an estimated one third of the residents of Tau village walked away from their ruined houses and migrated to the main island of Tutuila to live with relatives, and some left the islands entirely (Leui Malae interview, 1995). Today, on Tau Island, numerous houses that were wrecked by the hurricanes stand abandoned (Figure 7); many of their owners now live in or near the urban center of Pago Pago on Tutuila. These migrations add to the problems associated with increasing urbanization in Pago Pago, such as lack of economic opportunity, the inability of services to support increase demand, and crime.

The consequences of migration of ecological refugees, in terms of increased urbanization in main island population centers, have yet to be fully studied. Anecdotal evidence, along with historical precedents such as Kapingamarangi, point to vulnerability of atoll populations, in particular, and to intensified drought conditions such as those suggested by the climate model results.

The chief concern of island governments with regard to climate change has been long-term sea level rise due to thermal expansion of the oceans in a future warmer climate (Rappa *et al.*, 1995). However, problems associated with freshwater resource deficits on atolls have not yet been adequately recognized. This vulnerability could result in problems with regard to atoll habitability sooner than long-term sea level rise. Additionally, freshwater lens depletion would be exacerbated by sea level rise since increased sea level drives the freshwater lens higher and thus results in a reduction of the potential water holding capacity of the atoll islets (Rappa *et al.*, 1995). Climate change effects on freshwater resources should be taken into account by island governments in terms of planning adaptation strategies for coping with the possibility of near-term (50 years) enhanced drought conditions.

7. Conclusions

The ecological refugees from Kapingamarangi Atoll during the drought associated with the 1918 El Niño event point to the vulnerability of atoll populations to possible future intensified and prolonged drought conditions in certain regions of the tropical Pacific. The climate model results presented here and elsewhere (Knutson and Manabe, 1995) suggest that a warmer

Fig. 7. Ruined house in Tau village, Tau Island, American Samoa in April 1995.

climate in the tropical Pacific could be associated with greater warming of SSTs in the east than in the west, partly due to a combination of cloud-albedo feedback effects and attendant changes in surface energy balance and the large-scale circulation of atmosphere and ocean.

Because of the greater warming of SSTs in the eastern tropical Pacific compared to the western tropical Pacific in the model, the mean changes in Pacific region climate due to increased CO_2 resemble not only the climate anomalies associated with present-day El Niño events in many areas, but also the decadal time-scale climate anomalies observed during the 1980s. Recent observations of warming in the Pacific show that a mean warming of SSTs can occur with El Niño-like variability superimposed upon the warmer mean SSTs (Wang, 1995), as also occurs in the coupled climate model.

Though we cannot definitively attribute the recent warming in the Pacific (and associated global warming of the 1980s) to increased CO_2 in the atmosphere, the model results lead to the possibility that CO_2 climate change in the Pacific region could have this signature. There may also be decadal time-scale variability that has elements of this signal as well. Further clarification of these effects awaits more definitive cloud-albedo feedback observational and modeling studies, improved cloud formulations, and a better understanding of observed decadal time-scale climate fluctuations in the Pacific region.

There is evidence that if such enhanced drought conditions occur over wide regions of the tropical Pacific, atoll populations may migrate in significant numbers to main high-island population centers with more stable freshwater resources. This migration of ecological refugees, as has already been noted from anecdotal evidence to be the case from recent tropical cyclone devastation, contributes to urbanization of high-island population centers, with the attendant problems of increasing population density and diminishing economic ability to support such populations.

Island countries have not yet acknowledged the problems that could occur from climate change-related stress on freshwater resources, particularly on atolls, since the majority of attention in the Pacific has been focused on possible sea level rise associated with future climate change. However, climate model results indicate that significant alterations of precipitation patterns over the tropical Pacific could begin affecting atoll populations over the next 50 years.

An essential next step would be to conduct surveys of freshwater lenses on atolls to assess the freshwater storage in these emergency water supply resources. This would help quantify the vulnerability of atoll populations to possible future increases in strength or duration of droughts.

Acknowledgments

The author thanks Robert Dixon for his helpful comments on the manuscript. A portion of this study was supported by the Office of Health and Environmental Research, U.S. Department of Energy, as part of its Carbon Dioxide Research Program. A portion of the computations were performed under the auspices of the Model Evaluation Consortium for Climate Assessment (MECCA) and CRAY Research, Inc. The National Center for Atmospheric Research is sponsored by the National Science Foundation.

References

Boer G.J.: 1993, *Clim. Dyn.* **8**, 225.

Bottomly M., Folland C.K., Hsiung J., Newell R.E., and Parker D.E.: 1990, *Global Ocean Surface Temperature Atlas*, U.K. Met. Office, Bracknell, England.

Chen T.-C., van Loon H., Wu D.-D., and Yen M.-C.: 1992, *J. Meteorol. Soc. Japan* **70**, 1137.

Graham N.E.: 1994, *Clim. Dyn.* **10**, 135.

Graham N.: 1995, *Science* **267**, 666.

Houghton J.T., Callander B.A., and Varney S.K. (eds): 1992, *Climate Change 1992: The IPCC Scientific Assessment Supplementary Report*, Cambridge University Press, Cambridge, England.

Kiladis G.N. and van Loon H.: 1988, *Mon. Wea. Rev.* **116**, 120.

Knutson T.R. and Manabe S.: 1995, *J. Clim.* **8**, 2181.

Kumar A., Leetma A., and Ji M.: 1994, *Science* **266**, 632.

Meehl G.A.: 1987, *Mon. Wea. Rev.* **115**, 27.

Meehl G.A.: 1995, *Bulletin of the American Meteorological Society* **76**, 951.

Meehl G.A. and Washington W.M.: 1995, *Clim. Dyn.* **11**, 399.

Meehl G.A. and Washington W.M.: In press, *Nature*.

Meehl G.A., Branstator G.W., and Washington W.M.: 1993, *J. Clim.* **6**, 42.

Nitta T. and Kachi M.: 1994, *J. Meteorol. Soc. Japan* **72**, 823.

Nitta T. and Yamada S.: 1989, *J. Meteorol. Soc. Japan* **67**, 375.

Oberhuber J.M.: 1988, *An Atlas Based on the COADS Data Set: The Budgets of Heat, Buoyancy and Turbulent Kinetic Energy at the Surface of the Global Ocean*, Max Planck Institute, Hamburg, Germany.

Ramanathan V. and Collins W.: 1992, *Nature* **351**, 27.

Rappa P., Tomlinson A., and Ziegler S. (eds): 1995, *Climate Change Implications and Adaptation Strategies for the Indo-Pacific Island Nations: Workshop Proceedings*. U.S. Country Studies Management Team, Washington, DC,

Ropelewski C.F. and Halpert M.S.: 1987, *Mon. Wea. Rev.* **115**, 1606.

Salinger M.J., Basher R.E., Fitzharris B.B., Hay J.E., Jones P.D., Macveigh J.P., and Schmidely-Leleu I.: 1995, *Int. J. Climatol.* **15**, 285.

Senior C.A. and Mitchell J.F.B.: 1993, *J. Clim.* **6**, 393.

Shankman P.: 1993, Samoan Exodus, *Contemporary Pacific Societies*, Lockwood V., Harding T., and Wallace B.J. (eds), Prentice Hall, Englewood Cliffs, New Jersey.

Stanley D.: 1985, *Micronesia Handbook*, Moon Publications, Chico, California.

Trenberth K.E.: 1990, *Bull. Amer. Meteorol. Soc.* **71**, 988.

Trenberth K.E. and Hurrell J.W.: 1994, *Clim. Dyn.* **9**, 303.

van Loon H. and Madden R.A.: 1981, *Mon. Wea. Rev.* **109**, 1150.

Wang B.: 1995, *J. Clim.* **8**, 267.
Washington W.M. and Meehl G.A.: 1993, *Clim. Dyn.* **8**, 211.
Zhang G.J. and McPhaden M.J.: 1995, *J. Clim.* **8**, 589.

INTEGRATED MODEL SYSTEMS FOR NATIONAL ASSESSMENTS OF THE EFFECTS OF CLIMATE CHANGE: APPLICATIONS IN NEW ZEALAND AND BANGLADESH

R.A. WARRICK,[1] G.J. KENNY,[1] G.C. SIMS[1]
N.J. ERICKSEN,[1] Q.K. AHMAD,[2] and M.Q. MIRZA[1]
[1]*Centre for Environmental and Resources Studies (CEARS), University of Waikato*
Private Bag 3105, Hamilton, New Zealand
[2]*Bangladesh Unnayan Parishad (BUP), Dhaka, Bangladesh*

Abstract. To examine the sensitivity of environmental systems to climatic variability and change, integrated model systems for climate impact assessment are being developed for New Zealand (CLIMPACTS) and Bangladesh (BDCLIM). Features common to both model systems include a global climate model, regional modules for generating climate scenarios, and models for biophysical impact analyses. For CLIMPACTS, modified ecosystem models for horticultural crops, arable crops, and pasture production are being incorporated. For BDCLIM, the emphasis is on analysis of possible changes in agroclimatic zones and hydrology, including the risks of floods and droughts. The initial emphasis of both systems is on nationwide spatial analyses, using simplified models as much as possible. The development of integrated model systems supports the needs of the respective countries in assessing scientific uncertainties, evaluating vulnerabilities, and identifying adaptation options as a basis for international reporting requirements under the U.N. Framework Convention on Climate Change and for policy and planning at national and regional levels. The major advantage of such integrated model systems is that they can readily be updated as the science of climate change advances, thus providing an evolving tool for future reassessments of climate impacts.

Key words: New Zealand, Bangladesh, coastal resources, water resources, agroclimatic zones, hydrology, integrated model system, BDCLIM, CLIMPACTS

1. Introduction

Under the provisions of the U.N. Framework Convention on Climate Change (FCCC), signatory nations are obligated to report to the Conference of the Parties regarding their vulnerability and adaptation to climate (FCCC, 1992). To meet this obligation, a large number of national assessments of climate change effects have recently been conducted, stimulated by financial assistance provided to developing countries through international (e.g., the Global Environmental Facility) and national (e.g., the U.S. Country Studies Program) sources. Overall, such assessments represent a tremendous global investment of financial and human resources.

However, in relation to the investment, one questions the "shelf life" of such assessments. The scientific knowledge of climate change is continually expanding. How easily can improved scenarios of climate change, for example, be taken into account? How easily can improved scientific knowledge of crop responses or coastal processes be incorporated and vulnerability reassessed? In the majority of cases, the answer is: not very easily. National assessments tend to be rather rigid, one of a kind, result-oriented exercises. Future reassessments (which will be required under the FCCC) will often mean starting at the beginning. Is there a more flexible alternative?

The purpose of this paper is to present such an alternative — an integrated model-based system for assessing the effects of climate change. The paper begins by first briefly describing the methodology. The methodology is then illustrated by reporting its application in a developed country, the CLIMPACTS system for New Zealand, and in a developing country, the BDCLIM system for Bangladesh. Finally, some judgments are offered about the prospects for the methodology's wider application in other national situations.

Water, Air, and Soil Pollution **92:** 215–227.
© 1996 *Kluwer Academic Publishers. Printed in the Netherlands.*

2. A Methodological Framework

2.1 INTEGRATED ASSESSMENT APPROACHES — THE CONTEXT

The development of methods for assessing climate change impacts is well summarized in the Intergovernmental Panel on Climate Change (IPCC) technical guidelines for assessing climate change impacts and adaptations (Carter *et al.*, 1994). An early attempt at a partially integrated approach to impact assessment was a compilation of national case studies of impacts on agriculture carried out in the mid-1980s (Parry *et al.,* 1988a, b). Several studies directly followed from this work, and essentially became an integration of impacts, again with a focus on agriculture (Carter *et al.*, 1991; Kenny and Harrison, 1992; Kenny *et al.*, 1993a, b). A further evolution of this type of approach has been the linking of output from agricultural impact models to economic models (e.g., Rosenzweig and Parry, 1994).

In parallel with these developments was the emergence of computer-based, integrated model systems. The development of integrated systems for understanding the implications of climate and environmental change (Dowlatabadi and Morgan, 1993a; Nakicenovic and Toth, 1994), including policy implications, has been expanding. Integrated systems include IMAGE (Rotmans, 1990; Alcamo, 1994), ICAM (Dowlatabadi and Morgan, 1993b), and PAGE (Hope *et al.*, 1993). Regionally, such systems have been developed for Europe (the ESCAPE model: CRU & ERL, 1992a, b; Hulme *et al.,* 1994; Rotmans *et al.,* 1994) and China (Hulme *et al.*, 1992), for example. The latter systems incorporate simple global climate and methods for generating regional scenarios of climate change based on complex general circulation model (GCM) output. The work presented in this paper builds directly on these efforts.

2.2 AN INTEGRATED MODEL FRAMEWORK

Collaborative research programs have been implemented for New Zealand and Bangladesh which are aimed at the development of integrated, computer-based systems with the capacity for conducting comprehensive country-scale assessments over time and space. Central to these efforts is the integration of models and data in a manner that allows for quick, systematic assessments and reassessments of model sensitivities and uncertainties in climate change related impacts.

The model systems being developed differ substantially from methods employed in past assessments. In most past assessments, the approach has been to use specific sectoral models (e.g., process-based crop models; see, for example, Rosenzweig and Iglesias, 1994) and specific climate change scenarios. The results of such work, although valuable, can rapidly become outdated. In contrast, the integrated model system described here has the flexibility to allow the user to generate a vast number of climate change scenarios, to adjust model parameters for sensitivity analyses, and to readily update and supplement model components as the science advances. This is the crucial difference.

The conceptual framework for the model-based systems is shown in Figure 1. At the top end is an integration of models and data for producing time-dependent scenarios of regional changes in climate and associated environmental effects. This is accomplished by linking a global climate model, MAGICC (Model for the Assessment of Greenhouse-Gas Induced Climate Change; Wigley and Raper, 1992, 1993; Wigley, 1993; Osborn and Wigley, 1994), with regional patterns of climate change, as derived from various GCM results. The core of MAGICC is a one-dimensional, energy-balance, upwelling-diffusion climate model. The inputs

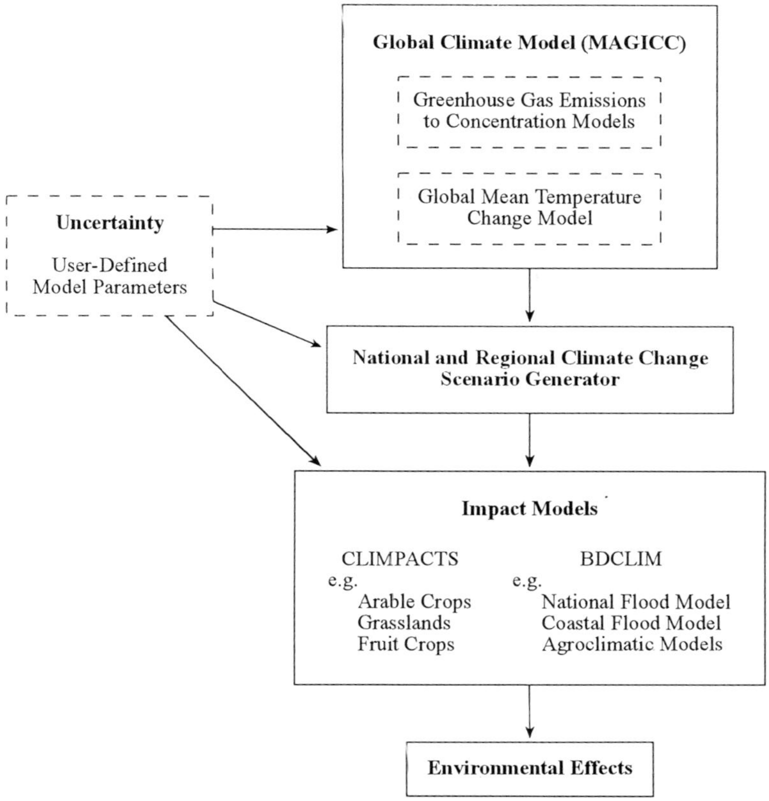

Fig. 1. Conceptual framework for model-based systems.

to MAGICC are emissions of greenhouse gases, and the outputs are global mean temperature and sea level changes. MAGICC includes models for converting emissions to atmospheric concentrations (e.g., a carbon cycle model) and for converting concentrations to changes in radiative forcing. The key to MAGICC is its simplicity and flexibility. The model can quickly generate time-dependent temperature changes for any scenario of greenhouse gas emissions to 2100. A variant of MAGICC was used by the IPCC (Houghton *et al.*, 1990, 1996) to make projections of global temperature and sea level changes.

Standardized regional patterns of temperature or precipitation change (expressed as the change per degree of global warming) are derived from the output of GCM experiments and incorporated into the model system. A diversity of GCM patterns is used to capture the range of uncertainty in global models. These standardized patterns are scaled by the projection of global-mean temperature change from MAGICC in order to generate time-dependent regional scenarios of climate change for user-selected years between 1990 and 2100. A principal assumption in the application of this method is that the large-scale patterns of greenhouse gas induced climate changes remain fixed with time.

A basic dataset of the integrated system is the baseline climatology. The range of climate variables and their spatial and temporal resolution in the system depend on climate data availability, the requirements of impact models, and computational demands. For simple spatial models, the minimum requirement is monthly temperature and rainfall for a standard 30-year period. Inclusion of the standard deviations, as well as the means, allows for analyses of the effects of climate variability and extremes as well as climate averages. The projected changes in climate are then used to perturb this baseline climatology.

Both the baseline and the perturbed climatologies are used to drive a set of sectoral impact models. The output from the integrated system currently emphasizes the spatial and temporal patterns and distributions of climatic and biophysical effects, but a longer-term goal is to incorporate social and economic components as well. The structure of the system also allows for incorporation of different management options to enable assessment of adaptation measures (e.g., changing cultivars, changing planting dates). The integrated system is PC-based and can be readily updated as the science advances.

The details of the application of the integrated model system to New Zealand and Bangladesh are discussed in the following sections.

3. CLIMPACTS: An Integrated Model System for New Zealand

3.1 THE CLIMPACTS SYSTEM

A description of preliminary development of the CLIMPACTS system is provided in Kenny *et al.* (1995). Key components of these developments, as well as more recent developments to the system and its application, are summarized here.

3.1.1 Reference Climatologies, Scenario Generation, and Land Use Data

The climate of New Zealand is dominated by the effects of the Southern Ocean, with prevailing westerly winds, and ranges from subtropical in the north to cool temperate in the south. The orography of New Zealand (approximately 75% of the land mass is 200 m or more above mean sea level) has a strong modifying influence on the climate, so that the western part of the country is generally much wetter than the east.

Reference climatologies for New Zealand were developed for monthly rainfall, temperature, and derived solar radiation data (1951-1980 normals), which were interpolated to a 0.05° latitude by 0.05° longitude grid for the New Zealand land surface. The interpolation was accomplished using methods described in detail by Mitchell (1991) and originally developed by Hutchinson (1989). The mean annual temperature, for example, is shown in Figure 2a. Daily weather data, covering the same 30-year period, are being incorporated for a minimum of 10 sites.

Patterns of climate change for New Zealand were derived from GCM output (Mullan and Salinger, 1994). The CSIRO4 model (Gordon and Hunt, 1991) and the GFDLQ model (Manabe and Wetherald, 1987) were used in the initial development of the CLIMPACTS system. A statistical approach, using principal components analysis and linear regression, was applied to infer local-scale changes from these GCMs. The change fields for temperature and precipitation for both GCMs were also interpolated to a 0.05° latitude by 0.05° longitude grid for the North and South Islands. The standardized patterns of change (i.e., the changes per degree of global warming) are scaled by the global temperature change from MAGICC to give regional patterns of climate change over time. These, in turn, are used to perturb the reference climatology and give a "new" climate for New Zealand (see also Section 2.2). An example of such a new climate by 2100, based on the IPCC IS92a emission scenario and the CSIRO4 GCM pattern, is shown in Figure 2b.

An important database component of the CLIMPACTS system is the New Zealand Land Use Capability (LUC) classification (Water and Soil Division, 1979). The data are held as part of

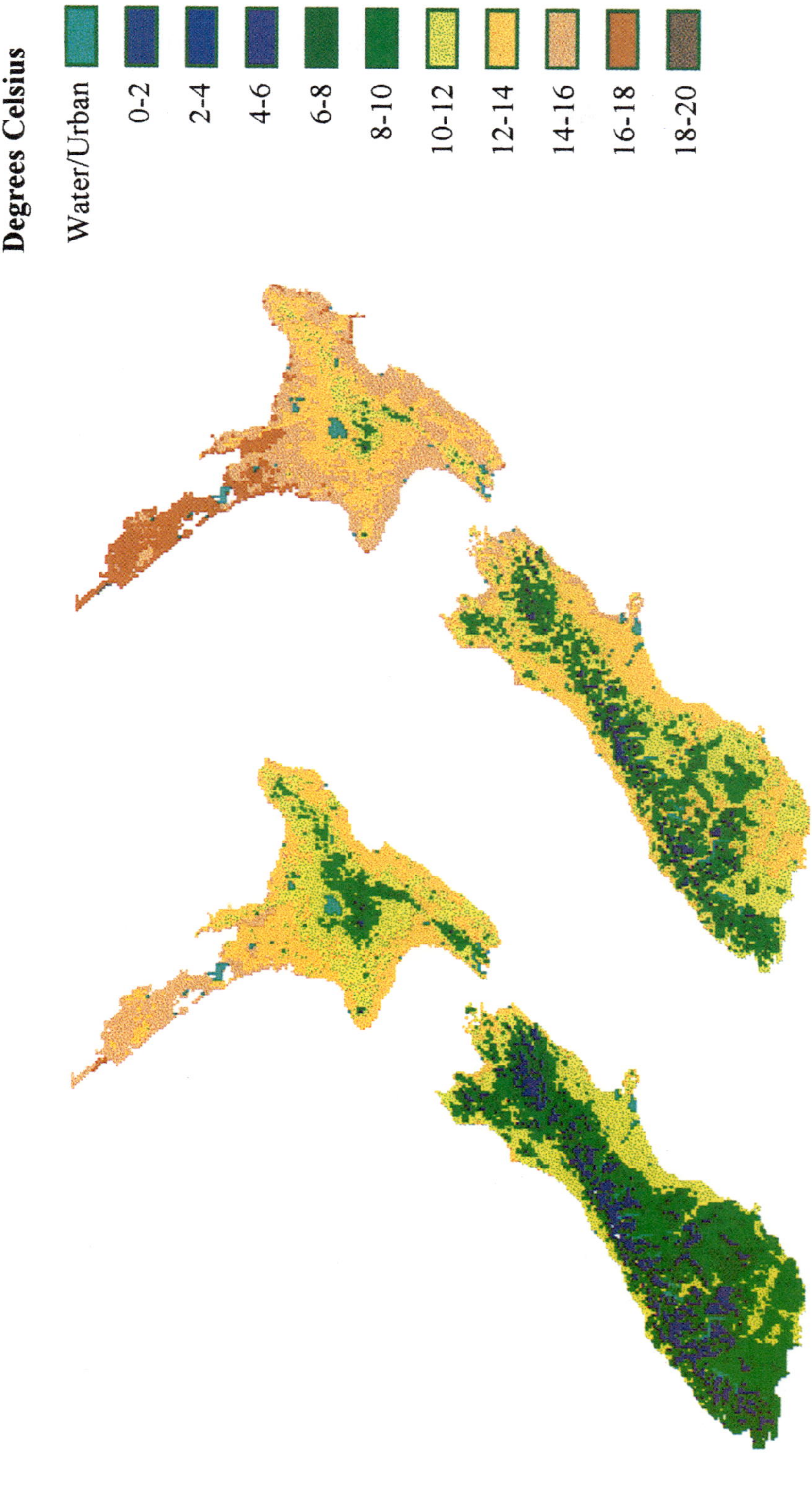

Fig. 2. (left) Mean annual temperature interpolated from data from 1951-1980; (right) climate in 2000.

the New Zealand Land Resource Inventory. Of the eight LUC classes, I to IV are mostly crop, dairy, and horticultural land, with I as the best arable land (flattest) and IV as generally marginal arable land. Classes V to VII are generally suitable for only pastoral agriculture or forestry, and Class VIII is catchment protection land. Inclusion of these data in the CLIMPACTS system substantially reduces the computational time (for example, the substantial area of nonarable land can be excluded from simulations for arable and horticultural crops) and, importantly, gives more realistic representations of crop suitability.

3.1.2 Sectoral Impacts Models

New Zealand's vulnerability to climate variability and change has grown with the rapid development of pastoral-based agriculture and other land-based industries over the past century. For a developed market economy, New Zealand has an unusually high dependence on agriculture. It has been estimated that losses to the agriculture sector due to climatic variability average about NZ$600 million (US$400 million) per year (Wratt and Matthews, 1992). For this reason, agriculture is a primary focus of CLIMPACTS.

Within the CLIMPACTS research program, climate change issues need to be addressed at the different scales of assessment, driven largely by the information needs of end users (Table I). A hierarchy of modeling approaches is thus being developed for application across a range of temporal and spatial scales. Initially, the focus has been on nationwide spatial assessments, in order to address questions concerning possible changes in area-wide yield or production, cropping areas, or production risks. This has involved the use of simple agroclimatic models for maize (Kenny *et al.*, 1996; Tate *et al.*, in press) and kiwifruit (Salinger and Kenny, 1995), and climate profiles for subtropical pasture species (Campbell and Mitchell, personal communications) based on the methods of Nix (1986) and Yee and Mitchell (1991).

To address detailed questions concerning physiological responses to environmental variations, more complex, physiologically based models are being adapted and refined for application at selected sites. These include models for wheat (Jamieson *et al.*, in press), sweet corn and maize (Muchow *et al.*, 1990; Wilson *et al.*, 1995), oats and peas (based

TABLE I
CLIMPACTS research strategy

	Stage 1 Develop CLIMPACTS System, NZ-Wide Applications	Stage 2 Develop Site-Specific Capacity	Stage 3 Develop Subregional Capacity	Stage 4 Develop Economic and Land Use Analysis Capability
Spatial Scale (Resolution)	North or South Island (~5 km × 5 km)	~10 sites (nonspatial)	Regional council scale (~1 km × 1 km)	Combination
Time Scale of Climate Forcing	Monthly/seasonal/ annual	Hourly/daily	Combinations	Model dependent
Environmental Modeling Approach	Simplified impact models and indices	Complex biophysical process models	Mix of models	Production, land-use models
Uses	Broad-scale sensitivity analyses; national impact assessments	Detailed analyses of biophysical responses; evaluation of management options	Sensitivity analyses at local level; resource management & planning	Sectoral sensitivity analyses; economic assessments and management
Potential Users	Government	Private sector; modelers	Local authorities; planners	Industry; local government

on Jamieson *et al.*, in press), pasture production (a modified version of a model developed by McCall, 1984), kiwifruit phenology (McPherson *et al.*, 1992; Hall and McPherson, in press), and kiwifruit and apple growth. Within the CLIMPACTS program, work has progressed on scaling up site-specific models for area-wide applications, including methods for simplifying site-specific models (e.g., Brisson *et al.*, 1992) and for spatially interpolating site-specific results.

3.2 AN EXAMPLE OF CLIMPACTS SYSTEM APPLICATION

Some applications of the CLIMPACTS system can be illustrated by focusing on kiwifruit. A model for mapping kiwifruit suitability in New Zealand has been developed by Salinger and Kenny (1995). Spatial suitability is determined by overlaying output from three indices. First, there is a winter chilling threshold. As a perennial, warm temperate crop, kiwifruit require a period of winter cold to promote budburst in the spring. Empirical evidence (Salinger *et al.*, 1993) suggests that the mean winter (May-July) temperature should be less than 11°C. Second, once fruit have set, a period of sufficient warmth over the summer is required for the fruit to develop to maturity; generally, a minimum temperature accumulation (thermal time or degree days) of 1100°C days above a base temperature of 10°C is required. Third, availability of water is important for fruit development, with an annual rainfall requirement of at least 1,250 mm under rainfed conditions. Where all three conditions are met, kiwifruit suitability is defined as optimal (Salinger and Kenny, 1995). It is possible, within CLIMPACTS, to conduct analyses of the spatial changes in kiwifruit suitability for an almost unlimited number of scenarios of climate change, and for any specified time in the future up to 2100. Such spatial analyses can help identify key areas where suitability may decline or new areas where it may increase.

For example, Figure 3 compares the model-produced spatial distributions of kiwifruit suitability for current climate (based on the 1951-1980 climatology) to a scenario of climate change. The scenario includes the following user-defined specifications:

- greenhouse gas emission scenario: IPCC IS92a
- climate model parameters: MAGICC best-estimate parameter values
- GCM climate change pattern: CSIRO4
- year: 2100
- land suitability: LUC Classes I-III.

As shown in Figure 3, this climate change scenario results in a significant reduction in the area of land climatically suitable for kiwifruit production in the northern half of the North Island, where actual production is currently centered. In these areas, the winters become too warm for sufficient winter chilling, resulting in suboptimal conditions for kiwifruit. Other areas become marginal for kiwifruit because of insufficient rainfall. There is limited southward expansion of optimal areas, largely because of the drier climate, particularly on the east coast of the South Island. Thus, under this scenario, there are large shifts in suitability for this particular horticultural activity. Sensitivity analyses can easily be performed to test the sensitivity of this pattern to different assumptions regarding, for example, greenhouse gas emission rates, climate responses, GCM patterns, and kiwifruit model parameters. Using CLIMPACTS, each such analysis takes about 15 to 30 minutes.

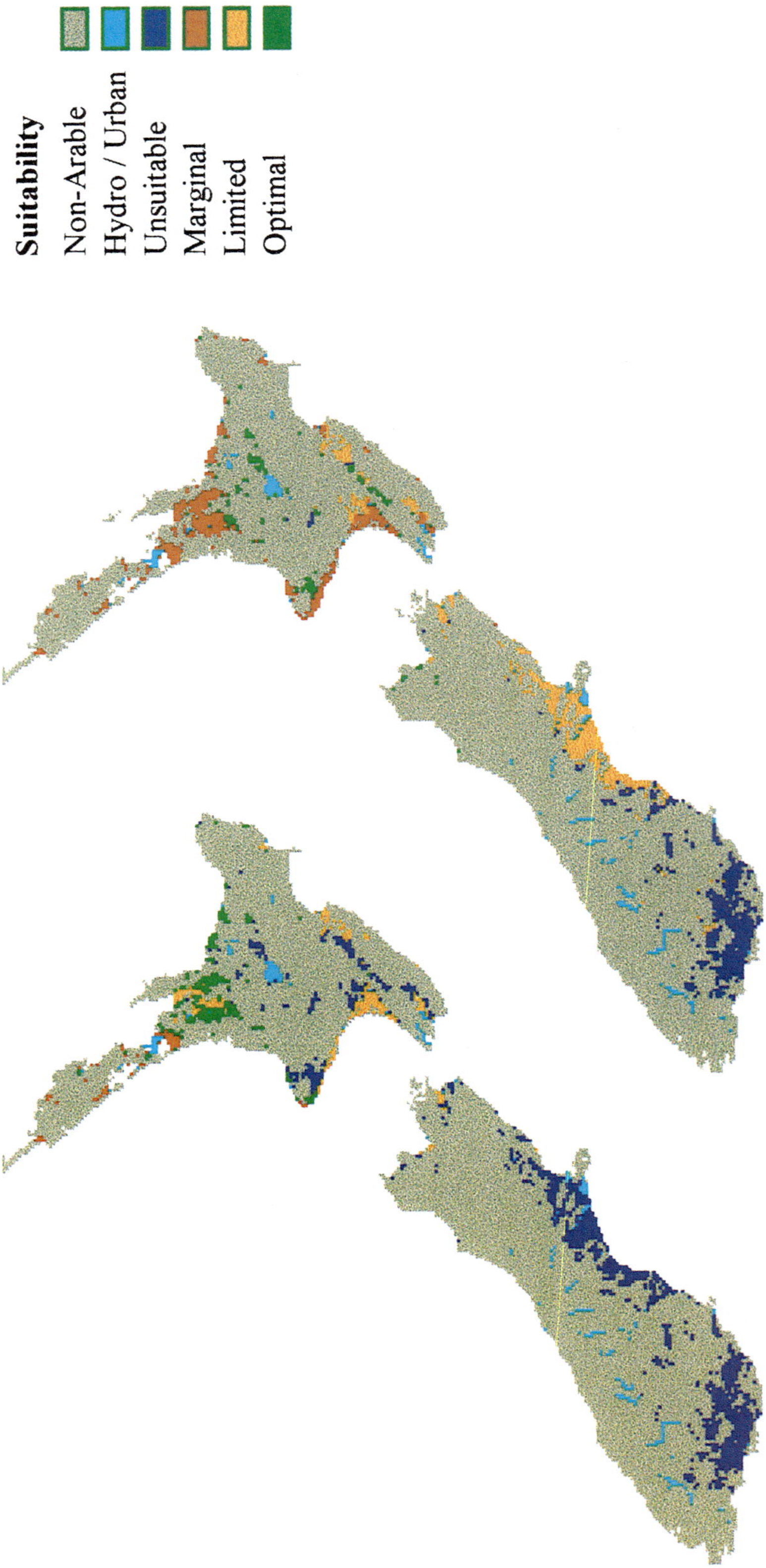

Fig. 3. Comparison of spatial distributions for kiwifruit suitability for (left) current and (right) future climate generated by CLIMPACTS.

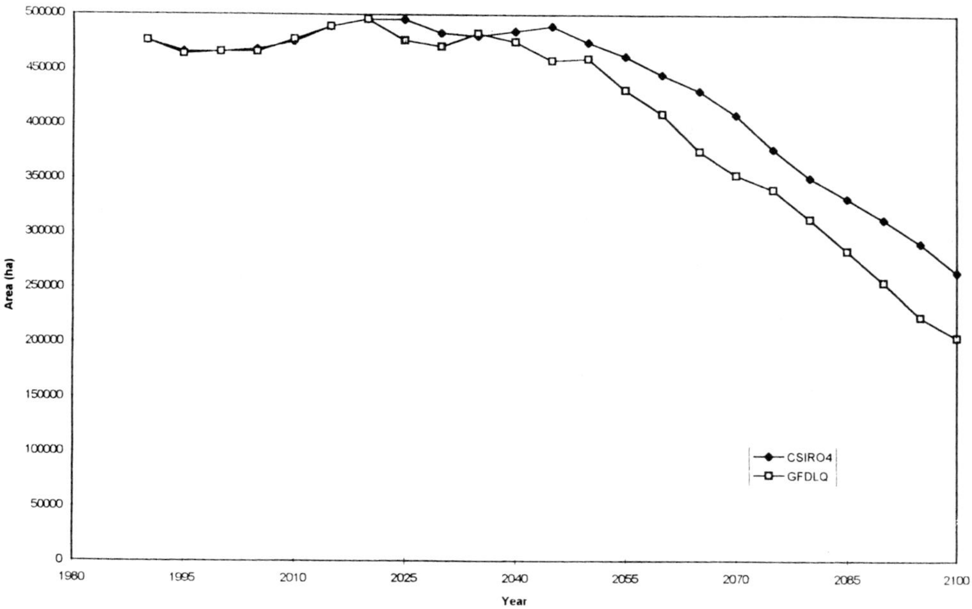

Fig. 4. Temporal pattern of suitability for kiwifruit in the North Island generated by CLIMPACTS.

It is also possible to examine the rates and thresholds of change over time. For example, Figure 4 illustrates the changing temporal pattern of suitability for kiwifruit in the North Island, with the same model parameters noted above, but run iteratively on five-year time steps (a more time-consuming analysis). Until about the middle of the 21st century, the decline in existing optimal areas (largely as a consequence of warmer winters and hence insufficient winter chilling) is compensated by an expansion of new optimal areas (as a consequence of warmer summers in cooler areas). At this point in time, however, a threshold is reached, and a steady decline in net area suitable for kiwifruit occurs thereafter.

4. BDCLIM: An Integrated Model System for Bangladesh

4.1 VULNERABILITY TO CLIMATE VARIABILITY

The climate of Bangladesh is dominated by the southwest and northeast monsoons. Temperature variations are relatively small, but important. There are significant rainfall gradients, which are driven largely by the pattern of the monsoon and, in some areas, by elevational gradients. Approximately 75% of the total land area is less than 30 m above mean sea level.

Like New Zealand, Bangladesh is heavily reliant on agriculture, and therefore strongly affected by natural variations in climate. In the larger Bengal region, rice culture is dominant. Vulnerability to climatic variations in this region is very much a function of physical location, topography, timing and magnitude of climatic events, nature of land use, the state of economic underdevelopment, and high population density and growth rates. If, as expected, the population of Bangladesh doubles in the next 30 years, enormous pressure will be placed on the already limited land resource; this can only exacerbate existing vulnerability to climate variability. Small changes in climate, therefore, could potentially have a much greater relative impact in

Bangladesh than in countries with low, stable populations like New Zealand (Ahmad *et al.*, 1996).

The three main weather-related phenomena that affect Bangladesh are the summer monsoonal rains and associated flooding, drought through the November-May dry period, and tropical cyclones in the Bay of Bengal. If rainfall increases with climate change, as many GCMs suggest for this region, there is the distinct possibility of a greater frequency of damaging floods. Changes in the dry period also need to be quantified as much as possible in order to determine possible future demands for water. Again, the uncertainties need to be expressed.

4.2 THE BDCLIM SYSTEM

A prototype integrated model system for Bangladesh (BDCLIM) is currently being developed. BDCLIM is conceptually and methodologically similar to CLIMPACTS. Thus, BDCLIM also includes MAGICC, the results of which are used to scale GCM output for the region to produce time-dependent climate change scenarios for Bangladesh. These scenarios in turn are used to perturb a baseline climatology of temperature and rainfall, developed from a 30-year (1961-1990) record of stations maintained by the Department of Meteorology. For precipitation, the spatial coverage is being extended using the records maintained by the Bangladesh Water Development Board.

It is evident that a key factor in the environmental and economic systems of Bangladesh is water. With respect to climate change, the implications of changes in flood and drought occurrence are paramount, because they play fundamental roles in determining the variability and long-term sustainability of land-use patterns and practices in Bangladesh. For this reason, the development of the prototype BDCLIM system is focusing on the following three water-related sectoral impact models as the foundation on which to build more detailed assessments of vulnerability and adaptation:

- river floods (encompassing the hydrological effects of climate change on the large river basins of the Ganges, Brahmaputra, and Meghna rivers)
- agro-ecological systems (as influenced by flood depth, duration, and frequency)
- coastal inundation (as influenced by sea level rise, with additional influences to be incorporated).

River discharge is a common element of each sector and an important link to climate change. For this reason, attention has focused on the development of a simple empirical model that can estimate changes in peak discharges at the border stations of the Ganges, Brahmaputra, and Meghna rivers from changes in basin-wide precipitation. The catchments of these rivers cover 1.5 million km^2 in India, Nepal, Tibet, Bhutan, and Bangladesh. Time-series data for precipitation and river discharge were obtained from a wide variety of sources. Principal components analysis was used for the regionalization of precipitation and the reduction of variables. Multiple regression models were then developed that can predict annual mean and peak discharge from precipitation amounts.

The impact models for BDCLIM are currently under development. The first impact model completed for BDCLIM will be a national flood model. The outputs of the flood peaks from the regional precipitation-discharge model will be transformed into flood depths using a GIS-based land elevation model. These spatial data will allow the development of discharge-flood depth relationships throughout the country. These relationships will provide the basis for BDCLIM to make country-wide estimates of changes in flood depth from scenarios of climate change, both for mean annual flood and selected extreme floods.

The next impact model to be developed will be a coastal inundation and flood model. In the first instance, a simple inundation model will be developed, combining a digital elevation model for Bangladesh with sea level rise data (output from MAGICC). A future development will involve estimations of coastal flood/inundation depth throughout the coastal zone, as a function of both river discharge and sea level rise. Generalized relationships based on these data will allow BDCLIM to estimate coastal flood/inundation depth for any specific scenario of river discharge change and sea level rise.

Finally, the information on flood depths, along with the national climate model, will be integrated with agroclimatic models in order to assess possible changes in crop calendars and cropping patterns in response to climate change. This is especially important for the aman rice crop — the staple crop of Bangladesh — since flooding during the monsoon has a substantial influence on its production. The spatial coverage of the floods will be linked with the agroclimatic models to assess the possible spatial impacts on aman rice crop.

The outputs of all these climate and sectoral models of BDCLIM are scenarios of impacts that aim to capture the spatial and temporal dimensions of change, and that are displayed in tabular or color map form, similar to a Geographic Information System environment. The prototype version of BDCLIM is scheduled for completion in 1997.

5. Summary and Conclusions

To make informed decisions, policy makers need tools that enable them to estimate the implications of climate change over a wide range of policy options, and that can provide a concise overview of the uncertainties surrounding global climate change. However, current assessment approaches, while often serving short-term needs quite adequately, tend to be cumbersome, inflexible, and not easily repeatable to test policy options, sensitivities to input assumptions, or new scientific knowledge.

In this paper we described an alternative approach — integrated model systems. The key advantages of the systems, because they are PC- and model-based, are the following:

- quick running
- flexible
- modular, and therefore easily updated
- promote sensitivity analyses of model inputs and assumptions
- allow examination of ranges of uncertainties
- provide spatial as well as temporal analyses.

However, there are disadvantages of integrated model systems as well. First, basic data needed to satisfy requirements for the models' spatial and temporal resolution, calibration, and validation can be difficult to acquire (either because they simply do not exist, or because of political restrictions or proprietary rights). Second, the intricacies of computer modeling may often exceed the expertise available in some countries. Third, relatively large initial investments are required, both in terms of time and commitment of interdisciplinary expertise; team collaboration is essential.

Finally, it should be emphasized that integrated model systems should not be viewed as a replacement for other approaches to vulnerability and adaptation assessment. Rather, they are complementary. It is only through such a combination of quantitative and qualitative, local and national, and scientific and policy analyses that the complete picture of climate change vulnerability and adaptive capacity will begin to emerge.

Acknowledgments

The authors gratefully acknowledge the research support provided by the Foundation for Research, Science and Technology (FRST contract UOW506) for the development of CLIMPACTS, and the Ford Foundation for the development of BDCLIM. Also acknowledged are the contributions of Brett Mullan and Jim Salinger (NIWA), Neil Mitchell (University of Auckland), Bruce Campbell (AgResearch), Hugh McPherson (HortResearch), Pete Jamieson (Crop & Food Research), and their colleagues to the development of CLIMPACTS.

References

Ahmad, Q.K., Warrick, R.A., Ericksen, N.J., and Mirza, M.Q.: 1996, The implications of climate change for Bangladesh: a synthesis, in: Warrick, R.A. and Ahmad, Q.K. (eds), *The Implications of Climate and Sea-Level Change for Bangladesh*, Kluwer Academic Publishers, Dordrecht, The Netherlands, pp, 1-34.

Alcamo J. (ed): 1994, *IMAGE 2.0: Integrated Modeling of Global Climate Change* (reprinted from *Water, Air and Soil Pollution* **76**), Kluwer, Dordrecht, The Netherlands.

Brisson N., King D., Nicoullaud B., Ruget F., Ripoche D., and Darthout R.: 1992, *Eur. J. Agron.* **1(3)**, 163.

Carter T.R., Parry M.L., and Porter J.H.: 1991, *International Journal of Climatology* **11**, 251.

Carter T.R., Parry M.L., Harasawa H., and Nishioka S.: 1994, *IPCC Technical Guidelines for Assessing Climate Change Impacts and Adaptations,* Department of Geography, University College London, United Kingdom and Center for Global Environmental Research, National Institute for Environmental Studies, Tsukuba, Japan.

CRU & ERL (Climatic Research Unit and Environmental Resources Ltd): 1992a, *Development of a Framework for the Evaluation of Policy Options to Deal with the Greenhouse Effect: A Scientific Description of the ESCAPE Model,* Report to the Commission of the European Communities.

CRU & ERL (Climatic Research Unit and Environmental Resources Ltd): 1992b, *Development of a Framework for the Evaluation of Policy Options to Deal with the Greenhouse Effect: A User Manual for the ESCAPE Software,* Report to the Commission of the European Communities.

Dowlatabadi H. and Morgan M.G.: 1993a, *Science* **259**, 1813.

Dowlatabadi H. and Morgan M.G.: 1993b, *Energy Policy* **21**, 209.

FCCC: 1992, *United Nations Framework Convention on Climate Change*, UNEP/WMO Information Unit on Climate Change (IUCC), Geneva, Switzerland.

Gordon H.B. and Hunt B.G.: 1991, *International Journal of Climatology* **11**, 347.

Hall A.J. and McPherson H.G.: In press, *Acta Horticulturae.*

Hope C.W., Anderson J., and Wenman P.: 1993, *Energy Policy* **21**, 327.

Houghton, J.T., Jenkins, G.J., and Ephramus, J.J. (eds): 1990, *Climate Change: The IPCC Scientific Assessment*, Cambridge University Press, Cambridge, United Kingdom.

Houghton, J.T., Meiro Filho, L.G., Callander, B.A., Harris, N., Kattenberg, A., and Maskell, K. (eds): 1996, *Climate Change 1995: The Science of Climate Change*, Cambridge University Press, Cambridge, United Kingdom.

Hulme M., Raper S.C.B., and Wigley T.M.L.: 1994, An integrated framework to address climate change (ESCAPE) and further developments of the global and regional climate modules (MAGICC), in Nakicenovic N. and Toth F. (eds), *Integrative Assessment of Mitigation, Impacts and Adaptation to Climate Change*, IIASA Collaborative Paper Series, Luxembourg.

Hulme M., Wigley T.M.L., Jiang T., Zhao Z-ci., Wang F., Ding Y., Leemans R., and Markham A.: 1992, *Climate Change due to the Greenhouse Effect and its Implications for China,* WWF, Gland, Switzerland.

Hutchinson M.F.: 1989, *CSIRO Division of Water Resources Tech. Memo.* **89/5**, 95.

Jamieson P.D., Semenov M.A., and Francis G.S.: In press, *Field Crops Research.*

Kates R.W.: 1985, The interaction of climate and society, in: Kates R.W., Ausubel J.H., and Berberian M. (eds), *Climate Impact Assessment: Studies of the Interaction of Climate and Society,* SCOPE 27, Wiley, Chichester, England.

Kenny G.J. and Harrison P.A.: 1992, *Climate Research* **2**, 113.

Kenny G.J., Harrison P.A., and Parry M.L. (eds): 1993a, The Effect of Climate Change on Agricultural and Horticultural Potential in Europe, Research Report No. 2, Environmental Change Unit: University of Oxford, England.

Kenny G.J., Harrison P.A., Parry M.L., and Porter J.R.: 1993b, *European Journal of Agronomy* **2(4)**, 243.

Kenny G.J., Warrick R.A., Mitchell N.D., Mullan A.B., and Salinger M.J.: 1995, *Journal of Biogeography* **22(4/5)**, 883.

Manabe S. and Wetherald R.T.: 1987, *Journal of Atmospheric Science* **44**, 1211.

McCall D.G.: 1984, A Systems Approach to Research Planning for North Island Hill Country, PhD thesis, Massey University, Palmerston North.

McPherson H.G., Hall A.J., and Stanley C.J.: 1992, *Journal of Horticultural Science* **67(4)**, 509.

Mitchell N.D.: 1991, *Journal of the Royal Society of New Zealand* **21(1)**, 13.

Muchow R.C., Sinclair T.R., and Bennett J.M.: 1990, *Agronomy Journal* **82**, 338.

Mullan A.B. and Salinger M.J.: 1994, CLIMPACTS 1993/94*: Climate Change Scenarios from GCM Output and Palaeoclimate Data,* NIWA Report, February.

Nakicenovic and Toth F. (eds): 1994, *Integrative Assessment of Mitigation, Impacts and Adaptation to Climate Change,* IIASA Collaborative Paper Series, Laxenburg, Austria.

Nix H.A.: 1986, A biogeographic analysis of Australian Elapid snakes, in Longmore, R. (ed) *Atlas of Elapid Snakes of Australia,* Australian Government Publishing Services, Canberra, Australia, pp. 4-15.

Osborn T.J. and Wigley T.M.L.: 1994, *Climate Dynamics* **9**, 181.

Parry M.L., Carter T.R., and Konijn N.T. (eds): 1988a, *The Impact of Climatic Variations on Agriculture. Volume 1. Assessments in Cool Temperate and Cold Regions,* Kluwer, Dordrecht, The Netherlands.

Parry M.L., Carter T.R., and Konijn N.T. (eds): 1988b, *The Impact of Climatic Variations on Agriculture. Volume 2. Assessments in Semi-Arid Regions,* Kluwer, Dordrecht, The Netherlands.

Rosenzweig C. and Iglesias A. (eds): 1994, *Implications of Climate Change for International Agriculture: Crop Modeling Study,* U.S. Environmental Protection Agency, Office of Policy, Planning, and Evaluation, Climate Change Division, Adaptation Branch, Washington, DC.

Rosenzweig C. and Parry M.L.: 1994, *Nature* **367**, 133.

Rotmans J.: 1990, IMAGE: *An Integrated Model to Assess the Greenhouse Effect,* Kluwer, Dordrecht, The Netherlands.

Rotmans J., Hulme M., and Downing T.E.: 1994, *Global Environmental Change* **4(2)**, 97.

Salinger M.J. and Kenny G.J.: 1995, *New Zealand Journal of Crop and Horticultural Science* **23**, 173.

Salinger M.J., Kenny G.J., and Morley-Bunker M.J.: 1993, *New Zealand Journal of Crop and Horticultural Science* **21**, 235.

Tate K.R., Giltrap D.J., Parshotam A., Hewitt A.E., Ross D.J., Kenny G.J., and Warrick R.A.: In press, Impacts of climate change on soils and land systems in New Zealand, in: Pearman G. and Manning M.R. (eds), *Greenhouse* 94, CSIRO.

Water and Soil Division: 1979, *Our Land Resources,* A bulletin to accompany New Zealand land resource worksheets, Ministry of Works and Development, Wellington, New Zealand.

Wigley T.M.L.: 1993, *Tellus* **45B**, 409.

Wigley T.M.L. and Raper S.C.B.: 1992, *Nature* **357**, 293.

Wigley T.M.L. and Raper S.C.B.: 1993, Future changes in global-mean temperature and sea level, in: Warrick R.A., Barrow E.M., and Wigley T.M.L. (eds), *Climate and Sea Level Change: Observations, Projections, and Implications,* Cambridge University Press, Cambridge, United Kingdom, pp. 111-133.

Wilson D.R., Muchow R.C., and Murgatroyd C.J.: 1995, *Field Crops Res.* **43,** 1.

Wratt D.S. and Matthews W.A.: 1992, *Submission for Funding from Output Class 33: Atmosphere and Climate Research,* Ministry for Research, Science and Technology, Wellington, New Zealand.

Yee T.W. and Mitchell N.D.: 1991, *Journal of Vegetation Science* **2**, 587.

A PROCESS FOR EVALUATING ANTICIPATORY ADAPTATION MEASURES FOR CLIMATE CHANGE

J.B. SMITH, S.E. RAGLAND, G.J. PITTS

Hagler Bailly Consulting, Inc., P.O. Drawer O, Boulder, CO 80306, U.S.A.

Abstract. Many countries are preparing national climate change action plans that describe specific measures they are taking to mitigate greenhouse gas emissions and adapt to the potential effects of climate change. Among the reasons for preparing such plans are that climate change is likely to occur, and many anticipatory measures that would be taken in response to climate change are "no regret" measures that will produce benefits even if climate does not change. Additionally, these plans can serve as communications required by the U.N. Framework Convention on Climate Change. We propose here an assessment process for anticipatory adaptation measures that will enable countries to identify and select measures to adapt to climate change. These measures anticipate potential climate changes and are flexible enough to meet objectives under a wide variety of future climate conditions. The process builds on assessments of vulnerability by focusing on adaptation measures for the most sensitive regions, or populations, within a country. Potential anticipatory adaptation measures are identified, and two or three are chosen based on expert judgment and analysis regarding which measures would produce the greatest benefits and be easiest to implement. Analytic techniques are used to assess the benefits and costs of each of the measures and evaluate barriers to implementation. The measure that is most cost-effective and is easiest to implement is selected. We illustrate the application of the process by examining a hypothetical forest threatened by climate change.

Key words: adaptation, benefit-cost analysis, cost-effectiveness analysis, multicriteria analysis

1. Introduction

Many countries are preparing national climate change action plans that describe the specific measures they are taking to mitigate greenhouse gas emissions and adapt to the potential effects of climate change. Among the reasons for preparing such plans are that climate change is likely to occur (IPCC, 1995), and many measures that would be taken in response to climate change are "no regret" measures, which means that they will produce benefits even if climate does not change.

In addition, many developing countries and countries with economies in transition face numerous opportunities in the near future to incorporate climate change in their infrastructure and institutions. Many infrastructure projects for managing climate-sensitive resources will be built in these countries. These projects include reservoirs, hydropower facilities, coastal development, and irrigation systems. In addition, institutions and plans for managing climate sensitive resources, such as agriculture extension services, coastal zone management plans, sustainable development plans, and others, are being developed. The design of these infrastructure projects, institutions, and plans can and should incorporate the potential for climate change to significantly affect the natural resources being addressed.

A few articles have been published on how governments could address adaptation to climate change (e.g., Titus, 1990; Goklany, 1995; Fankhauser, 1996; Smith, in press a). Carter *et al.* (1994) described general steps that can be taken to assess adaptation. This paper describes specific techniques that can be applied by governments to assess adaptation needs, identify anticipatory adaptation measures, and select those measures that are most cost-effective and can be most readily implemented. In describing this process, a hypothetical example is offered to

Water, Air, and Soil Pollution **92:** 229–238.
© 1996 *Kluwer Academic Publishers. Printed in the Netherlands.*

present and illustrate the specific analytic techniques. (This process is also described in Smith, in press b, although the example is for water resources.)

For this example, we assume we are examining a tropical forest region that is highly fragmented as a result of past and ongoing development. Furthermore, we assume that these fragments contain a large number of locally endemic species, including some relic species that are especially sensitive to climate change and that the forest supports an indigenous population. We assume that annual precipitation is usually high, but that changes in climate may decrease precipitation and increase average temperature.

2. Selection of Vulnerable Sectoral Components

The first step in the anticipatory adaptation assessment process is to select a vulnerable region, population, or other sectoral component (e.g., natural reserves) to focus on. This should begin with an assessment of the vulnerability to climate change of regions, populations, or other sectoral components. These assessments should fully consider baseline changes in socioeconomic and environmental conditions, biophysical and socioeconomic impacts, and the capacity for systems to react to climate change through autonomous adaptation (Carter *et al.*, 1994; Benioff *et al.*, in press). In this selection process, governments should consider not just the magnitude of potential climate change impacts (e.g., where forests may change the most), but also the value (e.g., some forests may have more highly valued uses than others), the irreversibility of impacts (e.g., species extinction is irreversible), and the uniqueness of the resource (e.g., some areas such as the Monteverde Cloud Rain Forest in Costa Rica are considered unique habitats).

If quantitative vulnerability assessments have not been conducted, expert judgment may be used to identify particularly vulnerable regions, populations, or other sectoral components. For example, there is extensive literature on potential climate change impacts on forests upon which to base judgments of the relative vulnerability of these resources. Studies of vulnerability at regional levels include Tegart *et al.* (1990), Smith *et al.* (1992), World Wildlife Fund (1992), Melillo *et al.* (1993), Morse *et al.* (1993), Asian Development Bank (1994), Markham and Malcolm (1996), and IPCC (in press).

Examples of characteristics indicating forest ecosystems that are the most sensitive to climate change are:

- forests/species with limited geographic range
- forest/species that are drought/heat intolerant
- forests/species at boundaries of compatible climate regions, particularly at their heat tolerant or drought tolerant limits of range
- species with limited seed dispersal/migration capabilities
- highly fragmented forests or species populations
- regions where estimated changes in climate are relatively large

- regions that are vulnerable to sea level rise impacts
- regions that are currently experiencing stress from climate, use, and pests.

Based on expert judgment, team members could identify the most vulnerable forests or ecosystems.

Our hypothetical example exhibits the following vulnerability characteristics: it is fragmented and perhaps under stress from use; it contains relic species that are extremely rare; and estimated climate changes are relatively large.

3. Selection of Measures to Be Evaluated

The next step is to identify adaptation measures to be considered. This list should include only measures to *anticipate* climate change (e.g., planting drought-tolerant trees) because this exercise addresses what can be done now to adapt to climate change. Anticipatory measures increase the flexibility of systems to perform under a wide range of potential climate change conditions as well as under current climate. For example, since trees do not reach maturity for decades, planting decisions now should consider future climate conditions. Reactive adaptations (measures taken as or after climate changes) should not be on the list. Crops are planted for the current year's climate, not for future years. Thus, farmers should also change to drought- or heat-tolerant crops as climate changes. Anticipatory adaptations can take effect now and also generally make it easier to implement reactive adaptations. Note that these adaptations may be justified by considering the risks of climate change or even without considering the risks of climate change. Potential forest measures adapted from Smith and Lenhart (1996) are listed below.

- *Encourage mixed use strategies.* Ecosystems that support multiple uses tend to be more resilient to change. For example, forests managed for timber supply alone will probably be more sensitive to climate change than forests that are also used for recreation, nontimber harvesting, watershed management, and biodiversity preservation. Another example of a mixed land use strategy is game farming, which enhances biodiversity by incorporating wildlife in traditional agricultural management practices (Markham and Malcolm, 1996).

- *Plant a variety of tree species.* Planting a greater variety of tree species may provide a buffer against the uncertainties of climate change. One example of how this policy has been implemented is the planting of trees with greater resistance to heat and drought on the equatorial range of managed forest boundaries, if the trees can survive in the current climate (OTA, 1993).

- *Establish flexible criteria for intervention.* Policies should be in place that establish flexible criteria for the use of existing intervention management practices. The use of management practices such as salvage harvests, silvicultural management, insect and fire control, and restoration activities should be allowed to change as conditions change.

- *Reduce habitat fragmentation and promote development of migration corridors and buffer zones.* Geographic fragmentation may threaten the ability of species to migrate or adapt to changing climate. Forest fragmentation may be reduced through incentive programs for multiple-use management that balance preservation and use within a single parcel, or through the negotiation of conservation easements that protect geographically important land parcels from development. Buffer zones around current reserve areas that include different altitudes and ecosystems increase the adaptive potential for species within the preserve. A graded system of management may be implemented, where the innermost areas receive the greatest protection and more uses are allowed in the outer buffer zones or corridors (NAS, 1992; Lillieholm, 1993; OTA, 1993; Markham and Malcolm, 1996).

- *Use integrated ecosystem planning and management.* Integrated planning and management along watershed and ecosystem lines reduces the institutional fragmentation in the management of natural areas and focuses on protecting a variety of species and natural systems. Because the impacts of climate change are difficult to predict, the preservation of a variety of species in a healthy ecosystem may be the most effective way to protect those species that will be able to adapt to climate change (OTA, 1993).

- *Enhance methods to protect biodiversity off-site.* Preservation efforts for threatened or endangered species can be enhanced by off-site strategies such as breeding programs. These methods must be in place before climate changes to avoid the irreversible loss of species extinction, however, they should not be considered a substitute for in situ biodiversity (NAS, 1992).

- *Enhance seed banks and develop new plant varieties.* Seed banks that maintain a variety of seed types preserve biological diversity and provide foresters with an opportunity to diversify, allowing them to both counter the threat of climate change and develop a profitable specialization. Development of more and better heat- and drought-resistant tree and plant varieties will benefit areas that become warmer or drier over time.

Other measures may be appropriate, depending on particular national, cultural, geographical, or other circumstances.

4. Screening of Measures

There are many more potential anticipatory adaptation measures than can be efficiently analyzed as part of the adaptation assessment process. However, we suggest screening the adaptation measures based on the following criteria to identify several measures upon which to focus further analysis:

1. The measure addresses high priority adaptation (Smith, in press a). High priority adaptations address the following:

   a. *Irreversible or catastrophic consequences of climate change.* Biodiversity losses are caused by extinction; cultural losses are caused by dislocation.

   b. *Long-term decisions.* Seedlings may take decades to reach maturity.

   c. *Unfavorable trends.* Forests are being converted to other uses, which may impair flexible responses to climate change.

2. The adaptation measure addresses targets of opportunity. In particular, are decisions being made now on the following:

   a. *Land purchases or development.*

   b. *Plans being revised* (e.g., national environmental action plans, contingency plans, sustainable development plans).

   c. *Research and development investments or priority setting.*

3. The measure generates other benefits to the economy or environment. In particular, the measure is justified under current climate conditions. Many of the adaptation measures listed in Table I, such as reducing forest fragmentation and establishing migration corridors, are justified even without considering climate change.

4. The measure will be inexpensive to implement. Adaptation measures generally should have minimal or low cost. Given the long time frame and uncertainty about climate change, it is difficult to justify significant costs that provide benefits only if climate changes. Large costs could be justified more easily based on other benefits independent of climate change.

5. The measure is feasible: it can gain support for adoption. There are no significant barriers to implementation. Barriers can be:

   a. *Institutional/legal.* Migration corridors and ecosystem preservation may require international cooperation when migration routes or ecosystems cross international borders; access to ecosystem management funds may be limited.

   b. *Social and cultural.* Existing land use practices such as shifting agriculture may be difficult to change.

   c. *Market (e.g., pricing, availability of capital).* Forest management and policy changes may affect domestic and international wood product markets.

   d. *Technological (existence, access).* Access to technologies such as mechanized harvest technologies, or drought or disease resistant tree cultivars may be limited.

6. The measure is consistent with mitigation measures and adaptation measures in other sectors. The measure makes it less difficult to adapt to changes in other sectors. It makes it less difficult to mitigate greenhouse gas emissions or achieve other goals.

TABLE I
Example: Screening adaptation options

Adaptation Option	High Priority	Target of Opportunity	Other Benefits	Low Costs	Low Barriers	Consistent
Reduce Fragmentation	Yes	Yes	Yes	No	No	Yes
Plant Mix of Species	Yes	Yes	Yes	Yes	Yes	Yes
Protect Biodiversity Off-Site	Yes	No	Yes	No	No	Yes

The screening process should use expert judgment to address these points. Answers to these screening criteria could be arrayed in a table such as Table I. Those measures with the most "yes" answers should be the ones that are analyzed in the next step.

In the hypothetical example, we consider three adaptation options, which are displayed in the first column of Table I. The first two options, reduce fragmentation and plant mixed species, have more yeses than the third option, protect biodiversity off-site. So, the first two options are evaluated below.

5. Evaluation of Measures

The adaptation measures considered for each vulnerable region should be analyzed in detail. The goal of the analysis should be to rank the measures based on their effectiveness (or cost-effectiveness) in fulfilling policy goals and their relative ease of implementation. The analysis of fulfilling policy goals is called effectiveness analysis.

5.1 IDENTIFY POLICY OBJECTIVES

The effectiveness analysis should start with the identification of policy objectives. Policy objectives are selected based on the socioeconomic role that the resource fulfills. For example, forests can be managed to provide valuable wood products, to protect biodiversity, and to provide recreational opportunities.

5.2 SELECT METHOD FOR EVALUATING MEASURES

Having identified the policy objectives, the next step is to select a method to examine how well current policy and the adaptation measures perform in fulfilling the objectives under a variety of climate change scenarios. The current policy is analyzed to determine if it is more cost-effective in meeting policy objectives than the adaptation alternatives. If so, no change in policy need be made.

Quantitative or at least ordinal measures of how well these objectives are satisfied are preferable to nonquantitative analysis. Ideally, the same metric, such as monetary value, is used to measure this success. This will make it easier to compare results of the analysis across policy objectives. Whatever the method used, the process should build a consensus among decision makers and key stakeholders. Workshops or meetings should be held in which the decision makers and key stakeholders discuss the vulnerability of the natural resources and evaluate the various adaptation measures.

A number of different approaches can be used in these meetings to help assess the effectiveness of the policy measures. Among the approaches are benefit-cost analysis, cost-effectiveness analysis, and multicriteria assessment. It is important that the analysis considers multiple climate change scenarios within a plausible range (e.g., hot and wet, hot and dry, mild and wet, mild and dry) in order to assess the flexibility of adaptation measures to meet policy objectives under a wide range of potential climate change conditions. The evaluation should also consider the discounted benefits and costs of the measures to assess their economic efficiency.

5.2.1 Benefit-Cost Analysis

Benefit-cost analysis relies on monetizing all the benefits and costs of policy options in present value terms (Gramlich, 1981). Fankhauser (1996) discusses the application of benefit-cost analysis to evaluate climate change adaptation measures. The advantage of this type of analysis is that it uses a common, monetary measure to value all of the benefits and costs. This enables users to determine whether policy options will produce benefits that exceed costs. The disadvantage is that it is often difficult to apply monetary values to all benefits and costs. Although techniques exist for determining the value of nonmarket goods, for example, the value of a statistical life or an ecosystem, such techniques are still not yet broadly accepted.

5.2.2 Cost-Effectiveness Analysis

Cost-effectiveness analysis identifies the least-cost measure for reaching a specific goal. Rather than valuing benefits, it helps determine the least expensive alternative for reaching a certain level of benefits. For example, cost-effectiveness analysis may seek to identify the option with the fewest "dollars per life saved." An example of this type of analysis is a decision matrix that analyzes the cost-effectiveness of adaptation measures by comparing costs of adaptation measures to the benefits of the measures measured in a common metric (not necessarily dollars). Such measurements can be added up across the different policy objectives (and weighted based on relative importance) and compared to costs to determine cost-effectiveness (e.g., cost per incremental unit of benefit) and rank measures.

Table II displays the use of a decision matrix to examine adaptation measures for the hypothetical forest example. The objective is to manage the forest to provide wood products, protect biodiversity, and provide recreation. The current policy is compared to planting a mix of species and reducing fragmentation. The current policy is included in the decision matrix to ensure that the adaptation measures are superior to the current policy and that they increase benefits. In the example, expert judgment based on quantitative assessments or the knowledge of experts is used to apply an ordinal ranking (e.g., on a 1 to 5 scale) of how well objectives are fulfilled. Note that moving away from high-valued wood species to a more diverse set of species results in a lower score on wood products. Planting a mixed species scores higher for the other two objectives compared to the current policy. The two policy alternatives have higher scores, indicating that they produce more benefits than the current policy. Reducing habitat fragmentation has a much higher total score than planting a mix of species. The total scores can be compared to costs to determine cost-effectiveness and rank measures. Reducing habitat fragmentation costs much more than planting a mix of species. However, since the former results in a much larger increase in benefits than the latter, it turns out to be more cost-effective (US$0.77 million per unit of benefit compared to US$1 million per unit of benefit for planting a mix of species).

5.2.3 Multicriteria Analysis

Multicriteria analysis can use nonquantitative approaches for assessing how well adaptation measures address multiple policy objectives (e.g., Smith and Chu, 1994). Unlike the prior two analytic approaches, it does not rely on quantifying benefits in dollars or another common metric. Decision makers are asked to identify policy objectives and subjectively determine how well the adaptation measures may meet the objectives. The advantage of applying multicriteria analysis is that it is relatively easy for decision makers to use. Since it shows which options are superior in meeting particular objectives, but does not resolve situations where different options are superior in meeting different objectives, it may not be able to rank options in all cases.

TABLE II
Example: Forests decision matrix using 1-5[§] scale

		Objectives							
		Wood Products	Bio-diversity	Recreation					
	Weights	4	3	2					
Measures	Scenario				Score	Total Score[†]	Score Increase[‡]	Cost of Measure (US$M)	Cost Effective-ness[¥]
Current Policy	Wet	3	2	2	22				
	Dry	2	1	1	13	35	N/A	N/A	N/A
Plant Mix of Species	Wet	2	3	3	23				
	Dry	1	2	2	14	37	2	$2	$1.00
Reduce Fragmentation	Wet	4	4	3	34				
	Dry	3	3	3	27	61	26	$20	$0.77

[§] The 1-5 scale in this matrix is an arbitrary ranking of how well objectives are met under different scenarios and policies. Five is the highest score, implying the greatest benefits, and one is the lowest score, implying the lowest benefits.

[†] In this example, we added the wet and dry scores, assuming each scenario has the same probability of occurrence. Probabilities for individual scenarios can be adjusted.

[‡] Increase from current policy.

[¥] Cost-effectiveness equals cost (US$M) per incremental unit of benefit.

5.3 ANALYZE EASE OF IMPLEMENTATION

A parallel analysis to the effectiveness of policies is the identification of any barriers to implementation, and an evaluation of how difficult or easy it may be to overcome these barriers. Matrices can be used to identify barriers, and to evaluate the degree of difficulty associated with overcoming the barriers based on the required time and financial resources. The results from the analysis of barriers can be used to adjust the rankings from the results of benefit-cost analysis or the decision matrix. If the barriers, however, are only a matter of cost, the costs of overcoming the barriers can be entered into the cost calculations under benefit-cost analysis or the decision matrix.

In Tables III and IV, matrices are displayed showing the barriers to implementation for each adaptation measure in our example. The barriers are analyzed with respect to the response to overcoming the barrier, the time, costs, and overall level of difficulty. The "X's" are used to assess overall difficulty, with more "X's" indicating greater difficulty in overcoming the barrier. In our example, planting a mix of species has more barriers than reducing habitat fragmentation, although resistance by property owners to habitat fragmentation reduction may be harder to overcome than the barriers to planting a mix of species.

TABLE III
Example: Overcoming barriers to implementation of planting a mix of species

Barriers	Response	Time	Costs	Difficulty
Technical — forest managers need training in techniques	Training	6 months	US$100,000	X
Cultural — indigenous populations resistant to change in forest	Education	1 year	US$500,000	XX

TABLE IV
Example: Overcoming barriers to implementation of reducing fragmentation

Barriers	Response	Time	Costs	Difficulty
Legal — property owners resist land purchases	Court cases	2 years	US$500,000	XXX

5.4 SELECT MEASURES

Having analyzed the measures for their relative effectiveness and ease of implementation, the final step of the evaluation process is to select one or more measures for implementation. If the effectiveness and barrier analyses show the same measure as being superior, the choice is easy. If they differ, then it is up to decision makers to select the measure they think is best. Our advice would be to put more weight on barriers to implementation. Selecting a measure that has lower barriers increases the likelihood of its being adopted.

In our example, reducing habitat fragmentation is more cost-effective, but has more difficult barriers to implementation. Since more weight should be placed on barriers to implementation, we would recommend planting a mix of species.

6. Evaluation of Success

The evaluation of the success of the adaptation may take a long time because benefits of some adaptation measures may not be realized until climate changes significantly. On the other hand, adaptation measures that have benefits if climate does not change (e.g., no regret measures) should show benefits in the near term. Evaluations should be based on measuring the benefits against the policy objectives used in the effectiveness analysis. Quantitative measures of success should be used when available.

7. Conclusions

A systematic approach is suggested here for countries to use in assessing needs for adapting to climate change and for identifying, evaluating, and selecting anticipatory adaptation measures. It is not necessary for those addressing adaptation to climate change to follow our process by rote. However, it is important that they use a clear, systematic, and replicable process for arriving at decisions on the selection and implementation of climate change adaptation measures.

Acknowledgments

The work that led to this article was funded by the U.S. Country Studies Program through the Energy Efficiency Project contract with the U.S. Agency for International Development. We wish to thank Ron Benioff and John Warren from the U.S. Country Studies Program for their thoughtful guidance and comments on this work; Jay Malcolm for reviewing the policy options and example; Barrie Pittock, Saleemul Huq, and Joe Wisniewski for providing very useful comments on the manuscript; Chris Thomas for editing the manuscript; and Lisa Foos for assisting us in the production of the paper.

References

Asian Development Bank: 1994, *Climate Change in Asia: Thematic Overview,* Asian Development Bank, Manila.

Benioff R., Guill S., and Lee J. (eds): In press, *Vulnerability and Adaptation Assessments: An International Guidebook,* Kluwer Academic Publishers, Dordrecht, The Netherlands.

Carter T.R., Parry M.L., Harasawa H., and Nishioka S.: 1994, *IPCC Technical Guidelines for Assessing Climate Change Impacts and Adaptations,* Department of Geography, University College London, London.

Fankhauser, S.: 1996, The potential costs of climate change adaptation, in: *Adaptation to Climate Change: Assessment and Issues,* Smith J., Bhatti N., Menzhulin G., Benioff R., Campos M., Jallow B., and Rijsberman R. (eds), Springer-Verlag, New York.

Goklany I.M.: 1995, *Climatic Change* **30**, 427-449.

Gramlich E.M.: 1981, *Benefit-Cost Analysis of Government Programs,* Prentice-Hall, Englewood, New Jersey.

(IPCC) Intergovernmental Panel on Climate Change: 1995, *Climate Change 1995: The Science of Climate Change, Contribution of Working Group I to the Second Assessment Report of the Intergovernmental Panel on Climate Change,* Houghton J.T., Meira Filho L.G., Callander B.A., Harris N., Kattenberg A., and Maskell K. (eds), Cambridge University Press, New York.

(IPCC) Intergovernmental Panel on Climate Change: In press, *Scientific-Technical Analyses of Impacts, Adaptations, and Mitigation of Climate Change,* Cambridge University Press, New York.

Lillieholm R.J.: 1993, Preserves at risk: An investigation of resource management strategies, implications and opportunities, in: *Preparing for an Uncertain Climate, Vol. 2,* OTA-O-567, Office of Technology Assessment, Washington, DC.

Markham A. and Malcolm J.: 1996, Biodiversity and wildlife conservation: Adaptation to climate change, in: *Adaptation to Climate Change: Assessment and Issues,* Smith J., Bhatti N., Menzhulin G., Benioff R., Campos M., Jallow B., and Rijsberman R. (eds), Springer-Verlag, New York.

Melillo J.M., McGuire A.D., Kicklighter D.W., Moore III D., Vorosmarty C.J., and Schloss A.L.: 1993, *Nature* **363**, 234.

Morse L.E., Kutner L.S., Maddox G.D., Kartesz J.T., Honey L.L., Thurman C.M., and Chaplin S.J.: 1993, The Potential Effects of Climate Change on the Native Vascular Flora of North America North of Mexico: A Preliminary Climate-Envelopes Analysis, Prepared by The Nature Conservancy, Arlington, Virginia, for the Electric Power Research Institute, Palo Alto, California.

(NAS) National Academy of Sciences: 1992, *Policy Implications of Greenhouse Warming,* National Academy Press, Washington, DC.

(OTA) U.S. Congress Office of Technology Assessment: 1993, *Preparing for an Uncertain Climate — Volume 2,* U.S. Government Printing Office, Washington, DC.

Smith A. and Quan Chu H.: 1994, *A Multi-Criteria Approach for Assessing Strategies for Anticipatory Adaptation to Climate Change,* Decision Focus, Inc., Washington, DC, Unpublished Manuscript.

Smith J.B.: In press a, *The Environment Professional.*

Smith J.B.: In press b, *Water Resources Development.*

Smith J.B. and Lenhart S.: 1996, *Climate Research* **6**, 193-201.

Smith T.M., Leemans R., and Shugart H.H.: 1992, *Climatic Change* **21**, 367-84.

Tegart W.J. McG., Sheldon G.W., and Griffiths D.C.: 1990, *Climate Change — The IPCC Impacts Assessment,* WMO/UNEP Intergovernmental Panel on Climate Change, Australian Government Publishing Service, Canberra.

Titus J.G.: 1990, *APA Journal* **311**, Summer.

(UNEP/WMO) United Nations Environment Programme/World Meteorological Organization: Undated, *United Nations Framework Convention on Climate Change,* UNEP/WMO Information Unit on Climate Change (IUCC), Geneva, Switzerland.

World Wildlife Fund: 1992, *Can Nature Survive Global Warming?* Discussion Paper, World Wildlife Fund, Washington, DC.

WORKSHOP PARTICIPANTS

REGIONAL WORKSHOP ON CLIMATE CHANGE VULNERABILITY AND ADAPTATION FOR ASIA AND PACIFIC

15 to 19 January 1996
Manila, Philippines

Amrita N. Achanta
Tata Energy Research Institute
Darbari Seth Block, Habitat Place
Lodhi Road
New Delhi 110003
India
Telephone: 919-689-4663/91-11-462-2246
Fax: 919-962-5206/91-11-426-1770
E-mail: aachanta.dcrp@.mhs.unc.edu

Anwar Ali
Bangladesh Space Research & Remote Sensing Organization (SPARRSO)
Agargaon, Sher-e-Bangla Nagar
Dhaka
Bangladesh
Telephone: 880-2-313329/327913/323981
Fax: 880-2-813080

Leoncio Amadore
PAGASA
1424 Quezon Avenue
Quezon City 1100
Philippines
Telephone: 632-922-8401
Fax: 632-922-8401/9291
E-mail: raq@dostmis.dost.gov.ph

Istiqlal Amien
Centre for Soil Research and Agroclimate
Jalan M.H. Juanda 98
Bogor 16123
Indonesia
Telephone: 62-251-323012
Fax: 62-251-311256

Emmanuel Anglo
Department of Meteorology and Oceanography, College of Science University of the Philippines
Diliman
Quezon City
Philippines
Telephone: 632-920-6096/6097
Fax: 632-920-6096
E-mail: eanglo@pusit.admu.edu.ph

Liu Bin
China Country Study Office
Energy Science Building
Tsinghua University
Beijing 100084
China
Telephone: 8610-259-4828
Fax: 8610-256-4177
E-mail: thinet@bec2.ihep.ac.cn

William Bolhofer
NOAA/NWS-W/IA, SSMC2
Room 13378
1325 East West Highway
Silver Spring, MD 20910
U.S.A.
Telephone: 301-713-1611 Ext. 148
Fax: 301-713-1255
E-mail: wbolhofer@smtpgate.
ssmc.noaa

Bold Bolortsetseg
Hydro Research Institute
Khudaldaany Gudamj - 5
Ullanbaatar - 11
Mongolia
Telephone: 976-1-321523
Fax: 976-1-321401

Kansri Boonpragob
Thailand Environment Institute
210 Sukhumvit 64, Bangkok
Refinery Building 4
Bangkok 10400
Thailand
Telephone: 662-331-0047/0060
Fax: 662-332-4873
E-mail: Kansri@ram1.ru.ac.th

Rodito D. Buan
PAGASA
1424 Quezon Avenue
Quezon City 1100
Philippines
Telephone: 632-922-1992
Fax: 632-922-8401

Water, Air, and Soil Pollution **92:** 239–246.
© 1996 *Kluwer Academic Publishers. Printed in the Netherlands.*

Leandro V. Buendia
International Rice Research Institute
P.O. Box 933
Manila 1099
Philippines
Telephone: 632-818-1926
Fax: 632-891-1292
E-mail: lbuendia@irri.cgnet.com

Ernesto C. Bumatay
Department of Forestry, VISCA
Baybay
Leyte
Philippines
Telephone: 632-58-8692

John A.T. Bye
Flinders University of South Australia
GPO Box 2100
Adelaide 5001
Australia
Telephone: 08-201-2113
Fax: 08-201-2676
E-mail: mojatb@earthsci.es.flinders.edu.au

Michael Cabalda
Mines & Geosciences Bureau
North Avenue
Diliman, Quezon City
Philippines
Telephone: 632-926-0935
Fax: 632-920-1635

Peter Cabral
University of the Philippines
Los Baños College
Laguna
Philippines

Efren P. Carandang
Oceanography Division, NAMRIA
421 Barraca St., San Nicolas
Manila 1010
Philippines
Telephone: 632-241-3494 to 98
Fax: 632-242-2090

Nathaniel Cruz
PAGASA
1424 Quezon Avenue
Quezon City 1100
Philippines
Telephone: 632-922-7282
Fax: 632-922-7813
E-mail: nac@dostmis.dost.gov.ph

Rex Victor Cruz
University of the Philippines at Los Baños
FDC College of Forestry College Laguna 4031
Philippines
Telephone: 63-94-2341
Fax: 63-94-3206
E-mail: rvoc@mudspring.uplb.edu.ph

A.H. Dhanapala
Forest Department
Participatory Forest Project
Battaramulla
Sri Lanka
Fax: 941-8666-33

Robert Dixon
Country Studies Program
CSMT (PO-63)
1000 Independence Avenue SW
Washington, DC 20585
U.S.A.
Telephone: 202-425-1464
Fax: 202-426-1540
E-mail: csmt@igc.apc.org

Anwest P. Eleas
Environmental Protection Authority
P.O. Box 1322
Majuro
Republic of the Marshalls 9696
Telephone: 692-625-3035
Fax: 692-625-5202

Michael Ellis
Hagler Bailly Consulting
G/F Suite B, 675 Ayala Avenue
Makati City, Metro Manila
Philippines
Telephone: 632-844-4421
Fax: 632-844-5563
E-mail: 7662.2473@compuserve.c

Lin Erda
Agrometeorology Institute
30 Baishiqiao Road
Beijing 100081
China
Telephone: 8610-217-4433, ext. 2883 (o)
218-6101 (h)
Fax: 8610-217-4142 or 217-5320
E-mail: lined@sun.ihep.ac.cn

Perfecto Evangelista
Bureau of Soils & Water Management
Cor. Visayas Avenue, Diliman
Quezon City
Philippines
Telephone: 632-920-4378
Fax: 632-923-0454
E-mail: bswm@phil.gn.apc.org

Renato Feir
NAMRIA
421 Barraca Street
San Nicolas, Manila
Philippines
Telephone: 632-242-2955
Fax: 632-242-2953

Raquel V. Francisco
PAGASA
1424 Quezon Avenue
Quezon City 1100
Philippines
Telephone: 632-922-7282
Fax: 632-922-7813
E-mail: raq@dostmis.dost.gov.ph

Carol Freysinger
Camber Corporation
601 13th Street NW, Suite 350 N
Washington, DC 20005
U.S.A.
Telephone: 202-393-1648
Fax: 202-628-8498
E-mail: camberdc@access.digex.net

Erlinda Gonzales
Environmental Quality Division
Environmental Management Bureau
Topaz Building, Kamias Street
Diliman, Quezon City
Philippines
Telephone: 632-928-3373
Fax: 632-924-7540

Sandy Guill
Country Studies Program CSMT (PO-63)
1000 Independence Avenue
Washington, DC 20585
U.S.A.
Telephone: 202-426-1464
Fax: 202-426-1540
E-mail: csmt@igc.apc.org

Tara Man Gurung
Department of Irrigation
P.O. Box 406, Babar Maha
Kathmandu
Nepal
Telephone: 977-1-215975
Fax: 977-1-224648

Hideo Harasawa
National Institute for Environmental Studies
16-2 Onogawa, Tsukuba
Ibaraki 305
Japan
Telephone: 81-298-50-2507
Fax: 81-298-58-2645
E-mail: harasawa@nies.go.jp

Flaviana Hilario
PAGASA
1424 Quezon Avenue
Quezon City 1100
Philippines
Telephone: 632-922-7282
Fax: 632-922-7813
E-mail: fdh@dostmis.dost.gov.ph

Nguyen Hoang Tri
SRES, Vietnam National University
91 Nguyen Khuyen
Hanong
Vietnam

Joseph Huang
Country Studies Program, Asian Nations
CSMT (PO-63)
1000 Independence Avenue
Washington, DC 20585
U.S.A.
Telephone: 202-426-1673
Fax: 202-426-1540
E-mail: csmt@igc.apc.org

Saleemul Huq
Bangladesh Centre for Advanced Studies
620 Road 10A (New) Dhanmondi
Dhaka 1209
Bangladesh
Telephone: 880-2-313682
Fax: 880-2-811344
E-mail: saleem@pradeshta.net

Annie Idanan
Environment Division, ADB
ADB Avenue, Ortigas Center
Pasig City, Metro Manila
Philippines
Telephone: 632-632-5779
Fax: 632-636-2195
E-mail: aidanan@mail.asiandevbank.org

Ana Iglesias
CIFOR-INIA
Carretera de la Coruña Km. 7
Madrid 28040
Spain
Telephone: 34-1-347-6859
Fax: 34-1-357-2293
E-mail: iglesia@cit.inia.es

Kazi Jalal
Asian Development Bank
ADB Avenue, Ortigas Center
Pasig City, Metro Manila
Philippines
Telephone: 632-632-6717
Fax: 632-741-7961
E-mail: kjalal@mail.asiandevbank.org

Dr. Aida M. Jose
PAGASA
1424 Quezon Avenue
Quezon City 1100
Philippines
Telephone: 632-922-7282
Fax: 632-922-7813
E-mail: amj@dostmis.dost.gov.ph

Edna L. Juanillo
PAGASA
1424 Quezon Avenue
Quezon City 1100
Philippines
Telephone: 632-922-8408
Fax: 632-922-7813

Dr. Zahurul Karim
Bangladesh Agriculture Research Corporation
Farmgate, New Airport Road
Dhaka 1215
Bangladesh
Telephone: 880-2-311432
Fax: 880-2-813032

Hillarius Kludze
International Rice Research Institute
P.O. Box 933
Manila 1099
Philippines
Telephone: 632-818-1926
Fax: 632-818-1292

Dr. K.I. Kobak
State Hydrological Institute
23, Second Line
St. Petersburg 199053
Russia
Telephone: 7-812-213-3517
Fax: 7-812-213-1028

Rhoda Lantin
IRRI Soil & Water Sciences
P.O. Box 933
Manila 1099
Philippines
Telephone: 632-818-1926/812-7686 loc. 29
Fax: 632-891-1292/817-8470
E-mail: rlantin@irri.ognet.com

Dr. Stephen Leatherman
Laboratory for Coastal Research
Department of Geography
University of Maryland
1113 LeFrak Hall
College Park, MD 20742
U.S.A.
Telephone: 301-405-4059
Fax: 301-314-9299
E-mail: sl9@umail.umd.edu

Dr. Louis Lebel
CSIRO Division of Wildlife & Ecology
GCTE Core Project Office
P.O. Box 84
Lyneham ACT 2602
Australia
Telephone: 616-242-1619
Fax: 616-241-2362
E-mail: louis.lebel@dwe.csiro.au

Stephanie Lenhart
Hagler Bailly Consulting
P.O. Drawer O
Boulder, CO 80306-1906
U.S.A.
Telephone: 303-449-5515
Fax: 303-443-5684
E-mail: slenhart@habaco.com

Weihong Lin
International Rice Research Institute
P.O. Box 933
Manila 1099
Philippines
Telephone: 632-818-1926/762-0127
Fax: 632-761-2404/2406
E-mail: wlin@irri.cgnet.com

Bindu N. Lohani
Asian Development Bank
ADB Avenue, Ortigas Center
Pasig City, Metro Manila
Philippines
Telephone: 632-632-6883
Fax: 632-632-2195
E-mail: blohani@mail.asiandevbank.org

Cynthia Lowry
USAID/Energy Program Manager
1680 Roxas Boulevard
Manila
Philippines
Telephone: 632-522-4411 Ext. 4467
Fax: 632-522-2512
E-mail: clowry@usaid.gov.

Mochtar Machful
State Ministry for Research and Technology
U. Veteran III
Jakarta
Indonesia
Telephone: 021-384-8905
Fax: 021-344-6733

Amado R. Maglinao
PCARRD
Los Baños, Laguna 4020
Philippines
Telephone: 630-945-0015
Fax: 630-945-0016

Resi Mariñas
Office of Senator Alvarez
12th Floor, Gotesco Tower Building Concepcion Street
Manila
Philippines

Dr. D.T. Mathes
Coconut Research Institute
Biometry Division
Lunuwila
Sri Lanka
Telephone: 94-1-45-3795
Fax: 94-1-31-7195

Gerald Meehl
National Center for Atmospheric Research
P.O. Box 3000
Boulder, CO 80307-3000
U.S.A.
Telephone: 303-497-1331
Fax: 303-497-1333
E-mail: meehl@ncar.ucar.edu

Riyad M. Mistry
RMI Environmental Protection Authority
P.O. Box 1322
Majuro, MH 96960
Marshall Islands
Telephone: 692-625-5203
Fax: 692-625-5202
E-mail: mistry_r@marshall.usp.ac.fj

Tolentino B. Moya
IRRI Agronomy, Plant Physiology, and Agroecology Division
P.O. Box 933
Manila 1099
Philippines
Telephone: 632-818-1926
Fax: 632-891-1292
E-mail: tmoya@irri.cgnet.com

Shishir Mukherjee
Asian Development Bank
ADB Avenue, Ortigas Center
Pasig City, Metro Manila
Philippines
Telephone: 632-632-5779
Fax: 632-632-2195

Ofelia S. Namuco
International Rice Research Institute
P.O. Box 933
Manila 1099
Philippines
Telephone: 632-818-1926/762-0127
Fax: 632-761-2404/2406
E-mail: o.namuco@cgnet.com

Betty Neal
Hagler Bailly Consulting
P.O. Drawer O
Boulder, CO 80306-1906
U.S.A.
Telephone: 303-449-5515
Fax: 303-443-5684
E-mail: bneal@habaco.com

Edna Ona
National Research Council of the Philippines
Bicutan, Taguig, MM
Manila
Philippines
Telephone: 632-837-0409
Fax: 632-837-0409

Raymond Ordonez
International Rice Research Institute
P.O. Box 933
Manila 1099
Philippines
Telephone: 632-818-1926
Fax: 632-891-1292

Bonifacio G. Pajuelas
PAGASA
1424 Quezon Avenue
Quezon City 1100
Philippines
Telephone: 632-922-7736
Fax: 632-922-8401

Sangsant Panich
International Environmental Institute
210 Sukhumvit 64, Bangkok Refinery Building 4
Bangkok 10400
Thailand
Telephone: 662-311-1708
Fax: 662-311-1708

Melinda Perez
National Research Council of the Philippines
Bicutan, Taguig, MM.
Manila
Philippines
Telephone: 632-837-0409
Fax: 632-837-0409

Dr. Rosa T. Perez
PAGASA, V&A Team
1424 Quezon Avenue
Quezon City 1100
Philippines
Telephone: 632-922-1992
Fax: 632-922-8490

Dr. A. Barrie Pittock
CSIRO, Atmospheric Research
PMB 1
Aspendale 3195
Australia
Telephone: 613-9239-4527
Fax: 613-9293-4444
E-mail: abp@dar.csiro.au

Poerwoto
KPPL.OKI Jakarta
Jakarta Capital City
Indonesia
Telephone: 021-520-9665
Fax: 021-520-9643

Marisol Porta
GEOSPHERE Technologies
Philippines

Anil Premaratne
Coast Conservation Department
4th Floor New Sec. Building
Maligawatte
Colombo 10
Sri Lanka
Telephone: 94-1-449756
Fax: 94-1-438005

L. Heidi Primo
FSM Office of the President
P.O. Box PS-121
Palikir, Pohnpei FM 96941
Federated States of Micronesia
Telephone: 691-320-2228
Fax: 691-320-2785

K.L. Narasimha Rao
S.V. University
Department of Geology
Tirupati, AP, 517 502
India
Telephone: 91-8574-24766 Ext. 251
Fax: 91-8574-24111
E-mail: nicnet!svims!root@uunet.uu.net

Sang G. Shim
Korean Institute of Science and Technology
P.O. Box 131
Cheongryan, Seoul
Korea
Telephone: 82-2-958-5815
Fax: 82-2-958-5805
E-mail: sgshim@hp710.kist.re.kr

Madan Shrestha
Department of Hydrology & Meteorology
P.O. Box 406, Babar Mahal
Kathmandu
Nepal
Telephone: 977-1-215979
Fax: 977-1-224648

Graham Sims
CEARS, University of Waikato
Private Bag 3105
Hamilton
New Zealand
Telephone: 64-7-838-4427
Fax: 64-7-838-4289
E-mail: gsims@waikato.ac.nz

Hotman Silaen
Environmental Bureau
Jakarta
Indonesia

Dr. K.U. Sirinanda
Department of Geography, University of Brunei Darussalam
B.S.B. 2028
Darussalam
Brunei
Telephone: 673-2-421528
Fax: 673-2-249003

Joel Smith
Hagler Bailly Consulting
P.O. Drawer O
Boulder, CO 80306-1906
U.S.A.
Telephone: 303-449-5515
Fax: 303-443-5684
E-mail: jsmith@habaco.com

Bernardo Soriano, Jr.
PAGASA
1424 Quezon Avenue
Quezon City 1100
Philippines
Telephone: 632-922-8411
Fax: 632-922-8411

Dr. R.T.M. Sutamihardja
Ministry of Environment Indonesia
J1, Medan Merdeka Barat
Jakarta 10110
Indonesia
Telephone: 62-21-374307
Fax: 62-21-374307

Sobar Sutisna
National Coordination Agency for Surveying and Mapping
JL. Raya Bogor Room 46
Cibinong 16911
Indonesia
Telephone: 62-21-8754654
Fax: 62-21-8752064

Amien Suwarto
Department of Transportation
Merdeka Borat Jakarta
Toman Agri Blok E7/12
Indonesia
Telephone: 361308 ext. 1403
Fax: 384-9795

Wanna Tanunchaiwatana
Office of Environmental Policy & Planning
60/1 Soi Phibun-Wattana, Rama 6 Road
Payathai, Bangkok 10400
Thailand
Telephone: 662-272-3054
Fax: 662-271-4322/3226

Lourdes Tibig
PAGASA
1424 Quezon Avenue
Quezon City 1100
Philippines
Telephone: 632-922-7282/8407
Fax: 632-922-7813/8401

Saksit Tridech
Office of Environmental Policy & Planning
60/1 Soi Phibun-Wattana, Rama 6 Road
Payathai, Bangkok 10400
Thailand
Telephone: 662-279-0130
Fax: 662-271-3226

Vanchindor Ulziisaikhan
Research Center for Climate Change
Khudaldraany Str. 5
Ulaanbaatar 210646
Mongolia
Telephone: 976-1-326603
Fax: 976-1-321401

Imelda Valeroso
PAGASA
1424 Quezon Avenue
Quezon City 1100
Philippines
Telephone: 632-922-1992
Fax: 632-922-9291

Lando Velasco
Climate Action Network SEA
Rm. 403 Cabrera Building
Timog Avenue
Quezon City
Philippines
Telephone: 632-96-5362
Fax: 632-96-5362
E-mail: cansea@phil.gn.apc.org

Jonathan Villasper
Haribon Foundation
340 Villamor Street
San Juan, Metro Manila 1500
Philippines
Telephone: 632-78-4179
Fax: 632-70-4316
E-mail: haribon@phil.ag.apc.org

Jinghua Wang
Agrometeorology Institute
30 Baishiqiao Road
Beijing 100081
China
Telephone: 8610-217-4433
Fax: 8610-217-4142/5320

Richard Warrick
CEARS, University of Waikato
Private Bag 3105
Hamilton
New Zealand
Telephone: 64-7-838-4276
Fax: 64-7-838-4289
E-mail: cearsr@waikato.ac.nz

R. Wassman
International Rice Research Institute
P.O. Box 933
Manila 1099
Philippines
Telephone: 632-818-1926
Fax: 632-818-1292
E-mail: rwassman@irri.cgnet.com

Upali Weerakkody
University of Ruhuna
Matara
Sri Lanka
Telephone: 94-1-041-22681
Fax: 94-1--502566

M.A. Wijeratne
Tea Research Institute
Research Advisory and Extension Centre
Ratnapura
Sri Lanka
Telephone: 94-45-2428
Fax: 94-45-3262

Yin Yanfang
Agrometeorology Institute
30, Baishiqiao Road
Beijing 100081
China
Telephone: 8610-217-4433
Fax: 8610-217-4142
E-mail: lined@sum.ihep.ac.cn

W.W.S. Yim
The University of Hong Kong
Department of Earth Sciences
Pokfulam Road
Hong Kong
Telephone: 852-2859-2829
Fax: 852-2517-6912
E-mail: wwsyim@hkucc.hku.hk

AUTHOR INDEX

Ahmad, Q.K. [215]
Ahmed, M. [53]
Ali, A. [171]
Amadore, L. [1]
Amien, I. [29]

Bolhofer, W.C. [1]
Bolortsetseg, B. [95]
Boonpragob, K. [107]
Buan, R.D. [41]

Carandang, E. [137]
Cruz ,N.A. [191]
Cruz, R.V. [1]

Dhanapala, A.H. [129]

Ericksen, N.J. [215]
Evangelista, P.P. [41]

Feir, R.B. [1], [137]
Freysinger, C.A. [1]

Gonzalez, E.B. [137]
Guill, S. [1]

Hussain, S.G. [53]

Iglesias, A. [1], [13]

Jalal, K.F. [1]
Jose, A.M. [1], [191]

Karim, Z. [53]
Kenny ,G.J. [215]
Kobak, K.I. [119]
Koch, H. [119]
Kondrasheva, N.Yu. [119]

Leatherman, S.P. [1], [149]
Lenhart, S. [1]
Lin Erda [13], [63], [75]

Maglinao, A.R. [41]
Mcinnes, K. [159]
Meehl, G.A. [203]
Mirza, M.Q. [215]
Mukherjee, S. [1]

Pajuelas, B.G. [41]
Perez, R.T. [137]
Pittock, A.B. [159]
Pitts, G.J. [229]
Pramudia, A. [29]

Ragland, S.E. [229]
Rejekiningrum, P. [29]
Rosenzweig, C. [13]

Santisirisomboon, J. [107]
Schulze, E.-D. [119]
Schulze, W. [119]
Sims, G.C. [215]
Smith, J.B. [1], [229]
Somaratne, S. [129]
Sosa, L.M. [191]
Susanti, E. [29]

Turchinovich, I.Ye. [119]
Tuvaansuren, G. [95]

Vygodskaya, N.N. [119]

Walsh, K. [159]
Wang Jinghua [75]
Warrick, R.A. [215]
Wijeratne, M.A. [87]
Wisniewski, J. [1]
Wyss W.-S. Yim [181]

KEY WORD INDEX*

adaptation [1], [13], [229]
agriculture [1], [13], [29], [41], [53], [63], [75], [87]
agroclimatic zones [215]
Asia [1], [149]
Australasia [203]
Australia [159]
AVVA [137]

Bangladesh [53], [171], [215]
BDCLIM [215]
beach nourishment [149]
benefit-cost analysis [229]

cattle production [95]
CCCM [41], [53], [95], [129], [191]
CERES-Corn [41]
CERES-Rice [41], [51]
CERES-Wheat [51]
China [63], [75]
CLIMPACTS [215]
coastal erosion [149]
coastal resources [1], [137], [149], [159], [171], [181], [215]
coastal zone management [137]
corn [41]
cost-effectiveness analysis [229]

drought [29], [203]

El Niño [13], [29], [203]
floods [181]
forest [1], [107], [119], [129]

GFDL [29], [41], [53], [63], [75], [95], [129], [191]
GISS [29], [41], [95], [107]
grasslands/livestock [95]

Holdridge Life Zone Classification [107], [129]
Hong Kong [181]
hydrology [215]

Indonesia [29]
integrated model systems [215]

Java [29]

landslides [181]
limited area models [159]

maize [63], [75]
Manila Bay [137]
Mongolia [95]
MPI [63], [75]
multicriteria analysis [229]

national action plans [1]
New Zealand [215]

Pacific island nations [1], [149], [203]
pasture production [95]
Philippines [41], [137], [191]

rice [29], [41], [53], [63]

sea level rise [1], [137], [149], [181]
Siberia [119]
SPUR2 [95]
Sri Lanka [87], [129]
storm surges [159], [171], [181]

tea [87]
Thailand [107]
tropical cyclones [159], [171]
typhoons [181]

UK89 [107]
UKMO [29], [41], [63], [75], [95], [107], [191]

WATBAL [191]
water resources [1], [191], [203], [215]
wheat [53], [63]

*Only first page of each paper addressing a subject is included in the key word index.